Studies in the History of Civil Engineering

Volume 11

Early Reinforced Concrete

STUDIES IN THE HISTORY OF CIVIL ENGINEERING

General Editor: Joyce Brown

with the assistance of M.M. Chrimes, A.W. Skempton, N.A.F. Smith and R.J.M. Sutherland

---

Twelve volumes are to appear from the Autumn of 1997:

1. **The Engineering of Medieval Cathedrals**
   Edited by LYNN T. COURTENAY, University of Wisconsin

2. **Masonry Bridges, Viaducts and Aqueducts**
   Edited by TED RUDDOCK, University of Edinburgh

3. **Land Drainage and Irrigation**
   Edited by SALVATORE CIRIACONO, University of Padova

4. **Dams**
   Edited by DONALD C. JACKSON, Lafayette College

5. **Water-Supply and Public Health Engineering**
   Edited by DENIS SMITH, formerly University of East London

6. **Port and Harbour Engineering**
   Edited by ADRIAN JARVIS, Merseyside Maritime Museum and Liverpool University

7. **The Civil Engineering of Canals and Railways before 1850**
   Edited by MIKE CHRIMES, Institution of Civil Engineers

8. **The Development of Timber as a Structural Material**
   Edited by DAVID T. YEOMANS, Liverpool University

9. **Structural Iron, 1750–1850**
   Edited by R.J.M. SUTHERLAND, Consultant, Harris and Sutherland, London

10. **Structural Iron and Steel, 1850–1900**
    Edited by ROBERT THORNE, Editor, *Construction History Journal*

11. **Early Reinforced Concrete**
    Edited by FRANK NEWBY, formerly F.J. Samuely and Partners, London

12. **Structural and Civil Engineering Design**
    Edited by WILLIAM ADDIS, University of Reading

Studies in the History of Civil Engineering
General Editor: Joyce Brown

Volume 11

Early Reinforced Concrete

edited by
Frank Newby

Ashgate
VARIORUM

Aldershot • Burlington USA • Singapore • Sydney

Published in the series **Studies in the History of Civil Engineering** by

Ashgate Publishing Limited
Gower House, Croft Road
Aldershot, Hampshire GU11 3HR
Great Britain

Ashgate Publishing Company
131 Main Street
Burlington, Vermont 05401–5600
USA

Ashgate website: http://www.ashgate.com

ISBN 0–86078–760–5

British Library CIP data
Early Reinforced Concrete. – (Studies in the History of Civil Engineering; 11).
1. Reinforced concrete–History. 2. Civil engineering–History
I. Newby, Frank.
624. 1' 8341

Library of Congress Card Number: 00–109453

This book is printed on acid free paper.

Printed in Great Britain by St Edmundsbury Press Ltd,
Bury St Edmunds, Suffolk

Contents

PUBLISHER'S NOTE

As this book was going to press, the General Editor and Publishers learned with deep regret of the death of Frank Newby. It is a sad duty to present this work following his death and it is especially sad that Frank Newby did not see this work in print, although he committed much time to its careful completion just prior to his death.

Acknowledgements

The chapters in this volume are taken from the sources listed below, for which the editors and publishers wish to thank their authors, original publishers or other copyright holders for permission to use their material as follows:

Chapter 1: J. Mordaunt Crook, 'Sir Robert Smirke: a pioneer of concrete construction', *Transactions of the Newcomen Society*, 38 (1965), pp. 5–20. Copyright © 1965 by the Newcomen Society, London.

Chapter 2: B.L. Hurst, 'Concrete and the structural use of cements in England before 1890', *Proceedings of the Institution of Civil Engineers: Structures and Buildings*, Special Issue, *Historic Concrete*, 116, issue nos. 3–4 (August/November 1996), pp. 283–294. Copyright © 1996 by B.L. Hurst and the Institution of Civil Engineers, London.

Chapter 3: Joyce M. Brown, 'W.B. Wilkinson (1819–1902) and his place in the history of reinforced concrete', *Transactions of the Newcomen Society*, 39 (1966–67), pp. 129–142. Copyright © 1966–67 by the Newcomen Society, London.

Chapter 4: A.W. Skempton, 'Portland cements, 1843–1887', *Transactions of the Newcomen Society*, 35 (1962–63), pp. 117–151. Copyright © 1967 by the Newcomen Society, London.

Chapter 5: John W. de Courcy, 'The emergence of reinforced concrete, 1750–1910', *The Structural Engineer*, 65A, no. 9 (September 1987), pp. 315–322. Copyright © 1987 by the Institution of Structural Engineers, London.

Chapter 6: Cyrille Simonnet, 'The origins of reinforced concrete', *Rassegna*, 49, no. 1 (March 1992), pp. 6–14. Copyright © 1992 by Editrice Compositori, s.r.l., Bologna.

Chapter 7: Gwenaël Delhumeau, 'Hennebique and building in reinforced concrete around 1900', *Rassegna*, 49, no. 1 (March 1992), pp. 15–25. Copyright © 1992 by Editrice Compositori, s.r.l., Bologna.

Chapter 8: Patricia Cusack, 'Agents of change: Hennebique, Mouchel and ferro-concrete in Britain, 1897–1908', *Construction History*, 3 (1987), pp. 61–74. Copyright © 1987 by the Construction History Society, Ascot, Berks.

Chapter 9: G.J. Edgell, 'The remarkable structures of Paul Cottancin', *The Structural Engineer*, 63A, no. 7 (July 1985), pp. 201–207. Copyright © 1987 by the Institution of Structural Engineers, London.

Chapter 10: M.N. Bussell, 'The era of the proprietary reinforcing systems', *Proceedings of the Institution of Civil Engineers, Structures and Buildings*, Special Issue, *Historic Concrete*, 116, nos. 3–4 (August/November 1996), pp. 295–316. Copyright © 1996 by M.N. Bussell and the Institution of Civil Engineers, London.

Chapter 11: Patricia Cusack, 'Architects and the reinforced concrete specialist in Britain, 1905–1908', *Architectural History*, 29 (1986), pp. 183–196. Copyright © 1986 by the Society of Architectural Historians of Great Britain.

Chapter 12: W.K. Hyatt, 'Genesis of reinforced concrete construction', *American Concrete Institute, Proceedings of the Twelfth Annual Convention* (1916), pp. 21–39. Copyright © 1916 by W.K. Hyatt and the American Concrete Institute, Detroit.

Chapter 13: Ellen W. Kramer and Aly A. Raafat, 'The Ward House: a pioneer structure of reinforced concrete', *Journal of the Society of Architectural Historians*, 20, no. 1 (March 1961), pp. 34–37. Copyright © 1961 by the Society of Architectural Historians, Philadelphia.

Chapter 14: Carl W. Condit, 'The first reinforced-concrete skyscraper: the Ingalls Building in Cincinnati and its place in structural history', *Technology and Culture*, 9, no. 1 (January 1968), pp. 1–33. Copyright © 1968 by the Society for the History of Technology, Chicago. Reproduced by permission of the Johns Hopkins University Press.

Chapter 15: John Snyder and Steve Mikell, 'The consulting engineer and early concrete bridges in California', *Concrete International* (May 1994), pp. 38–44. Copyright © 1994 by the American Concrete Institute, Detroit.

Chapter 16: D.J. Fraser, 'Early reinforced concrete structures in New South Wales, 1895–1915', *Transactions of the Multi-Disciplinary Engineering, Institution of Engineers, Australia*, GE9 (1985), pp. 82–91. Copyright © 1985 by the Institution of Engineers, Barton, ACT

Chapters 1, 3, and 4 are reproduced from the *Transactions of the Newcomen Society* by permission of the Society's Council. The Newcomen Society promotes, encourages and co-ordinates the study of the history of all types of engineering and technology from the earliest times to the present day. Details from the

Executive Secretary, The Newcomen Society, The Science Museum, London SW7 2DD, Tel/Fax 0207–589–1793.

Every effort has been made to trace all the copyright holders, but if any have been inadvertently overlooked the publishers will be pleased to make the necessary arrangement at the first opportunity.

***

Frank Newby would personally like to thank Julia Elton for her encouragement during the preparation of this book, and for her help in the writing of the introductory essay and for her German translations. Without her typing and editing, the task would have been more difficult.

General Editor's Preface

Joyce Brown

Civil engineering has a long history. Although practitioners did not begin to describe themselves as 'civil engineers' until the eighteenth century, the origins of their work lie in the construction skills of the ancient world and in the works of military engineers. Shortly after British civil engineers had formed themselves into a professional body, the Institution of Civil Engineers in London, their role was described in 1828 by Thomas Tredgold in these words:

> Civil engineering is the art of directing the great sources of power in Nature for the use and convenience of man.[1]

He thus neatly encapsulated what the civil engineer's job is about – the provision of the means to give society what it wants for its survival.

As time has gone by, the civil engineer's skills have had to diffuse over an ever-widening field and at the same time become more specialised. In all periods he has had to be the ingenious 'fixer', but more than a mere artisan – a point observed in the Institution's motto, *scientia et ingenio*.

The history of civil engineering is a fascinating one. Certainly 'the great sources of power in Nature' are not always directed without a certain amount of opposition by them, and the provision of the basic requirements – habitation, water supply, main drainage, harbours, bridges, places of worship, transport systems by land and water – has called for inventiveness and sometimes for daring and courage.

Yet interest in the history of civil engineering is not of long standing, and it seems fair to state that most of the important work has been done in the last forty years. With a few notable exceptions, most of the major books on the subject have appeared in that era; many specialist societies with their own transactions have been founded; and many of the journals in which articles might be published have come into existence. While archaeologists and architectural historians have long made their contribution to this subject, their ranks have been swelled by a large number of serious scholars – including engineers themselves – now interested in recording and evaluating the engineering achievements of earlier times. In particular, there is a lively interest in the study of our industrial past, at least in part because some of the evidences of it are already disappearing from the landscape.

Our intention in creating the series of which this volume forms part is to provide, through the reproduction of important contributions to this subject, an

---

[1] *Minutes of Council, Institution of Civil Engineers*, 4 January 1828.

invaluable reference collection for its study. The series encompasses many different branches of engineering from early times to the beginning of the twentieth century; its perspective is global; and the chosen articles have an international authorship, to the extent that this can be achieved for an essentially English-language series. The introduction to each volume has allowed the volume editor to set the pages selected in the context of the whole history of the subject and its historiography, while the provision of references and a bibliography will enable the reader to go further into the study of the topic.

This series will have succeeded if it gives civil engineers some unexpected insights into their own craft and other readers a new way of looking at engineering structures.

***

The volume editors were chosen for their particular knowledge and expertise in the field. They have worked with enthusiasm and good will to put together collections of articles intended to be both informative and stimulating, and I am grateful to them for their efforts. Where possible, they have incorporated the work of the best scholars, while at the same time giving preference to articles often obscurely published and not readily available. The learned journals are represented, as also are conference proceedings and essay chapters contributed to books.

In all of this, I have been much assisted by my Advisory Panel. Emeritus Professor Sir Alec Skempton (Imperial College) has made himself freely available to me for consultation on many aspects of this venture. Mike Chrimes (Librarian of the Institution of Civil Engineers), Dr. Norman Smith (Emeritus Reader, Imperial College) and James Sutherland (formerly a Senior Partner in Harris and Sutherland, London, consulting engineers) have all contributed valuable advice based on their extensive knowledge of particular fields. The idea for the series was developed in discussion with Dr John Smedley of Ashgate, and Publisher of the Variorum Collected Studies Series, who has given me much personal support and the benefit of his experience.

I am grateful also for the help given by the staff of several libraries, notably Kay Crooks and Susannah Parry in the library of the Department of Civil Engineering, Imperial College, and Mike Chrimes and staff of the library of the Institution of Civil Engineers, London. Many friends have given me encouragement and useful advice; in particular, I want to thank Paula Kahn and Mark Baldwin.

Joyce Brown,
Formerly Department of Civil Engineering,
Imperial College, London

Introduction

Frank Newby

Pier Luigi Nervi has described reinforced concrete as 'the most fruitful and generous of all building materials'.[1] Over the last hundred years this remarkable man-made material has transformed the built environment, giving a new scale to architecture and civil engineering. It has a long and fascinating history, beginning with the use of mass concrete when the building industry came to terms with the idea that an initially wet mix of ingredients could slowly harden into a stone-like mass. The addition of metal reinforcement to provide tensile strength, which only occurred in the last quarter of the nineteenth century, produced a new structural material of greater versatility than wrought iron, steel or timber, which, like cast iron, could be moulded into a variety of shapes, but for the first time could also form a continuous monolithic structure of almost limitless size.

Sixteen papers have been selected for reproduction in this volume to give the reader a picture of the emergence of reinforced concrete as a new structural material. Some of the papers naturally overlap in subject while other aspects of reinforced concrete, such as the chemistry of cement, the control of materials and construction on site, or structural failure are not covered. However, it is hoped that the bibliography will supply a list of sources for further study.

Binders, Mortars and Concrete

Concrete is an agglomerate of sand and a coarse material such as gravel or crushed stone which, when mixed with water and a binder, slowly hardens in the atmosphere to form an artificial stone. The earliest binder to be widely used was lime, and concrete made with it is known as lime concrete; lime mortar is a mix of sand and lime and has been used for centuries of masonry construction. Lime concrete and mortar only gain strength slowly, a factor which influences the rate of building, but its great disadvantage is that it will not set under water.

This was the problem encountered by John Smeaton, engineer and scientist, when he was commissioned in January 1756 at the age of 32 to rebuild the Eddystone lighthouse. He had to have an hydraulic mortar, that is one which would set under water, to bind the foundation blocks together. He knew that some

---

[1] Nervi, *Structures*, 49. Nervi, engineer and contractor, was one of the greatest twentieth century designers in reinforced concrete. This is a translation of his 1955 book, *Construire Correttamente*, his most important publication, giving his views on the essential qualities of reinforced concrete obtained over many years of practical experience.

limes had hydraulic qualities and that the addition of Dutch tarass (or trass) to limes would render them hydraulic, so he carried out a series of simple tests on mortars with tarass and with different limes, using both fresh and sea water. He consulted the few available books on engineering and found in Belidor a reference to a statement by Vitruvius that pozzolana, a ground volcanic rock from Naples, also gave hydraulic qualities to lime.[2] It was widely used by the Romans from the first century BC, most spectacularly for the 43-metre diameter dome of the Pantheon in Rome in 128 AD.[3] Fortuitously, Smeaton was able to acquire a quantity of pozzolana and further tests showed him that it was a much better material than tarass. His great discovery was that lime made from Aberthaw limestone, which was already known to be hydraulic, had this property because of a significant clay content.[4] He stopped short of analysing the clay chemically and it was left to Vicat to complete the explanation. The first stones of the lighthouse were bedded on 12th June 1757, some seventeen months after his appointment, but he did not publish details and results of his tests until 1791.[5]

In 1796 James Parker discovered that by calcining particular stones, found on the beaches of Kent and in chalk formations, and grinding the residue to powder, he was able to produce a natural hydraulic binder, which he named 'Roman cement'. Its only drawback was that it set very quickly and was therefore unsuitable for large volumes of concrete, but for mortar it was ideal and was widely used.

The next major step in the understanding of the nature of a binder or cement, as it became known, was carried out by the French engineer, L.J. Vicat, who published his first book on the subject in 1818.[6] His second book of 1828 was translated into English in 1837 by Capt. J.T. Smith,[7] who greatly added to the original in the light of his own experimental work, which also enabled him to explain and evaluate Vicat's conclusions. Smith comments in his preface that, 'It is not merely in the accuracy of the details of his valuable invention that M. Vicat has done the most service to the profession by the publication of his work [but] the variety of other processes which he has illustrated has increased the resources of the practical engineer in every situation'. Vicat's main contribution was to show that the presence of silica was essential to produce an hydraulic

---

[2] B.F. de Belidor, *Architecture hydraulique*, Vol. 2, part 2 (Paris, 1737–53), 86.

[3] Rowland J.Mainstone, *Developments in Structural Form* (London, 1975), 115–18.

[4] Norman Smith, 'Scientific Work', in A.W. Skempton (ed.), *John Smeaton FRS* (London, 1981), chapter 11, 54–7.

[5] Smeaton, *A Narrative of the building...*, 103–23.

[6] Vicat, *Recherches expérimentales...*

[7] Vicat, *A practical and scientific treatise...translated by Captain J.T. Smith.* This translation with its extensive commentary gives a focussed picture of the rapid rise of interest in the make up and properties of various binders or cements at the beginning of the nineteenth century. A key historical reference book.

lime, but so too was aluminium provided silica was already present. Vicat also found that pozzolana was a silicate of aluminium which reacts with lime in the mortar and that tarass was a baked clay of volcanic origin with similar properties to pozzolana. He later went on to produce artificial pozzolana and to publish a book on the subject. By 1840 the French had begun to surpass the British in the quality of lime and cement and produced geological maps to identify sources of suitable materials.[8] In Britain Col. C.W. Pasley of the Corps of Royal Engineers also undertook comprehensive research and tests on various limes, mortars, pozzolanas, and cements including those of Frost, Ranger, Francis and White. He published the results in his authoritative study of 1838.[9]

The first artificial hydraulic binder was produced and patented in 1824 by Aspdin, who called it 'Portland cement'. Davis describes the original method of manufacture and the subsequent efforts over many years to increase its strength.[10] John Grant carried out extensive tests on Portland cement from 1859 before it was used in the construction of Bazalgette's London Outfall Sewer system,[11] which led to its widespread use in this country.

As the use of concrete in England and France grew, the manufacture of cement was changing from a craft to a science-based industry. In chapter 4 A.W. Skempton charts the history of cement, emphasizing strength and methods of testing. Tension tests were carried out on standard briquettes, and 6-inch cubes were used in compression tests, while chemical tests indicated composition. In time an understanding of the nature of cement emerged in England, France and Germany and national standard specifications appeared. Skempton suggests that there were three periods in the development of the strength of cement. The first occurred in England up to 1866, the second took place in Germany from 1875 to 1900, where chemical control and fine grinding were introduced, thus doubling the strength, while the third period (1910–50) was international, when rotary kilns and cement chemistry took strength ever further. It could also be said that the first period related to unreinforced concrete and the exploitation, by architects in particular, of load-bearing concrete walls and fireproof floors, while the second phase covers the development of early reinforced concrete, primarily by engineers, and the third, and possibly the most exciting phase, the innovative use and artistic interpretation of the essential qualities of reinforced concrete by both civil engineers and architects.[12]

---

[8] Guillerme, 'From lime to cement...'.

[9] Pasley, '*Observations on limes*...'. The best of all the early British sources for the history of cements and concrete. Also covers virtually all the foreign contemporary work, including that of Vicat, some of whose findings he disagrees with. In some cases Pasley provides the only known information on the foundations of some major London buildings.

[10] Davis, *A Hundred Years*...

[11] Grant, *Experiments on the Strength of Cement*...

[12] Newby, 'The innovative uses of concrete...'.

The Development of Reinforced Concrete in Europe

The first use in England of mass concrete for building came in 1817 when the foundations of the newly-built Millbank Penitentiary, which were beginning to settle, had to be underpinned. In chapter 1 J.M. Crook describes the investigations into the problem and the views put forward by leading civil engineers. A proposal to use a grouted concrete foundation came from the civil engineer, John Rennie, before Sir Robert Smirke was appointed as architect for the remedial work, but he too came to the conclusion that the walls should be underpinned with concrete using a Dorking hydraulic lime. Smirke subsequently used mass concrete for the foundations of all his major works. According to Pasley, the method of construction varied between building up the concrete in layers of 6 inches to pouring it into the shuttering from a height of 6 feet.[13]

Apart from taking sole responsibility for the structure and foundations of buildings, architects were also concerned with the provision of a fireproof construction. The development of fireproof construction in England is well documented by B.L. Hurst in chapter 2. He outlines the early use of mortar and concrete in floor construction, illustrating systems where concrete slabs were supported on inverted cast-iron Tee-sections or wrought-iron I-sections. In particular he mentions early work by Foulston for suspended floors in Bodmin asylum in 1818, which eventually led to the first patented system by Fox and Barrett in 1844.

Meanwhile, William Boutland Wilkinson of Newcastle-upon-Tyne, a plasterer by trade, was taking out in 1854 the first ever patent for a system of reinforced concrete with iron bars and cables acting as tension members within the concrete. His drawings indicate that he had a clear appreciation of the structural action of reinforced concrete. Joyce Brown in chapter 3 illustrates the patent and gives results of his tests on reinforced beams. One of his small buildings, which survived until 1955, was a cottage built in the Newcastle area in about 1865, which Cassie describes as having 6½ inch deep beams, 26 inches apart, and spanning 14 to 16 feet, with moulded plaster panels forming permanent shuttering for the concrete slab.[14] Unfortunately, Wilkinson was not able to market his system successfully and nothing came of the patent. Joyce Brown concludes, 'Perhaps, like most innovators, Wilkinson was a little before his time and was unable to grasp the full significance of his own discovery.'[15]

In France, the work of François Coignet is of particular interest and is well described by Collins.[16] In 1852, Coignet attempted to build a factory with exposed

[13] Pasley, *Observations on limes...*, 16–18.

[14] Cassie, W. Fisher, 'Early reinforced concrete in Newcastle-upon-Tyne', *The Structural Engineer,* 33 (1955), 137.

[15] Brown, chapter 3, 141.

[16] Collins, *Concrete,* 27–35. Collins' book is a brilliant study of the history and development of concrete and its role in architecture, which has never been superseded.

concrete walls following the ancient construction system known as pisé, in which earth was rammed between temporary timber panels.[17] After overcoming difficulties with mixes and water content on this, his first encounter with concrete, Coignet commissioned an architect to design a sophisticated four-storey house in St Denis, Paris, which imitated stonework. Because of its success he took out a British patent in 1855 entitled *Emploi de Béton*, which gave details of his method of construction and contained an addendum emphasising the complete sufficiency of concrete. 'Concrete walls need no facing materials such as stone etc. The hollow part of the mould in which the concrete is poured should have the form to be given to the mass whether the walls be plain or with projections such as cornices, string-courses...or any kind of ornament'.[18] In a submission to the organising committee of the Paris Exposition of 1855, where he wished to build a concrete house, he makes a prophetic remark, 'The reign of stone in building construction seems to have come to an end. Cement, concrete and iron are destined to replace it.'[19] In 1861 in his book, *Béton Aggloméré,* he wrote, 'Building cheap vaults of unusual spans for great public spaces...will not be a servile imitation of Roman works but will far surpass them in daring, elegance and economy.'[20] He also suggests the use of concrete with an armature of iron rods, but there is no evidence that he carried out any experiments.

Whether or not Coignet had seen, at the 1855 Exposition, Lambot's boat, which was made from a mesh of iron rods encased in a thin layer of hydraulic cement, is not recorded. Joseph-Louis Lambot had built the boat in 1848, and in 1855 he applied for a patent in Brussels covering 'iron cement as a substitute for timber' to be used both in shipbuilding and in structures having to withstand moisture, such as water tanks and tubs for orange trees. The patent also included a method of producing a beam or plank. A further boat built by Lambot was discovered in 1975 and it can be seen in Paris.[21]

The first system of reinforced concrete to progress from a simple idea to a viable system of construction was that of Joseph Monier (1823–1906), who in 1849 made flower pots using a mesh of iron rods enclosed within a mortar in a similar manner to that of Lambot. Monier took out a patent in 1867 and extended it the following year to cover reinforced concrete tanks, the first of which he built in 1872. Monier's first patent for the civil engineering use of reinforced concrete came in 1873 and covered the manufacture of arched bridges and foot bridges of all sizes; the patent included a drawing of an arched bridge with open spandrels. A patent covering beams appeared in 1877, but it lacked any real substance.

---

[17] Collins, ibid., 21.
[18] Translated by Collins, ibid., 28.
[19] Translated by Collins, ibid., 29.
[20] Translated by Collins, ibid., 33.
[21] *Concrete Journal* (February 1977), 24–5.

The development of Monier's patents into a workable and outstandingly successful building system only came about with the involvement of Conrad Freytag, who bought the patent rights for South Germany in 1884. The history of this development as outlined here (which differs from information given by other writers) comes from information produced by Freytag's firm and published in its fiftieth anniversary *Festschrift*.[22]

Freytag's firm had been founded in 1875 in Neustadt an der Haardt under the name, Freytag & Heidschuch, as merchants for building materials, but Freytag increasingly specialized in cement and concrete construction work. On a chance visit to Trier in 1884, he saw a concrete water tank being built on the Bahnhofstrasse on Monier's instructions in order to demonstrate his system in Germany. Freytag, despite his lack of formal technical training, immediately recognized the potential of the system and barely three weeks later, in September 1884, he went to Paris with his friend, Josseaux, partner in the Offenbach building firm of Martenstein & Josseaux, to visit Monier . They saw various structures such as further water tanks and vaults and quickly agreed with him that his German patent would be acquired by Freytag & Heidschuch and by Martenstein & Josseaux. However, Josseaux was not as convinced as Freytag as to the merits of the system and only took the rights for Frankfurt am Main and the surrounding area, leaving the South German rights to Freytag & Heidschuch. The two firms also secured the first refusal for the rights in the remaining parts of Germany.[23] Immediately after the acquisition of the patent Freytag built a dog-kennel (which still exists) to satisfy himself that such a building method was indeed feasible, but the first major structures built were six bleaching tanks for a chemical works in Württemberg, which established the suitability of the material for industrial use.[24]

The next step came with the involvement of G.A. Wayss, who had received a formal engineering education at the Technische Hochschule in Stuttgart. Wayss set up his office in Frankfurt am Main and learnt of the Monier system when Josseaux began building with it there. He approached Freytag & Heidschuch about acquiring the North German rights, which they passed over to him free of charge as they themselves were overwhelmed with work in South Germany. Wayss then acquired the remaining rights for the rest of Germany and transferred his office to Berlin, a more advantageous centre from which to develop and exploit the system. He sold on the rights for Posen and Silesia to the firm of Huber in Breslau. Thus four firms between them now owned the patent rights for the whole of Germany and seem to have worked in close co-operation with one another.[25]

[22] Wayss & Freytag, *Festschrift...1875–1925*, 5–46.
[23] Wayss & Freytag, ibid., 5–6.
[24] Wayss & Freytag, ibid., 9–10.
[25] Wayss & Freytag, ibid., 11–12.

During the winter of 1886–87, Wayss with Freytag & Heidschuch carried out fire tests, reinforcement pull-out tests and load tests in Berlin to prove to the authorities and to the professional world that this was a reliable building method. At the same time Matthias Koenen, a Prussian State architect, produced a structural theory for the performance of Monier floors and vaults. In 1887 his work, together with the results of the tests carried out in Berlin, appeared in *Das System Monier*,[26] which was published by Wayss simultaneously in Berlin and Vienna. It revealed the commercial possibilities of the system, showing multi-storey buildings with a framework of cast iron encased within concrete and with arched floors. Also shown were arched bridges, the thinness of which, when compared with the equivalent stone arches, was remarkable for such a date. The age of modern reinforced concrete construction could be said to begin with the publication of this book.

In 1889 Wayss' firm became a joint stock company, changing its name to 'Aktiengesellschaft für Monierbauten', and in the same year Freytag & Heidschuch issued the first catalogue of executed Monier structures,[27] with an appendix giving the Monier works carried out by Wayss and by Huber. Its appearance heralded a period of expansion, which saw the acquisition of Freytag & Heidschuch by the Aktiengesellschaft and the setting up of satellite companies in Stuttgart and Neustadt, run by Freytag & Heidschuch. In 1892, after a series of technical errors caused by too rapid a growth, Wayss resigned and the company folded. Freytag then bought back the satellite firms and in 1893, together with Wayss, he formed a new firm, Wayss & Freytag, Heidschuch having died in 1891.[28]

Let us now turn to the life and work of François Hennebique (1842–1921), whose achievements are often considered to be superior to those of Monier. He was self-taught, becoming an apprentice stonemason to a builder in Arras. At the age of 25 he set up in business on his own account. The business flourished and he gained experience in carpentry and masonry, particularly in the repair of churches, and became a public works contractor. Collins reports that in 1879 Hennebique used reinforced concrete for the first time for a building in Brussels (though unfortunately he does not give the source of the information).[29] Instead of timber joists, Hennebique used precast concrete beams reinforced with round iron rods to achieve a fireproof construction. From this date onwards it seems that Hennebique, in collaboration with the chief engineer for Belgian roads and bridges, began his research into the design of reinforced concrete beams, and in 1892 he patented his monolithic reinforced concrete system in Belgium and in

[26] Wayss, *Das System Monier.*

[27] Freytag & Heidschuch, *Monier und Beton-Bauwerke.*

[28] Wayss & Freytag, *Festschrift...*, 15.

[29] Collins, *Concrete*, 64.

France, presumably having satisfied himself that under fire conditions the reinforcement and concrete would stay bonded. Certainly his anchoring of the main reinforcement bars by dovetailing their ends and providing stirrups to tie the upper and lower bars in the beam shows a clear understanding of practical problems. Hennebique's firm was originally based in Brussels, but as the success of his system grew he moved to Paris. He ceased to be a contractor and proceeded to open agencies throughout Europe staffed with engineers and architects and associated contractors specially trained by him. His success was phenomenal. For instance, in 1896 alone he was responsible for over 800 building and civil engineering projects throughout Europe.

Wayss & Freytag learned of the Hennebique system during the 1890s when Martenstein & Josseaux built a large factory for the Adler Works in Frankfurt am Main using Hennebique's monolithic framed reinforced concrete construction. Wayss immediately grasped the potential of this type of construction and went to Paris in order to negotiate with Hennebique for the building rights, but was neither prepared to pay a royalty to him nor to allow his firm to become one of Hennebique's concessionaries. Instead, Wayss & Freytag exploited a loophole in the method of calculation in Hennebique's system, which enabled them to evade his patent and to build monolithic structures despite it. Their first major building was a very large eight-storey warehouse in Strasbourg constructed in 1899, a job for which they had been direct competition with Hennebique's Strasbourg agent, Züblin.[30]

While the Monier and Hennebique systems were being developed in Germany and France, Edmond Coignet (1850–1915), the son of François, who trained as an engineer, took out various patents, first in 1890 for reinforced concrete pipes and aqueducts and then in 1892 for beams and later piles. That same year Coignet was awarded the contract to build the Paris main drainage system in reinforced concrete, thereby giving credence to this new material, as Grant had earlier done to Portland cement with the construction of the London main drainage system.

Coignet was appointed as contractor for the showpiece of the 1900 Paris Exposition, the Chateau d'Eau, a huge water sculpture over 150 feet high and 100 feet wide made of reinforced concrete, and he received a Grand Prix and a gold medal. The architect, Boileau, was entrusted with the design of a pavilion which was to represent 'Hennebique reinforced concrete' at the Exposition. Hennebique himself was consultant for the reinforced concrete floors of the Grand Palais and the Petit Palais and was also responsible for the construction of two spiral slab staircases designed by the architect, Girault, which connected the ground and first-floor levels. These were structures of great elegance and

[30] Wayss & Freytag, *Festschrift...*, 20–22.

beautifully expressed the spatial forms made possible by this material. Hennebique himself received a gold medal for his civil engineering structures. The Exposition, which was visited by such engineers as Robert Maillart, revealed for the first time to a wide international audience the great potential of reinforced concrete both in architecture and in construction.[31]

The success of Monier, Coignet and Hennebique inspired many others to take out patents, which usually covered only the geometry of the reinforcing bars. According to Christophe, over 100 different systems were patented and he illustrates them in his book of 1902.[32] Although some of these were on show at the Exposition, only a few of them survived very far into the twentieth century.

In chapter 6 Cyrille Simonnet gives an interesting overview of the sequence of events which led to the development of reinforced concrete as a viable building material. Because it was an entirely new material, the calculation, detailing of reinforcement, specification of materials, design of formwork, mixing of concrete and the supervision of the construction works were beyond the normal experience of an architect, engineer or builder, and Simonnet suggests that these operations could only be managed by new bodies of expertise, which became known in France as *bureaux d'études*. The leading specialist contracting firms, such as Wayss & Freytag and Hennebique's agencies, each had their own *bureau d'études*. Architects could suggest structural forms, but it was the *bureaux d'études* which determined details of the final design. Simonnet also mentions the role of the various periodicals such as *Le Béton Armé* and *Beton und Eisen* in the widespread dissemination of research and practical experience. He finally emphasises the importance of the introduction in 1897 of lectures on reinforced concrete by Charles Rabut at the Ecole des Ponts et Chaussées.

Further information on Hennebique is given in chapter 7 where Gwenaël Delhumeau, who has studied the Company's archives in Paris, suggests that Hennebique set up his journal, *Le Béton Armé*, to keep his agencies and contractors as well as his clients informed of new projects, loading tests, failures, fire tests and related developments, but also to emphasize the superior expertise of his agencies over independent engineers and architects.

After its transition to a Hennebique type of monolithic structure, Wayss & Freytag also expanded enormously and from 1901 established further satellite companies in all the major German cities and as far away as St Petersburg and Buenos Aires. These were staffed by engineers who had first been trained by Emil Mörsch, who in early 1901 had been engaged as head of the technical department, at head office in Neustadt.[33]

---

[31] Exposition Universelle, *Revue technique...*

[32] Christophe, *Le Béton Armé...*

[33] Wayss & Freytag, *Festschrift...*, 30.

In 1902 Wayss & Freytag published Mörsch's book, *Der Eisenbetonbau, seine Anwendung und Theorie*,[34] which gave his structural theory of reinforced concrete and made public all his experimental work for the firm. It went into many updated editions and became the standard work on the subject in the German-speaking world. By 1904, Wayss & Freytag had prepared provisional regulations for the Verband Deutscher Architekten- und Ingenieurvereine, which became the official Prussian regulations. The firm went on to acquire German patent rights for the Considère system (1904), and for Simplex piles (1907) from the Concrete Piling Co. of Philadelphia.

The Development of Reinforced Concrete in the USA

The work carried out in England, France and Germany on cement and concrete was closely observed by engineers in the USA. Jasper O. Draffin reports that cement was first required on a large scale for the construction of canals, notably the Erie Canal of 1817.[35] Natural cements were found in Madison County, NY, and the earliest cement works was established in 1828 near the route of the Delaware and Hudson Canal. Portland cement was first shipped from England in about 1871. Its import increased to a peak in 1896, then fell off as local production was increasingly able to supply the requirements of an expanding construction industry at a much-reduced cost per barrel. As in Europe, the USA prepared its own regulations for the production of cement and concrete.

In the early 1870s, William E. Ward (1821–1900), who had become interested in the possible use of reinforced concrete for fireproof construction, carried out tests before constructing a fireproof concrete house in 1876 in Port Chester with the architect, Robert Mock. The building, which was test loaded, is still extant and details of it are given in chapter 13 by Ellen W. Kramer and Aly A. Raafat. In 1883 Ward was persuaded to give a paper to the American Society of Mechanical Engineers on his work, which he concludes by hoping that reinforced concrete would lead to a system of fireproof construction which would merit public confidence and be universally accepted.[36]

In 1875, while Monier was still developing his system, Thaddeus Hyatt (1816–1901) went to London from New York to promote his iron and glass block side-walk construction. While in London, he designed a building for which he devised a fireproof system consisting of upright wrought-iron plates with holes

---

[34] Mörsch, *Der Eisenbetonbau...*

[35] Jasper O. Draffin, 'A brief history of lime, cement, concrete, and reinforced concrete' (1943), 14–47. Reproduced in Newlon (see note 41 below).

[36] W.E. Ward, 'Béton in combination with iron as a building material' (1882–83). Reproduced in Newlon (see note 41 below).

through which lateral round bars were placed. The whole framework was then enclosed within a thin concrete slab. This so-called 'slab ceiling' spanned between the bottom flanges of I-section wrought-iron beams, which themselves had a separate cover of concrete. He next changed the I-section beam to an inverted Tee-section, using concrete to carry the beam compression. The inverted Tee-section was subsequently replaced by round bars, and thus reinforced concrete beams were born.

Hyatt carried out tests to confirm that iron, steel and concrete had compatible coefficients of expansion so that under fire conditions the material would not disintegrate. Tests were carried out by David Kirkaldy in London on 5ft span beams with different forms of reinforcement and the results were included in Hyatt's book of 1877.[37] Hyatt became over-confident in his forecast for the future of the material, offering to build reinforced concrete domes from 2ft. to 200ft. diameter which incorporated his glass lenses. Alas, he failed to obtain building permission and returned to New York. He retired from his work with reinforced concrete after patenting his system in 1878.

The first major builder in reinforced concrete in the United States was Ernest Ransome (1844–1917). He went from England to the USA in 1870 to become superintendent at the Pacific Stone Co. in San Francisco, where he stayed for four years before setting up as an independent contractor. Ransome gives a personal reminiscence in his book of 1912[38] and details his experience in the construction of some of the earliest reinforced concrete buildings in the USA. He patented the twisted square reinforcing bar in 1884, and in 1889 introduced the economic ribbed floor construction which is still used today. Carl Condit in chapter 14 provides many details of Ransome's buildings and gives many *Engineering Record* references to Ransome's work.

In chapter 15 John Snyder and Steve Mikesell suggest that the work of Hyatt and Ransome influenced the use of reinforced concrete in California. They refer to the earliest buildings designed by Ransome with the architect, George Percy, which introduced reinforced concrete and fireproof construction to the local building industry. They also illustrate the first reinforced concrete bridge in America, designed by Ransome in 1889, which provided the model for many other small-span bridges, which continued to be built long after Ransome had moved his office to Chicago in the mid-1890s.

Ransome remained the most prominent contractor until the late 1890s, by which time the Monier and Hennebique systems had become established. According to Draffin,[39] Hennebique's system was the most widely used in

[37] Hyatt, *An Account of some Experiments...*

[38] Ransome and Saurbrey, '*Reinforced Concrete Buildings*'. The first chapter, headed 'Personal reminiscence', details the life and early work of Ransome and is reproduced in Newlon (see note 41 below).

[39] Draffin (see note 35), 41.

America. However, the tightly-controlled single system as experienced in Europe seems to have had only a brief life in America. The structure of the 16-storey Ingalls Building of 1902, the subject of Condit's paper, was made up from a combination of the systems of Ransome, Monier, Hennebique and Considère. The Ferro-Concrete Co. of Cincinnati were the independent engineers and contractors while the architect was local.

The American Concrete Institute was set up in 1905 in order to control the growth of construction in reinforced concrete but primarily to formulate regulations. Condit comments that, 'The primary document in the history of concrete construction during the twentieth century is the *Report* of the Joint Committee of the American Concrete Institute and the American Society of Civil Engineers (1909) on the fundamental standards of reinforced concrete design. This report became the basis of all American metropolitan building codes and of much subsequent development in the theory of elasticity of reinforced concrete and the stress analysis of concrete structures.'[40]

The next significant system of reinforced concrete construction was developed in America as against Europe. The Ingalls building had the standard two-way framework of beams and columns which echoed iron and steel construction, but in 1903 A.V. Turner concluded, at least theoretically, that he could do away with beams and use a flat slab supported directly on columns. He applied for a patent in 1905, building his first example in 1906. However, in 1902 Orland Norcross, an engineer from Boston, had patented and built a similar flat-slab system and Turner had to overcome the problem of infringement. Turner's patent was accepted in 1908 and he proceeded to develop his flared column capital system. Details of later tests by Arthur Lord on flat-slab construction in 1911 and the subsequent discussion on a structural design method are both included in an American Concrete Institute volume reprinting a selection of historic papers.[41]

The Structural Design of Reinforced Concrete

In chapter 5 De Courcy discusses the work of Wilkinson and Coignet, as the first patentees of the idea of reinforced concrete, although neither suggested any method of structural design. He also mentions the fireproof house built by William Ward, and comments on Ward's understanding of the fundamentals of reinforced concrete although he too did not provide any calculations. De Courcy is more admiring of Hyatt's work and his proposals for the structural design of beams based on an ultimate load method, though in fact these proved incorrect as he placed the neutral axis (the line or surface of zero stress) at mid-height of the

[40] Condit, *American Building Art*, 346.

[41] Newlon (ed.), *A Selection of Historic American Papers...*

beam. Hatt in chapter 12 comments similarly on Hyatt's work. However, both De Courcy and Hatt agree that the first structural design method was that of Matthias Koenen, which had appeared in 1887 in Wayss's *Das System Monier*, as mentioned earlier. Koenen also put the neutral axis at the mid-height of a beam, but assumed a triangular distribution of compression stress above the neutral axis with all tension forces taken by the reinforcement. De Courcy (chapter 5) illustrates Koenen's method, which was apparently used by designers for many years, as does Hatt in chapter 12. Hatt also illustrates the method used by Hennebique, which he considers to be less correct than that of Koenen. However, the factors of safety were so high that these errors in calculations would not lead to failure. What this shows is that there was then very little understanding of how this, the first composite structural material, really acted in bending.

During the 1890s Paris had become an active centre of research and many structural design proposals were produced. Hamilton illustrates eight rival theories, including one from Edmond Coignet and Napoléon de Tédesco, on the distribution of stress across the section of a reinforced concrete beam, most taking into account that concrete could carry some tension and others assuming parabolic or non-linear compression distribution above the neutral axis.[42]

In 1894 Coignet and Tédesco, who was co-director of Coignet's *bureau d'études*, had correctly proposed that the moduli of elasticity of both steel and concrete should be taken into account to position the neutral axis, that a triangular distribution of compression stress be assumed above the neutral axis and that no tension stresses were taken by the concrete.[43] This method of calculation became generally accepted as giving a good approximate solution.

The extent of the research work can be seen in Charles Marsh's book of 1904, *Reinforced Concrete,* which was the first textbook to appear in England. A much enlarged and rewritten third edition came out in 1906.[44] It includes the work of Considère with his patented system of spiral reinforcement for compression members such as columns and upper chords of trusses, and there is an extensive section dealing with calculation, which shows that the discussion of the distribution of stress through a beam was still very much alive in 1906.

Marsh also gives the Hennebique method of calculation,[45] which could well be the source of Hatt's information given in chapter 12. If Marsh is correct, it is surprising that by 1906 Hennebique had not amended his method to tie in with

---

[42] Hamilton, *A Note on the History of Reinforced Concrete...* This is the first and best study of the history of reinforced concrete. Hamilton covers the period up to 1900 outside Britain, and in Britain up to 1944.

[43] Coignet and Tédesco, *Du calcul des ouvrages...,* 282–363.

[44] Marsh and Dunn, *Reinforced concrete*, 3rd edn. (London, 1906). A study of the technical changes between this and the first edition of 1904. It highlights the research and building works being carried out in the early 1900s.

[45] Marsh and Dunn, ibid., 383–6.

the Coignet and Tédesco proposal of 1894. There seems to have been a certain amount of secrecy surrounding Hennebique's tests and method of design, though it is possible that information relating to this important aspect of his work might exist somewhere in the Hennebique archives in Paris.

In chapter 7 Delhumeau describes how the information gained from the building and demolition of the various reinforced concrete systems used at the 1900 Paris Exposition gave the Commission du Ciment Armé, chaired by Armand Considère, sufficient data to produce their official regulations of 1906, making it possible for independent engineers to design in reinforced concrete. Collins refers to the regulations of 1906 thus, 'By this act, the long efforts to create a new building material were given their final seal of recognition and from then onwards concrete was no longer a hazardous novelty, but an acknowledged medium of architectural design in the modern world.'[46] Hatt, who carried out a great deal of research on reinforced concrete from the early 1900s at Purdue University, rounds off his paper (see chapter 12), written in 1918, with an accolade on the early constructors. 'They showed a general knowledge of the strains to be resisted and a mechanical skill in meeting these strains with economical design that cannot but excite the admiration of present day constructors.'[47]

However, the structures referred to in the various building regulations were of a simple nature, namely beams, slabs and columns. The original research work in France and Germany was augmented by work in the USA and, as previously mentioned, it was there that a new structural form, the flat slab, appeared, and where methods of structural design were later established. With the growing understanding of the characteristics of reinforced concrete, national standards were, and still are, being constantly updated.

Early Use of Reinforced Concrete in Australia

In order to illustrate the extraordinary rapidity with which reinforced concrete construction reached distant parts of the world, I have included D.J. Fraser's 'Early reinforced concrete in New South Wales (1895–1915)', chapter 16. The Monier system was introduced to Australia in the 1890s by a draughtsman/engineer by the name of W.J. Baltzer, who had been educated in Germany and who had made a report on the use of concrete there, taking photographs of Monier arched bridges, during a visit home. When he returned to Sydney he was able to arouse the interest of his employers, the contractors Carter Gummer & Co., who subsequently became agents for the Monier system. In order to test it they built a small experimental culvert, which was loaded to failure. They went

[46] Collins, *Concrete,* 75.
[47] Hatt, chapter 12, 34.

on to build two sewer aqueducts near Sydney in 1897, for which they had to pay a royalty of 15 per cent. However, these aqueducts were so successful, both practically and financially, that the Monier system thrived, becoming the prime method of building culverts, arched bridges and water tanks, first in New South Wales and then in other parts of Australia.

In 1906 A.V. Turner applied for a patent covering flat-slab construction with mushroom heads in Australia and carried out his first contract there in 1909, while Truscon, again from the USA, introduced the Kahn bar. The innovative use of reinforced concrete can be seen in the design of the wool store in Geelong of 1911–12 by E.G. Stone, engineer, where large latticed arched roof trusses span about 180ft.[48] The upper chords of the arches were designed on the Considère system and the roof was precast. Stone followed this in 1914–15 with a kind of Forth Bridge construction to carry the Barwon sewer aqueduct over the local estuary.[49] By 1915 reinforced concrete and the method of structural design were generally accepted and there was a ready market for a method of construction which at that time was cheaper than the traditional imported steel-frame construction.

The Early Use of Reinforced Concrete in Britain

Although Wilkinson had patented a system as early as 1854, reinforced concrete was not used again in Britain until 1897, when Hennebique's Nantes agency successfully tendered for the construction of a mill for Messrs. Weaver in Swansea, which was to have a fully-framed structure. Calculations and detailed drawings were produced in Nantes and all materials and some workmen came from there.

For this and later British contracts, Hennebique appointed L.G. Mouchel as his agent. In chapter 8 Patricia Cusack describes how Hennebique set up his British organization, training his staff and contractors as he had done in France. She notes that the majority of projects were for civil engineering or for industrial buildings, where it was not difficult to get building permission, or for railway structures, which were free of regulations. In 1909 Mouchel published a book on his organization which included a list of contracts carried out to that date.[50] He, like Hennebique, published his own journal, *Ferro-Concrete*, which first appeared in 1909. Similarly, the Yorkshire Hennebique Company in *ca.*1910 published details of their contracts, which included the outstanding Lion Chambers in Hope Street, Glasgow of 1908 and the 17-storey Royal Liver Building in

---

[48] Lewis, *Two Hundred Years of Concrete...*, 17–21.

[49] Lewis, ibid., 30–31.

[50] Mouchel & Partners, *Hennebique Ferro-Concrete...*

Liverpool, begun in 1909.[51] Cusack's main conclusion is similar to that of Simmonet (chapter 6) that 'the early controlled conditions of its employment in Britain, largely by Mouchel's company and Mouchel's personal dedication to the work, ensured that in a period of rudimentary knowledge about reinforced concrete, Mouchel-Hennebique design and construction was enabled to adhere to practices that worked'.[52]

Other French engineers such as Paul Cottancin, who had been engineer for the San Marino pavilion at the 1900 Paris Exposition, later introduced their systems into Britain. Cottancin's patent involved reinforcing thin brick walls and concrete slabs with a mesh of continuously woven wire, and in 1902, when the tender price for a new Methodist church in Exeter came in too high, an English architect resident in Paris suggested that Cottancin be invited to quote. He gave a satisfactory price, but rearranged the structure to suit his system and persuaded the Exeter authorities to accept this solution. He proceeded to produce a structure with exceptionally slender members, which is described by G.J. Edgell in chapter 9. Edgell also includes a description of another extraordinary church, St Jean de Montmartre in Paris, built in 1903. Both these buildings are still extant, though some renovation has been necessary. They are examples of structural forms which can be achieved using a new material before regulations impede innovation, and Edgell writes, 'What is to be learnt from Cottancin is encapsulated in a quotation from *The Builder* in 1905, referring to St Jean de Montmartre, "The church is certainly one that deserves personal inspection and although our readers may not be anxious to emulate the example presented by St Jean de Montmartre they may learn from this building how great is the adaptability of reinforced construction and how useful is the aid offered by the system to the architect who is bold enough to break away from stereotyped methods of realising his designs."' Further information on the chequered history of the design and construction of this church is to be found in Collins.[53]

In addition to the systems which came from France, Trussed Concrete (Truscon), which used the Kahn bar, and the Bonna or Columbian floor system came from the USA. These and many more are illustrated in detail in chapter 10 by Michael Bussell, who highlights the great variety of patented shapes of reinforcing bars. Bussell also draws attention to the appearance of hollow block floors, which became one of the standard economic methods of construction. As in France, the more bizarre systems quickly disappeared, particularly after simple round bars were shown to have a satisfactory bond with the concrete. However, all types of floors are found during refurbishment of early twentieth-century

[51] Yorkshire Hennebique Co., *Views of works executed in Hennebique's patent ferro concrete by the Yorkshire Hennebique Contracting Co. Ltd. Leeds* (Leeds, *ca.*1910).

[52] Cusack, chapter 8, 13.

[53] Collins, *Concrete*, 115–17.

construction and knowledge of their characteristics is essential for an assessment of their capability.

In another contribution to this subject, Patricia Cusack in chapter 11 considers the problem of architects having to work within the restraints of a system or to cope with the whims of the specialist engineer. She describes how the Royal Institution of British Architects set up committees to study the material and to propose recommendations for its use by architects, who were particularly enthusiastic about its fireproof qualities. Edwin Sachs, who established the British Fire Prevention Committee in 1897, was in the forefront of architects critical of the stranglehold imposed by specialist design and construction engineers. In 1906 Sachs became editor of the successful periodical *Concrete and Constructional Engineering*, which was the main source of information for the building industry in Britain. An initial lack of interest in reinforced concrete by the Institution of Civil Engineers resulted in the setting up in 1908 of the Concrete Institute as a venue for the various professionals to come together and discuss and promote reinforced concrete. Anita Witten has given further details of the Institute's composition and work,[54] noting for instance its major role in the compilation of the first regulations issued in 1915 by the London County Council, by which time reinforced concrete was finally established as a viable, economic and exciting form of construction.

Towards the Great Age of Concrete Engineering

The successful demonstration of the potential of reinforced concrete at the Paris Exposition of 1900 had a profound influence on architects, engineers, academics and on potential future clients. Two outstanding civil engineers emerge from this period, Robert Maillart (1872–1940), and Eugène Freyssinet (1879–1962). After graduating from the Eidgenossische Technische Hochschule in Zurich, and gaining experience in concrete arch building while working with various Swiss civil engineering firms, Maillart set up his own design and construction firm in 1902.[55] In 1905 he built the Tavanasa bridge, which spanned 170 feet and was the first to have all the design features of what became the typical Maillart bridge. It was a three-hinged arch made up from the curved lower slab, the upper road slab and two vertical infill walls to produce a partial box construction of minimum material and maximum elegance, quite different in appearance from the heavy arched construction of Monier and Hennebique. In 1909 he built and tested beamless floors quite independently and without knowledge of the similar work

[54] Anita Witten, 'The Concrete Institute' in *Historic Concrete. Proceedings of the Institution of Civil Engineers: Structures and Buildings* (August/November 1996), 470–80.

[55] Billington, *Robert Maillart's Bridges,* 16–29.

being carried out by Turner in the USA. However, his system of reinforcement, simpler than Turner's, showed his better understanding of the structural action of such floors. He went on to design many further innovatory structures as his appreciation of the material slowly matured.

Freyssinet studied under Rabat in Paris and at the age of 26 joined a provincial government department as a civil engineer concerned with bridge construction.[56] In 1910 he designed the Verdre bridge with a central span of 240 feet and two side spans each of 225 feet, each of three-hinged arch construction. The remarkable feature of this bridge was his use of horizontal hydraulic jacks, which were inserted at the crown of each arch and stressed in order to lift the bridge off its centring. However, after a few years, when he found that there was substantial settlement at the crowns, he simply reinserted the horizontal jacks in order to bring the bridge back to its original profile. Because of this, he realized that the concrete in the arches must have shortened under constant compression. This was the first definitive evidence of what became known as plasticity or creep, a phenomenon which had already been investigated but the importance of which had not been fully appreciated.[57]

Years later, in 1955, the great engineer, P.L. Nervi, despite all his achievements and experience in reinforced concrete, admitted that there was still much to be understood about this extraordinary material. He states this with elegance and clarity in the passage from which came the opening quotation of this introduction, 'If the technique of reinforced concrete is to solve the great problems of tomorrow, we must tackle and understand the subject of concrete plasticity and of its viscous behaviour under load and in time. What is today an element of uncertainty will then become an essential factor in the successful realization of those daring structures awaiting construction in the most fruitful and generous of all building materials.'[58]

Bibliography

Cement and Concrete

Collins, Peter, *Concrete. The Vision of a New Architecture. A Study of Auguste Perret and his Precursors* (London, 1959).
Davis, A.C., *A Hundred Years of Portland Cement, 1824–1924* (London, 1924).
Francis, A.J., *The Cement Industry, 1796–1914: A History* (Newton Abbot, 1977).

[56] Ordonez, *'Eugène Freyssinet'*, 20.
[57] Marsh and Dunn, *Reinforced Concrete...*, 223–6.
[58] Nervi, *Structures*, 49.

Gillmore, Q.A., *Practical treatises on Limes, Hydraulic Cements, and Mortars. Containing reports of numerous experiments conducted in New York City, during the years 1858–1861, inclusive* (New York, 1863).

Grant, John, *Experiments on the Strength of Cement, chiefly in reference to the Portland Cement used in the Southern Main Drainage works* (London, 1875).

Guillerme, André, 'From lime to cement: the industrial revolution in French civil engineering (1770–1850)', *History and Technology*, 3 (1986), 25–85.

Hassenfratz, J.H., *Traité théorique et pratique de l'art de calciner la pierre calcaire, et de fabriquer toutes sortes de mortiers, ciments, bétons, etc., soit à bras d'hommes, soit à l'aide de machines* (Paris, 1825).

Higgins, Bryan, *Experiments and observations made with the view of improving the art of composing and applying calcareous cements, and of preparing quick-lime: theory of these arts; and specification of the author's cheap and durable cement for building, incrustation, or stuccoing, and artificial stone* (London, 1780).

Pasley, C.W., *Observations on limes, calcareous cements, mortars, stuccos, and concrete, and on puzzolanas, natural and artificial; together with rules deduced from numerous experiments for making an artificial water cement, equal in efficiency to the best natural cements of England, improperly termed Roman cement; and an abstract of the opinions of former authors on the same subjects* (London, 1838).

Potter, Thomas, *Concrete: its use in building and the construction of Concrete Walls, Floors etc.* (London, 1877; followed by a 2nd edn. in 1891 and a 3rd edn. in 1908, each substantially rewritten to take account of new developments).

Smeaton, John, *A Narrative of the building and a description of the construction of the Edystone Lighthouse with stone* (London, for the Author 1791).

Vicat, L.J., *Recherches expérimentales sur les chaux de construction, les bétons et les mortiers ordinaires* (Paris, 1818).

—, *Résumé des connaissances positives actuelles sur les qualités, le choix et la convenance réciproque des matériaux propres à la fabrication des mortiers et ciments calcaires; suivi de notes et tableaux d'expériences justificatives* (Paris, 1828).

—, *A practical and scientific treatise on calcareous mortars and cements, artificial and natural...Translated with the addition of explanatory notes, embracing remarks upon the results of various new experiments by Captain J.T. Smith* (London, 1837).

Reinforced Concrete

Adams, Henry, and Matthews, Ernest R., *Reinforced Concrete Construction in Theory and Practice* (London, 1911).

Atlas Portland Cement Co., *Reinforced Concrete in Factory Construction* (The Atlas Portland Cement Co., New York, 1907).

Badovici, Jean, *Grandes Constructions. Béton Armé-Acier-Verre* (Paris, nd. but *ca.*1925).

Banham, Reyner, *A Concrete Atlantis* (London, 1986).

Bay, H., *Emil Mörsch* (Dusseldorf, 1990).

Bennett, T.P., *Architectural Design in Concrete* (London, 1927).

Berger, C., and Guillerme, V., *La Construction en ciment armé. Applications générales, théories et systèmes divers* (Paris, 1902).

Beton und Eisen (ed. F. von Emperger). Periodical commencing 1901.

Bill, Max, *Robert Maillart: Bridges and Constructions* (New York, 1969).

Billington, David, *Robert Maillart's Bridges. The Art of Engineering* (Princeton, 1979).

—, *Robert Maillart and the Art of Reinforced Concrete* (Cambridge, Mass., 1990).

Boitel, Ch., *Les Constructions en Béton Armé. Calculs de résistance – Détails d'exécution* (Paris, 1899).

British Reinforced Concrete Engineering Co., *Applications of the B.R.C. System of Reinforced Concrete Construction* (B.R.C., London, 1917).

Buel, A.W., and Hill, C.S., *Reinforced Concrete*, 2nd edn. (New York, 1906).

Christiani & Neilsen, *Twenty-five years of Civil Engineering, 1904–1927* (Copenhagen, 1929).

Christophe, Paul, *Le Béton Armé et ses Applications,* 2nd edn. (Paris, 1902).

Coignet, Edmond, and Tédesco, N. de, *Du calcul des ouvrages en ciment avec ossature métallique* (Paris, 1894).

Coignet, Edmond Ltd., *Reinforced Concrete Systems* (Coignet, London, nd. but *ca.*1911).

Coignet, Edmond & Co., *Reinforced Concrete Construction* (Coignet, London nd. but *ca.*1913).

Coignet, François, *Bétons agglomérés appliqués à l'art de construire* (Paris, 1861).

Collins, Peter, *Concrete. The Vision of a New Architecture. A Study of Auguste Perret and his Precursors* (London, 1958).

Commission du Ciment Armé, *Expériences, rapports et propositions. Instructions Ministérialles relatives à l'emploi du béton armé* (Paris, 1907).

[Commission du Ciment Armé], *The Properties and Design of Reinforced Concrete. Instructions, authorised methods of calculation, experimental results and reports by the French Government Commissions on reinforced concrete* (London, 1912).

Concrete and Constructional Engineering. Periodical commencing 1906.

Condit, Carl W., *American Building Art. The Twentieth Century* (New York, 1961).

Considère, Armand, *Experimental Researches on Reinforced Concrete,* 2nd edn. (New York, 1906).

—, *The Considère System of Reinforced Concrete Design* (Considère, London,1912).

Emperger, F. von, *Handbuch für Eisenbetonbau* (Berlin, 1908 and later editions).

Expanded Metal Co., *Some particulars concerning 'Expanded Metal,' its production and uses in fire-proof and other building constructions. Also for agricultural, farming, horticultural, hardware, and like purposes. Further embracing some extracts from reports, tests, opinions, tables, &c., relating to the employment of the metal for structural purposes* (The Expanded Metal Co., London, 1897).

Expanded Metal and Corrugated Bar Co., *Corrugated Bars for Reinforced Concrete* (Expanded Metal & Corrugated Bar Co., St Louis, Missouri, 1906).

Exposition Universelle (Paris), *Revue technique de l'Exposition Universelle de 1900 par un Comité d'Ingénieurs, d'Architectes, de Professeurs et de Constructeurs. Première Partie. Tome II. Architecture et Construction* (Paris, 1900).

Faber, Oscar, and Bowie, P.G., *Reinforced Concrete Design* (London, 1912).

Freytag & Heidschuch, *Monier und Beton-Bauwerke* (Vienna (?), 1889). With appendices of Wayss's and of Huber's works.

Guidi, Camillo, *Le Costruzioni in beton armato* (Turin, 1906).

Hagermann, G. et al., ed., *Vom Caementum zum Zement* (Wiesbaden, 1964).

Hamilton, S.B., *A Note on the History of Reinforced Concrete in Buildings* (London, 1956).

[Hennebique, François], *Le Béton Armé. Organe des agents et concessionnaires du système Hennebique* (Maison Hennebique, Paris, periodical commencing 1898).

Hoffman, Richard, *Die Eisenbetonliteratur bis Ende 1910* (Berlin, 1911).

Hyatt, Thaddeus, *An Account of some Experiments with Portland-Cement-Concrete combined with Iron, as a building material, with reference to economy of metal in construction, and for security against fire in the making of roofs, floors, and walking surfaces* (Printed for private circulation, London, 1877).

Institution of Civil Engineers, *Committee on Reinforced Concrete. Interim Report on Reinforced Concrete* (Institution of Civil Engineers, London, 1910).

—, *Historic Concrete. Proceedings of the Institution of Civil Engineers: Structures and Buildings* (Special issue, August/November 1996).

Lewis, Miles, *Two Hundred Years of Concrete in Australia* (Sydney, 1988).

London County Council, *Reinforced Concrete Regulations* (London County Council, London, 1915).

Marsh, Charles F., *Reinforced Concrete* (London, 1904).

Marsh, C.F., and Dunn, W., R*einforced Concrete,* 3rd edn. (London, 1906).

Melan, J., *Plain and Reinforced Concrete Arches* (New York, 1915).

Mörsch, Emil, *Der Eisenbetonbau. Seine Theorie und Anwendung* (Konrad Wittwer for Wayss & Freytag, Stuttgart, 1902 and later editions).

—, *Concrete-Steel Construction (Der Eisenbetonbau)* (New York and London, 1909; translated from the 3rd (1908) edn., revised and enlarged by E.P. Goodrich).

—, *Applications of Reinforced Concrete, Historical* (New York, 1910).

Mouchel, L.G. & Partners, *Mouchel-Hennebique Ferro-Concrete* (L.G. Mouchel, London, 4 editions, 1909–1921).

—, *Hennebique Ferro-Concrete Theory and Practice. A Handbook for Engineers and Architects* (L.G. Mouchel, London, 1909).

—, *Ferro-Concrete. A monthly review of concrete and steel construction in engineering and architectural practice* (London, periodical commencing July 1909).

—, *Mouchel-Hennebique Ferro-Concrete. List of Works 1897–1919* (L.G. Mouchel, London, 1920).

Nervi, Pier Luigi, *Structures* (New York, 1956).

Newby, Frank, 'The innovative uses of concrete by engineers and architects', in *Historic Concrete. Proceedings of the Institution of Civil Engineers: Structures and Buildings* (Special issue, August/November 1996), 264–82.

Newlon, Howard, ed., *A Selection of Historic American papers on Concrete 1876–1926* (American Concrete Institute, Detroit, 1976).

Onderdonk, Francis S., *The Ferro-Concrete Style: Reinforced Concrete in Modern Architecture* (New York, 1928).

Ordonez, J.A.F., *Eugène Freyssinet* (Barcelona, 1979).

Patent Indented Steel Bar Co., *Indented Steel Bars for Reinforced Concrete Construction* (Patent Indented Steel Bar Co., London, 1907).

Ransome, Ernest L., and Saurbrey, Alex, *Reinforced Concrete Buildings: A treatise on the history, patents, design and erection of the principal parts entering into a modern reinforced concrete building* (New York, 1912).

Rings, Frederick, *Reinforced Concrete Bridges* (London, 1913).

Royal Institute of British Architects, *Report of the Joint Committee on Reinforced Concrete, Journal of the Royal Institute of British Architects,* 3rd series,14 (London, 1907).

—, *Second Report of the Joint Committee on Reinforced Concrete* (Royal Institute of British Architects, London, 1911).

Schmidt, H., ed., *Zur Geschichte des Stahlbetonbaus – Die Anfange in Deutschland 1850 bis 1910* (Berlin, 1999).

Tédesco, N. de, and Forestier, Victor, *Recueil de types de ponts pour routes en ciment armé calculés conformément à la Circulaire Ministérielle du 20 Octobre 1906* (Paris, 1907).

—, *Manuel théorique & pratique du constructeur en ciment armé avec une note sur le calcul des arcs par M. Henry Lossier* (Paris, 1909).

Trussed Concrete Steel Co., *Handbook of Instructions regarding materials and workmanship for the Foreman and the Clerk of Works. The Kahn system of reinforced concrete* (Trussed Concrete Steel Co., London, 1912).

Turner, C.A.P., *Concrete Steel Construction: Part 1. Buildings. A practical treatise for the constructor and those commercially engaged in the industry* (Minneapolis, 1909).

Twelvetrees, W. Noble, *Concrete-Steel. A treatise on the theory and practice of reinforced concrete construction* (London, 1905).

Warren, F.D., *A Handbook on Reinforced Concrete for architects, engineers, and contractors* (New York, 1906).

Wayss, G.A., *Das System Monier (Eisengerippe mit Cementumhüllung) in seiner Anwendung auf das gesammte Bauwesen* (G.A.Wayss, Vienna and Berlin, 1887).

Wayss & Freytag, *Bauten in Stampfbeton, Monierbeton und Moniermauerung. Fabrikation von Monier- und Betonröhren* (Berlin, 1895).

—, *Festschrift aus Anlass des fünfzigjährigen Bestehens der Wayss & Freytag A.G., 1875–1925* (Stuttgart, 1925).

1

Sir Robert Smirke: a pioneer of concrete construction

J. Mordaunt Crook

Today Sir Robert Smirke (Plate V) is remembered as a Greek Revivalist, the architect of the British Museum. But his reputation as the most successful architect of the Regency period was due to his proven ability as business man and builder rather than to his popularity as a stylist. Born in 1780, knighted in 1832, he remained throughout a long life (he died in 1867) an essentially eighteenth century figure, independent of architectural competitions, basing his enormous practice both public and private on an interlocking network of Tory patronage. His most powerful patrons were the Earl of Lonsdale, Earl Bathurst and Sir Robert Peel. Thus, in 1815, he joined John Nash and (Sir) John Soane as a member of the triumvirate of government architects attached to the Office of Works: the three architect-princes of the Regency era. But besides being a fashionable stylist, an Establishment architect par excellence, Smirke was a pioneer of cast iron and concrete construction at a time when the functions of architect and engineer were not readily distinguishable. His status as an innovator rests not only on his use of concrete for foundations, cast iron for beams, girders and braces, and wrought iron for ties, joists and supports, but upon his interest in heating and ventilation,[1] his use of novel measuring techniques and his employment of quantity surveyors to ensure accurate accounting.[2] His importance as a constructor lay less in invention than in publicity: in his use of novel methods and materials in several widely publicised commissions. His influence on contemporary constructional technique, as in matters of style, was enormous. "Mr. Smirke," remarked one critic, "is pre-eminent in construction: in this respect he has not his superior in the United Kingdom."[3]

This Paper trespasses on dangerous ground: it is an historian's attempt to assess Smirke's significance as an engineer. Although it deals only with Smirke's use of concrete foundations, it could never have been written without generous help, notably that of Professor A. W. Skempton. Smirke may or may not have been the first person in this country to combine the use of concrete foundations with the use of cast iron beams. But if the data collected here has any value at all, he must surely be regarded as a key figure in the development of modern structural techniques in England. The evidence has been divided into four sections dealing with the construction of Millbank Penitentiary, the state of contemporary opinion as to the use of concrete, the development of the Savoy Precinct and Smirke's later use of concrete foundations.

---

[1] Smirke used several ephemeral patent processes: Perkins' system in the British Museum print room, bird room and reading rooms, in the Custom House Long Room, in his own office at 5 Stratford Place, and in his temporary Houses of Parliament. He used Price and Manby's system in the Elgin and Egyptian galleries, and Stratton and Oldham's at the General Post Office. (P.R.O. Works 4/26, f. 290; 1/23, f. 398; 1/24, f. 197; *Arch. Mag.*, 1835, II, 137; 1836, III, 91; 1837, IV, 161, 263). C. J. Richardson's *Popular Treatise on Warming and Ventilation* (1837) was dedicated to Smirke. For Perkins' system, cf. *Repertory for Patent Inventions*, March 1832; J. C. Loudon's *Gardener's Magazine*, VIII, p. 292.

[2] *Journal of the R.I.B.A.*, 1867, p. 205; *The Builder*, 1847, V, 23; W. Papworth (ed.), *Dictionary of Architecture* (1852–92): "Measuring."

[3] *The Athenaeum*, 1828, p. 29. "He was a first class *builder architect*, . . . no building of his ever showed flaw or failing and . . . he was often called upon to remedy the defects of his brother artists." (J. Fergusson, *Observations on the British Museum, National Gallery and National Record Office*, 1849, p. 10.)

SIR ROBERT SMIRKE: A PIONEER OF CONCRETE CONSTRUCTION

I. Millbank Penitentiary

The frequency with which Smirke was invited to underprop a building in a state of "rigor mortis" won him from J. W. Croker's lips the title of "the Dr. Baillie of architects." This was his function at the County Courts, Carlisle (1810–12), Millbank Penitentiary (1817–22), and the London Custom House (1825–27). At Carlisle he inherited a building already "in a falling state . . . under circumstances which made it out of the question to ask for an estimate." Designed by Telford in 1807, and incompetently executed by the county surveyors John Chisholme and Peter Nicholson, the piling foundations proved inadequate and had to be reconstructed by Smirke.[1] At this stage it seems unlikely that he used any form of concrete. It was his "rescue" operation at Millbank which first demonstrated, on a grand scale, the value of concrete foundations.

The Millbank Penitentiary (Plate VI) for "offenders of secondary turpitude"[2] was opened in 1816 after a prolonged and troublesome period of gestation. The project originated in the humanitarian ideals of Jeremy Bentham, John Howard, Sir William Blackstone and Sir George Onesiphorous Paul. In 1794 Bentham contracted for the erection of a "Panopticon" Penitentiary "as far as strength is concerned, solely of cast iron and wood, in which the cast iron will greatly predominate."[3] Acting as feoffee for the Crown he purchased a 54-acre site from the Marquess of Salisbury for £12,000.[4] But his inability to fulfil the contract for £36,000, plus a certain amount of official ill will, caused the scheme to languish until Romilly revived it in 1810. Bentham was awarded £23,000 to cover the cost of his expenses, and an Act of 1812 appointed three Supervisors, Lord Farnborough, George Holford and the Reverend Mr. Becher, to organise a public competition for new designs.[5] Prizes of £200, £100 and £50 were awarded to W. Williams, C. A. Busby and J. Harvey. Revised plans by Williams were then submitted to Thomas Hardwick who agreed to supervise the works for a commission of 2 per cent. on his estimate of £239,725 plus £42,690 for foundations.[6] Dissatisfied with this low remuneration, and pessimistic as to the suitability of the marshy site, Hardwick resigned in 1813 and was replaced by Harvey.

Hardwick's resignation was nicely timed: the foundations had already begun to sink. In 1812 Bentham had bored to a depth of 14 ft. and, on finding nothing but loose soil and peat, condemned the ground as unfit to support buildings. At first Hardwick was confident that these difficulties could be overcome by a "substruction" of wooden "planking" or sleepers. But the Supervisors thought differently. They discovered that at the "Bedlam" Hospital in St. George's Fields, James Lewis had used Yorkshire flags instead of sleepers. This prompted them to experiment with various materials tested by weights of 200–300 tons. Finally, disregarding a Blackheath architect named

---

[1] Cumberland County Records Office: Ac./1–6; D. and S. Lysons: *Magna Britannia*, IV, 1816, Plan; "Report from the Select Committee on the Office of Works and Public Buildings," Parliamentary Papers, 1828, IV, 391. Similarly, Smirke's foundations for Lincoln County Court (1823–30) required excavations to the "extraordinary depth" of 28 ft.; but concrete does not appear to have been used. (J. Mordaunt Crook, "The building of Lincoln County Court," *Lincs. Arch. and Archaeol. Soc. Reports and Papers*, 1962, IX, 151–7).

[2] J. Elmes, *Metropolitan Improvements*, 1827–9, II, p. 150; for the original rules of the prison, cf. *Gents. Mag.*, 1817, I, 511.

[3] P.R.O. LRRO. 60/675; J. Bentham, *The Panopticon or Inspection House*, 1791. The radiating plan may possibly derive from Bentham's brother General Sir Samuel Bentham.

[4] A. Griffiths, *Memorial of Millbank*, 1875, I, 24.

[5] 52 G III cap. 44; *Monthly Mag.*, 1815, XXXIX, 302–3. Among the designs consulted were those for which William Blackburn (1750–90) had been awarded first prize in 1782 by the Commissioners originally appointed under the Penitentiaries Act, 1779 (P.R.O. Works 4/17, 20.7.1792; Works 4/18, ff. 22, 40, 41, 60; Works 6/22, ff. 50, 134; Works 1/5, f. 18).

[6] Griffiths, *op. cit.*, pp. 31, 34–5, citing Supervisors' Minute Book, 1812. J. I. Hawkins, one of the 43 competitors, claimed that Spencer Perceval secretly agreed to appoint an architect who would use the winning designs. (*Arch. Mag.*, 1835, II, 374–5.) Certainly, apart from the entrance, Hardwick closely followed Williams' designs which in turn followed Bentham's. (*The Builder*, 1868, XXVI, 181; R. Ackermann, *Repository of Arts*, 1812, VIII, 157; 1817, III, 153.) Williams, of 11 Tichborne Street, Golden Square, had taught drawing at Sandhurst.

Alexander, who claimed knowledge of a secret and unique system of foundations,[1] they instructed Hardwick to seek advice from John Rennie, S. P. Cockerell, James Lewis and Robert Browne. This quartet recommended that the topsoil and alluvial substrata consisting of "vegetable earth, light clay and morelog" should be excavated and that upon the gravel bed should be laid "puddled walling" composed of "gravel or ballast only, having perpendicular sides, cast and mixed in lime-water, with a small quantity of sand".[2]

This report was dated 22nd January 1813, and Rennie's seems to have been the dominating voice. The terminology is confusing, but the ingredients of concrete seem to be there. Puddling was of course the crude method of packing suitable earth to make canals and ponds watertight.[3] However, Hardwick's lack of interest, Harvey's incompetence and the Prison Supervisors' desire for economy made sure that this report was never fully implemented. It was a report, not a specification, and its recommendations were only made with regard to the main body of the building: the central chapel, the surrounding hexagon and the six radiating pentagonal salients, each three storeys high with angle turrets and a central "Taskmaster's Tower."[4] (Plate VI (*b*)). Hardwick's brief tenure of office was almost entirely concerned with the construction of the outer walls and entrances. The Pentagon foundations were only begun in the summer of 1813, and Hardwick resigned in the autumn. The octagonal boundary wall was given a foundation of rammed rubble no more than two feet thick; and the outer lodge and gateway, with its massively rusticated archway designed by Hardwick, was founded partly on rubble and partly on piles.[5] A few weeks before Hardwick's resignation the foundations of both wall and gateway began to subside.

For the next three years the Penitentiary was the responsibility of John Harvey. No reputable architect would accept a stipend of only 2 per cent. His gross inefficiency had already lost him one official appointment, at the Office of Woods and Forests.[6] At Millbank he again proved gouty and incompetent. The Supervisors had been lulled into believing that with regard to foundations Harvey possessed "more experience... than many architects of greater general knowledge."[7] Harvey first proposed an additional foundation of strong flat stones and then finding stone difficult to procure, used three courses of brickwork laid in Parker's cement.[8] By May 1816 little more than one third of the building had been executed, £128,304 had been spent and £228,813 was still required.[9] Less than three months after their arrival in June, the first prisoners (36 women from Newgate) in Pentagons I and II were "under considerable fear and alarm," as cracks and fissures began to divide the walls. At this point Harvey was dismissed and Smirke was called in at the usual rate of 5 per cent. commission.[10]

[1] "Independent of piles, planking and brick piers as arches . . . not subject to alteration by floods, damp, heat, frost, or any failures by time... neither pervious to vermin, nor liable to be undermined by prisoners." (Griffiths, *op. cit.*, p. 36.)

[2] *Ibid.*, p. 37; *Proc. Inst. C.E.* (1856–57), XVI, 434.

[3] "Puddle is formed by taking a quantity of earth, and mixing it with water, gradually poured upon it, to the consistency of mortar . . . It is applied . . . in horizontal layers or courses of about 9 in. in depth, and is mixed by men treading on it with mud boots, and stirring it with shovels. Each layer . . . is left to dry, until a man may stand on it without sinking . . . Clay and garden mould, or vegetable earth, are not good for forming puddle; the best kind of earth for that purpose being a lightish loam, mixed with coarse sand or fine gravel, all particulars of which, exceeding the size of a musquet ball, should be rejected. Puddle, when suffered to become thoroughly dry, cracks . . . but when properly made and either brought into immediate use or kept moist until required, it is impervious to water." (C. W. Pasley, *A Course of Military Instruction* (1817), III, 692–3.)

[4] Plans given by G. Holford, *Account of the Penitentiary at Millbank* (1828).

[5] Engraving in *Monthly Magazine*, 1814, XXXVII, 5.

[6] P.R.O. Crest 8/3 passim.

[7] Holford, *op. cit.*, p. xxix.

[8] Griffiths, *op. cit.*, p. 39.

[9] *Ibid.*, p. 40; *Commons Journals*, 1816–17, LXXI, 394, 649.

[10] Griffiths, *op. cit.*, p. 42, citing the *Governor's Journal*, 21.9.1816: "all three angle towers [in Pentagon I] . . . sunk a little lower, which cracked the arches and wall in several places . . . Sent for Mr. Harvey . . . Talked to prisoners and assured them there was no danger and succeeded in pacifying them" (Holford, *op. cit.*, pp. xxxi-ii).

SIR ROBERT SMIRKE: A PIONEER OF CONCRETE CONSTRUCTION

Under Hardwick and Harvey the contractors had been Messrs. Want and Richardson and Messrs. Joliffe and Banks. Smirke employed Samuel Baker and Sons. They were responsible for the reconstruction of three towers and the strengthening of foundations in Pentagons I and II; alterations to the foundations and then the erection of Pentagons III to VI; and the building of water machines, cornmills, reservoirs and sheds.[1] Smirke's reconstruction was partly based on Rennie's advice. Together they investigated the foundations and decided that there were two chief reasons for collapse: the sewers had been inadequately constructed and the artificial foundations were insufficiently firm and insufficiently wide. Tidal water had seeped through the sewer walls and had liquefied the subsoil which surrounded the narrow foundation trenches.[2] "The puddle or substratum of gravel," wrote Rennie, "has not been carried to a sufficient extent beyond the base of the outer walls of the pentagons and towers, and . . . the substratum itself has not been carried up with lime water or what may more properly be termed thin Grout except for about 18 in. below the base of the wall, which 18 in. is as hard and solid as any part of the brickwork, and to this I may add that I fear the gravel of which the substratum has been formed has not been carefully selected." Smirke and Rennie therefore recommended that "the bank of gravel now under the Towers should be dug over and well puddled with thin lime grout adding such extent to its breadth as the depth may require."[3]

Smirke was given full responsibility for the rescue work. It was he who produced the specification. His method of strengthening the foundations was to draw the walls together with "very strong iron screw rods," and to widen and underpin the footings of the main walls by laying down ballast and gravel and covering it with a thick layer of lime concrete. This method differed from his later practice, which was to dispense with the layer of ballast; it also differed from the method later adopted by Sir John Soane. At the New Law Courts and House of Lords (1822–24), at the Board of Trade and Privy Council Office (1824–27) and at the State Paper Office (1829–33), Soane rammed home the rubble in layers, grouting each layer with mortar.[4]

Smirke's foundations at Millbank took the form of a concrete raft. He called it a stratum of grouted gravel. In his *Course of Practical Architecture* (Chatham, 1826), C. W. Pasley describes the ingredients as follows: one part Dorking lime, burnt, pounded but unslaked, to 5/7 parts clean gravel,

---

[1] Holford, *op. cit.*, p. xxxii. Want and Richardson contracted for the boundary wall; they had previously worked on the Military College, Sandhurst, and Covent Garden Theatre under Copeland. Joliffe and Banks contracted for the entrance lodge; they were later responsible for London Bridge. One of Baker's labourers records that in the Winter of 1821 flooding from the river was so serious that workmen used boats to ferry themselves across to the nearby public house. (Charles Newnham, *The Life and History of a Man of Kent*, II, ff. 61 and 116. MS. in the possession of Mrs. M. Brain.)

[2] However, Smirke and Rennie presented Hardwick and Harvey with an alibi in the shape of economy. "We think it . . . just to add our opinion that the forming of an artificial foundation . . . for so extensive a building, observing at the same time the most rigid economy, was a work attended with no ordinary or inconsiderable difficulty, and we consider it to have failed in this instance principally from the desire of reducing its expence, to which only we can ascribe the attempt to give security to the buildings upon a foundation so contracted in width." (Library of the Institution of Civil Engineers, *Rennie MSS. Reports*, 1st Series IX, f. 56, 6.2.1817, Report signed "J. R." and "R. S.".)

[3] *Ibid.*, viii, ff. 428–9 (10 Dec. 1816) Rennie to Nathaniel Thorn, and ix, f. 56 (6 Feb. 1817), report signed "J. R." and "R. S." Smirke later recalled that "the foundations . . . which were giving way" were "composed of a bed of cleaned, washed gravel, rendered less fit . . . by having had all the sand sifted out of it. But this bed of gravel seemed to answer very well except where the pressure was near the edge of it under the external walls, and there it was seen to be giving way . . . the walls moving outwards with it." (Letter 16 Dec. 1837, from Smirke to Pasley, correcting Godwin's essay, sold (lot 265) at Sotheby's, 25 July 1962, and now in the possession of Mr. H. M. Colvin.)

[4] In 1836 George Bailey sent T. L. Donaldson Soane's specification for foundations at the Board of Trade and Privy Council Office: "the whole of the surface of the ground is to be excavated to the depth of 1 ft. 6 in. (or more if necessary) below the underside of the Yorkshire stone landings which are to be placed to receive the footings of the walls, and a general foundation is to be formed with layers of stone, broken to the size of an ordinary hen's egg, closely rammed with paviours' rammers, and grouted, every third layer or course, with fresh Dorking lime and sand in the porportion of two bushels of lime to one bushel of sand." (R.I.B.A. MSS. S.P.6, II, 14.1.1836.) Cf. also Wyatt Papworth (ed.), *Dictionary of Architecture* (1852–92): "Concrete."

plus water. "In the gravel, with which the lime is to be incorporated, no pieces larger than an egg, ought to be used, but there should be no limit in regard to the fineness of the minuter particles . . . . In fact, if only large gravel could be procured, it would be proper to mix a proportion of sand with it. This compound . . . forms by degrees a kind of Artificial Rock . . . The rule . . . is to spread the composition regularly throughout the whole foundation, and to raise it everywhere to the same general horizontal level, in courses of about 6 in. and in so doing labourers puddle it down with their feet . . . But Mr. Smirke does not in general recommend the grouting to be laid under the whole area of a building, as he has done at the Penitentiary. He is usually satisfied if it extends some feet each way beyond the footing of every wall; but [at Millbank] the apartments were so very small, that it would have been more troublesome and expensive to cut separate trenches for the foundations of the several walls than to excavate the whole space.[1]

Throughout his career Smirke used concrete as an artificial foundation, never as a building material. During the 1830s Thomas Cooper, G. L. Taylor and William Ranger produced concrete blocks for building purposes by using slaked lime, boiling water and wooden moulding boards. These patent processes were never very satisfactory. Before the invention of Portland Cement and modern techniques of reinforcement, they lacked the necessary cohesive ingredients. Both Smirke and Pasley realised that the friable texture of lime concrete made it unsuitable as a building material. But as a foundation it was cheaper, more durable and less troublesome than piles, sleepers, bricks or stone slabs.[2] At Millbank Smirke's foundations were crude in application and unnecessarily deep—as much as 18 ft. in places. Nevertheless, the early date (1817) and the importance of the commission make it a milestone in the development of modern structural techniques.

At the time Millbank was regarded less as a structural precedent than as a monument to official extravagance. The cost of additional accommodation, the cost of providing an embankment for the Thames frontage, the extravagance of experimenting with the latest patent locks, bedsteads, water closets and heating apparatus, and all the expense of underpinnng and rebuilding, resulted in the following progression of expenditure.[3]

1812: Hardwick's estimate : £259,725.
1816: Harvey's estimate: £350,000.
1819: Smirke's estimate: £380,288.
1822: Total expenditure: £450,310.

After a fever epidemic in 1823–24 a further £8,669 6s. 3d. was spent on alterations executed by Smirke on the advice of Sir Humphry Davy, and in 1828 the entrance lodge was rebuilt at a cost of

[1] C. W. Pasley, *Outline of a Course of Practical Architecture* (lithographed notes, Chatham, 1826; reprinted Chatham, 1862), pp. 13, 41–2. Smirke "examined and approved . . . and modified" Pasley's conclusions. (B.M. ADD. MSS. 41,696, f. 57A, Pasley to Sir John Rennie, 5.11.1838.)

[2] C. W. Pasley, *Observations on Limes, Calcareous Cements, Mortars, Stuccoes and Concrete* (2nd Ed., 1847), pp. viii, 18–20, 40; *Practical Architecture*, p. 45. Thomas Cooper, architect of Brighton Town Hall (1830–32), constructed the Sea Wall at East Cliff, Brighton, of concrete made with slaked lime, mixed in cold water and moulded into blocks by wooden frames. Ranger's artificial stone blocks followed the same principle, but were produced by mixing the concrete in boiling water. He used this method at Nos. 16 and 17 on the North side of Pall Mall, the College of Surgeons in Lincoln's Inn Fields (Barry's rebuilding 1835–37), the porticoes of the Wellington Barracks and the Guardhouse in Birdcage Walk; at Chatham and Woolwich Docks; and for a church and several houses in Brighton (*The Builder* (1847), V, 443. See also C. J. Richardson, *The Englishman's House* (1870), pp. 82 *et seq.*; *The Architect, Builder and Collector's Pocket Book* (1849), pp. 260–62.

[3] The accommodation for 300 males and 300 females was increased in 1816 to 400 males and 400 females and in 1819 to 600 and 400 respectively. (C. Knight, *London* (1871), V, 33.) For the embankment, cf. "Report from the Commissioners of Woods and Forests, for the Repair of the Embankment at Millbank, 1814," *Parliamentary Papers*, 1814–15, IX, 343–6; Additional Estimate, 1816, *Parliamentary Papers*, 1816, XII, 297. The estimate of the Commissioners of Woods was £9,500–£10,000. Harvey's accepted estimate was £8,984. For heating, etc., and final cost, *cf.* Griffiths, *op. cit.*, pp. 42, 45–6; *The Times*, 20 June 1816, p. 3; 26 June 1816, p. 3; 27 June 1817, p. 3.

SIR ROBERT SMIRKE: A PIONEER OF CONCRETE CONSTRUCTION

£5,180 9s. 7d.[1] "Erected in defiance of economy and taste,"[2] this labyrinthine prison survived until 1903. Built of sandstone from the Duke of Devonshire's quarries at Holt Stanwell Bridge, Derbyshire, it comprised cell passages two miles in length, 30 airing yards, 40 staircases, "miles of lead piping, hundreds of tons of iron, immense iron girders, gates in dozens—some of wrought iron, some of cast, flagstones without end, shiploads of timber, millions of bricks." Even so, it was popularly remarked of this penal iceberg that "there is more stuff below than above ground at Millbank."[3]

II. Contemporary Views on Concrete

Smirke's use of concrete on this occasion had been preceded by several Continental examples and by several British experiments. Concrete foundations were used in France at Toulon Harbour, *c.* 1748 (described by Bélidor, 1753); at the Avant Port Militaire in Cherbourg (1804); at the Pont d'Austerlitz (1802) and Pont d'Iéna (1806), both in Paris; at the bridge at Rouen (1814); at river locks on the Rhône-Rhine canal and at Saint-Valèry-sur-Somme (1813); and at the Souillac Bridge by Vicat (1818).[4]

Accounts of contemporary British experiments are confused both by imprecise terminology and by *parti pris*. Smirke himself never used the term 'concrete' before the late 1820s. Instead he used the word 'grouting,' which now carries a different meaning, namely a mixture of cement and water poured or pumped in to fill the voids in masonry, gravel or sand. It was not until the late 1830s that the word concrete was generally used in the modern sense of an intimate mixture of cement, inert aggregate and water.[5] Moreover, an influential article published in 1836[6] by George Godwin Jnr. (1815–88) perpetuated several legends. Godwin confused the *caementum* filling of the Romans[7]; the *Béton* of eighteenth century France; and the experimental use of gravel, rubble, and lime by George Semple in Dublin, by Smeaton at Calder Locks (1760) and by Ralph Walker at the East India Docks (1806), with the invention and use of concrete as we know it, that is concrete mixed in measured quantities for load-bearing foundations and other structural purposes. Sir John Rennie's adulation for his father proved another source of error. Sir John never ceased to claim that the elder Rennie's

[1] Penitentiary Expenses, *Parliamentary Papers*, 1830, XXIII, 395. For the prison's bad ventilation, *cf.* B.M. ADD. MS. 38295, ff. 275, 277.

[2] *A Letter to Rt. Hon. Sir Charles Long* (later Lord Farnborough) (1825), p. 21.

[3] Knight, *op. cit.*; M. E. C. Walcott, *Memorials of Westminster* (1851), pp. 335–6. In 1843 the name of the Penitentiary was changed to Millbank Prison (6 and 7 Vict. c. 26). It was superseded by Wormwood Scrubs in 1877 and closed in 1890. (H. B. Wheatley and C. Cunningham: *London Past and Present* (1891), II, 545.) Both the Panopticon plan and Smirke's concrete foundations were reproduced in Robert Abraham's Westminster Bridewell, Tothill Fields, 1830–34.

[4] I owe this information to Professor A. W. Skempton, who was kind enough to correct several important errors and omissions. During the 1790s in France, Cointeraux was also developing Béton from pisé. (Cf. F. Cointeraux, *Traité sur la construction des manufactures et des maisons de campagne* (Paris, 1791); P. Collins, *Concrete* (1959) passim.) *Bélidor's L'Architecture Hydraulique* was published in four parts in Paris, 1737–53; L. J. Vicat's *Récherches expérimentales sur les chaux de construction, les bétons et les mortiers ordinaires* was published in Paris in 1818. His *Mortiers et Ciments Calcaires* was published in Paris in 1828, and translated by Capt. C. J. Smith in 1837. For Henry Holland's use of pisé and cob, cf. Papworth, *Rural Residences* (1818), p. 15.

[5] "When a thick wall of rubble masonry, chiefly of small and irregular materials, is built in successive courses, with all the stones at first laid dry, excepting the outside ones of each course, or when long irregular stones in the form of an arch are laid dry over a wooden centering, and quick lime slaked with excess of water, either with or without sand is poured in a liquid state, into the mass of dry materials . . . the work is said to be *grouted*; and this process, which is of great antiquity, is so analogous to the mode of forming concrete foundations . . . that these were often called grouted foundations, when first introduced. It is proper however to make a distinction between the terms *Grout* and *Concrete*, because the best Architects are now in the habit of causing brick walls to be grouted (e.g. Smirke's British Museum) in a manner which... has no resemblance either to the ancient grouting or modern concrete." (Pasley, *Observations on Limes etc.* (1838), pp. 25–6.)

[6] G. Godwin, "The Nature and Properties of Concrete," *Transactions of the* (*R.*)*I.B.A.*, 1835–36.

[7] Ancient *caementum* normally contained a much higher percentage of rubble than modern concrete. For the difference between *Béton* and concrete, cf. W. Papworth (ed.), *Dictionary of Architecture:* "Béton."

use of concrete foundations antedated Smirke's by as much as a decade. These errors might never have gained credence if C. W. Pasley's *Course of Practical Architecture* (1826) had been published and distributed. Instead, Pasley's early work remained in the form of lithographed lecture notes, and is now excessively rare. Godwin's errors were eventually corrected in detail and buried in the footnotes and appendices of Pasley's *Observations on Limes, Calcareous Cements, Mortars, Stuccoes and Concrete etc.* (1838).[1]

General Sir Charles William Pasley, K.C.B., F.R.S., D.C.L., Royal Engineers (Plate VII), to give him his full and final title, was a Scotsman, born in 1780 and said to be descended from Robert III, King of Scotland. Long before his death in 1861 he had established himself as an authority of international stature on the subject of military engineering.[2] After serving in the Napoleonic War (he was with Sir John Moore at Corunna and was twice wounded at the siege of Flushing) Pasley became Director of a new course of military engineering at Chatham. There, between 1812 and 1841, he conducted a series of experiments with regard to fortifications, signalling, sapping, mining, pontooning, gunpowder, cement and concrete. Apart from a common interest in structural problems, his close connection with Smirke can possibly be tracted to the fact that General Sir John Malcolm, one of Smirke's leading Tory patrons, was also Pasley's cousin.[3] During his lifetime Pasley was famous as the man who cleared away the wreckage of the *Royal George* from the anchorage at Spithead (1839–41). Since his death, Pasley's discourse on practical architecture and his treatise on cement and concrete have been our most reliable sources for the history of artificial foundations in England.

Pasley began by examining Smeaton's investigations into the nature and properties of lime and cement. These discoveries he compared to the Baconian revolution which overturned Aristotelian principles of philosophy. By simply proving "that the real cause of the water-setting properties of limes and cements consisted in a combination of clay with the carbonate of lime," Smeaton upset every architectural theorist since Vitruvius.[4] Nevertheless, even in such ambitious projects as the Eddystone Lighthouse, he made no use of mixed concrete as a foundation; nor did Rennie and Stevenson at the Bell Rock Lighthouse (1807–10). But at Sheerness (1808–23), Chatham (1814) and London Docks (1802–05), Rennie seems to have experimented with a combination of piling, gravel and "grouted stones." And in 1810 the Martello Tower built "on the marsh near Walton Ferry opposite to Harwich," was given a foundation of grouted rubble plus a grating of elmwood sleepers.[5] Normally, however, foundations during this period took one of four forms according to the strength or weakness of the subsoil: either the brick footings were laid directly upon pounded earth and rubble (Fig. 1), or they were placed upon a combination of sleepers and piles (Fig. 2); or they were strengthened by the use of stone slabs and chain timbers (Fig. 3); or else they were raised upon a series of brick arches, like most of the Martello Towers (Fig. 4).

Smeaton and Rennie certainly experimented with combinations of lime and rubble, and must have been well aware of the potentialities of concrete as a building foundation. But it was Smirke who

[1] For Pasley's private letter to Godwin, correcting misapprehensions regarding Walker's work, cf. B.M. ADD. MS. 41969 f. 57b; 28.11.1836.
For Godwin's public apology to Pasley regarding Millbank, cf. *Civil Eng. & Archt's Jnl.*, 1839, II, 18–19, 26.11.1838.

[2] For details of Pasley's ancestry, career and publications, cf. his obituary, *Gentleman's Magazine*, 1861, LV, 698, and an unpublished MS. Biography by Col. J. C. Tyler (1929), B.M. ADD. MS. 41766.

[3] In 1832–3 Smirke made alterations to Malcolm's seat, Warfield Hall, Berks., *cf.* Sir J. Kaye, *Life of Sir John Malcolm* (1856); *Seats and Mansions of Berkshire* (pub. by Beard & Co., *c.* 1880).

[4] Pasley, *Observations on Limes etc.* (1847 ed.), pp. v and ix. For Smeaton's experiments *cf.* J. Smeaton, *Narrative of Eddystone Lighthouse* (1791); N. Davey, *A History of Building Materials* (1961), pp. 103–4.

[5] Pasley, *Practical Architecture*, pp. 13, 41. However, Pasley also gives a cross-section of a Martello Tower on the Suffolk coast which was raised on a conventional arched foundation (*ibid.*, p. 38). Ralph Dodd (1756–1822) is reputed to have experimented with a combination of cast iron and lime concrete as early as 1808 (C. M. Norrie, *Bridging the Years* (1956), p. 52.

SIR ROBERT SMIRKE: A PIONEER OF CONCRETE CONSTRUCTION

Fig. 1. Brick footings on earth.

Fig. 3. Stone slab and chain timbers

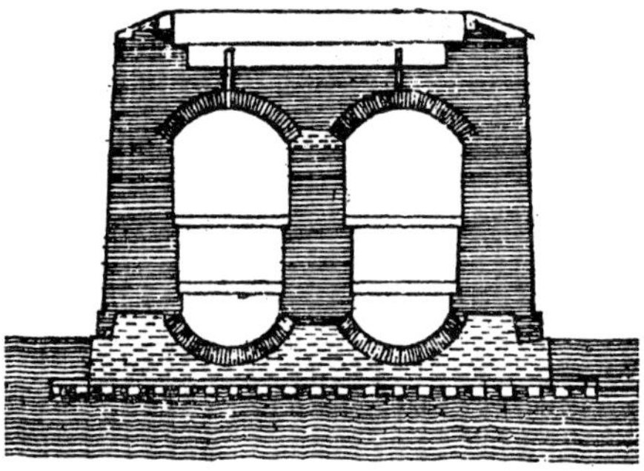

Fig. 4. Martello Tower on inverted brick arches, brickwork and sleepers

Fig. 5. Savoy wharf, 1819: footings on timber grid

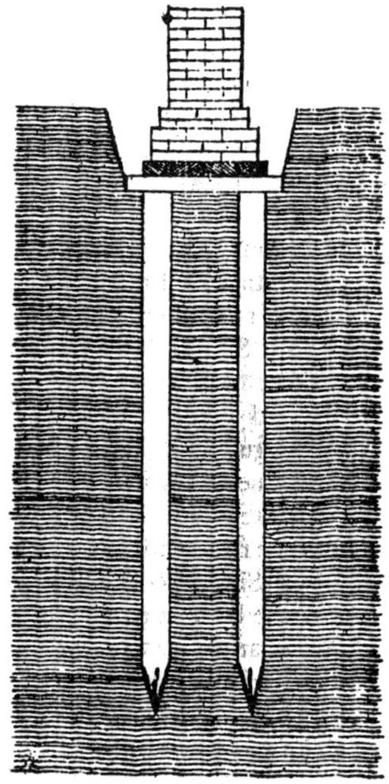

Fig. 2. Sleepers and piles

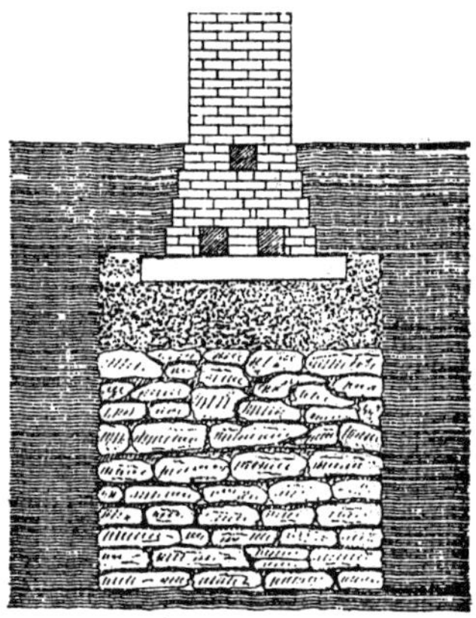

Fig. 6. Foundations at Lancaster Place comprising rammed rubble, mixed concrete and brick footings enclosing bonding timbers

Examples of foundations from C. W. Pasley's Practical Architecture (1826)

first used the new material on a large scale. Despite the assertion of his latest biographer, Rennie did not recommend concrete for the repair of Smeaton's Aberdeen Harbour in 1797.[1] "There is scarcely anything new under the sun," concluded Pasley, "but the merit of introducing this immense improvement systematically and generally into the practice of modern architecture is undoubtedly due to Sir Robert Smirke."[2]

For many years the Smirkes were connected with the Dobson family, architects of Newcastle, by ties of friendship and professional interest and eventually by marriage. In a biography of her father, John Dobson's daughter put Smirke's limited achievement in its proper place: "the French used concrete as a substitute for large baulks of timber, which was the usual mode of giving stability to walls on soft foundations, but Smirke was the first to introduce it into this country and apply it systematically for architectural purposes."[3]

Smirke's immediate inspiration, indeed the only precedent he agreed to acknowledge,[4] is sufficiently curious to deserve memorial. Apparently in 1812, during the construction of the cofferdams for Waterloo Bridge (1811–17), Rennie had discovered a large deposit of calcareous concrete. This deposit had been formed by the accidental combination of Thames shingle with a long-sunk load of hydraulic lime: the origin of modern concrete foundations. By a curious coincidence, George Semple's *Treatise on Building in Water* (1776) had cited a similar discovery at Dublin in 1757 during the building of Essex Bridge over the Liffey. In January 1838 we find Pasley asking Smirke whether this publication had influenced his choice of foundation at Millbank. "Did you read that book before you were employed at the Penitentiary? In his book Mr. Semple recommends much wilder schemes than even the fertile imaginations of Messrs. Taylor and Ranger ever attempted. But Mr. Semple *himself* never used *Concrete* at all, though he recommended it, and no one copied his ideas as he was an obscure writer of little authority . . . It can make no possible difference to your reputation as the first person who used concrete *usefully*, *extensively* and *successfully*, whether you ever read Mr. Semple's book or not. But as a matter of curiosity I should like to know." "I cannot recollect exactly," replied Smirke, "when I first saw Semple's work, but I think it was at Dublin long after the works at the Penitentiary were begun, and I certainly never heard then that anything like Concrete was suggested."[5]

This exchange of letters occurred just before the publication of Pasley's *Observations*. But within weeks of its publication Sir John Rennie stepped in to claim that his father deserved all the credit for the Millbank operations: he had made the discovery at Waterloo Bridge; he had recommended concrete for the Penitentiary; and only the incompetence of Hardwick and Harvey had made Smirke's intervention necessary.[6] "Sir John Rennie has written to me more than once," complained Pasley to Smirke, "and has since attacked me personally, to claim the entire merit of concrete for his father. I had always heard you admit that you took the hint from old Mr. Rennie, and have stated as much in

[1] C. T. G. Boucher, *John Rennie* (1963), p. 125. In fact Rennie's report recommended sheet piling, stones, wooden wedges and "rough lime rubbish" (Rennie MSS, Reports, 1st Series, I, f. 394). This preliminary stage in the development of mixed concrete foundations had been anticipated by George Dance as early as 1770 at Newgate where the foundations were 40 ft. deep: "he threw into the bog cartloads of whole and broken bricks and cartloads of mortar, in the proportion of one to four; and this rough concrete made an excellent foundation." (W. Papworth (ed.), *Dictionary of Architecture:* "Concrete.")

[2] Pasley, *Observations on Limes etc.*, p. 16.

[3] M. J. Dobson, *A Memoir of John Dobson* (1884), p. 53. Dobson's eldest daughter, Isabella, married Robert Smirke's younger brother Sydney, in 1841. (*Gentleman's Magazine*, 1841, XV, 91.)

[4] Letter (16.12.1837) from Smirke to Pasley on the subject of Godwin's essay, *op. cit.*

[5] R.I.B.A. MSS., Pasley to Smirke, 2.1.1838, annotated with Smirke's answers in pencil and misquoted in *Journal of the R.I.B.A.* 1929, XXXVI, p. 160. For Essex bridge, cf. Semple, *Building in Water* (1780 ed.), pp. 40, 75, 79–82, *The Builder* (1877), XXXV, 1217–9.

[6] B.M. ADD. MS. 41964, f. 219, Sir John Rennie to Pasley, 7.12.1837. Sir John Rennie's arguments on this occasion were spoilt by wild chronology, and on a later occasion by vague terminology, i.e. by failing to distinguish between grouting and mixed concrete. (*Proc. Inst. C.E.* (1856–57), XVI, 423–448; *The Builder*, 1857, XV, 288.)

SIR ROBERT SMIRKE: A PIONEER OF CONCRETE CONSTRUCTION

my Cement Book, but Sir John is not satisfied with what I have said . . . It appears to me that the Engineer of the Waterloo Bridge and of the Plymouth Breakwater does not want, in addition to his fame, to be also the engineer of the specification of the Penitentiary Foundations."[1] For all Sir John Rennie's accusations Smirke had an answer. Of course "old Mr. Rennie" had made the crucial discovery at Waterloo Bridge in 1812. But he was sceptical as to the value of this discovery, and his directions for Millbank in 1813 had never been properly executed. "It is quite true," admitted Smirke, "that if (Rennie) had not mentioned what he had observed . . . I should never have thought of applying the concrete as I did." But when Rennie and George Dance "examined by desire of the Penitentiary Committee what I was doing for the new foundations . . . their approbation was given in very guarded terms—and certainly not such as would have led anyone to suppose that either of them was practically acquainted with its effect."[2]

But there was really no need for Smirke to adopt this defensive posture. It was merely the Rennie-Stevenson controversy, with regard to Bell Rock Lighthouse, all over again.[3] Had the contestants but known it, "old Mr. Rennie" had set the record straight in his very first report on the collapse of the Penitentiary: "perhaps as I have had little or no practice in the support of buildings of this sort which have given way, it would be better to let this point (the method of reconstruction) stand over until I have had more time to consider the subject and had the assistance of some eminent Architect in whose line questions of this kind are more directly than in mine."[4] Rennie must be credited with the original idea; but Smirke put it into practice. He was coadjutor and sole executant, if not sole originator of the Millbank foundations.

In 1903, during the demolition of Millbank Penitentiary, a specimen of the now legendary concrete was placed, at the request of the German Ambassador, in the Munich Museum of Masterpieces of Natural Science and Technical Arts.[5] Apparently the Museum authorities credited Smirke with the rediscovery of concrete after the secret had been lost for centuries. Of course, this was both to simplify and to exaggerate his achievement. What he did was to actualise the experimental, and by publicising the new material in a major public building he revolutionised architectural practice.

III. The Savoy Precinct

In the development of artificial foundations a significant part was played by Smirke's contractor, Samuel Baker. Baker had married Smirke's sister Sarah in 1815.[6] Thereafter Baker's local business in Rochester became a metropolitan, even national, practice. The British Museum, the General Post Office, churches, mansions, clubhouses and provincial courts; Baker's tenders were generally accepted wherever Smirke was commissioned. He specialised in works involving structural

[1] R.I.B.A. MSS., Pasley to Smirke, 19.12.1838.

[2] B.M. ADD. MSS. 41964, f. 232, Smirke to Pasley, 26.10.1838. In actual fact the report by Dance and Rennie was brief but definite; "the foundations composed of gravel and lime . . . if allowed sufficient time to settle will form a solid basis upon which the building may be safely erected." (Rennie MSS., Reports 1st Series, IX, f. 209, 18.11.1817.)

[3] For this controversy, cf. Robert Stevenson, *Account of the Bell Rock Lighthouse* (1824); *Civil Engineer and Architect's Journal*, 1849, XII, 78, 137; Robert Louis Stevenson, *A Family of Lighthouse Builders* (1892); C. T. G. Boucher, *John Rennie* (1963), pp. 52–61.

[4] Rennie MSS., Reports 1st Series, VIII, f. 430, Rennie to Nathaniel Thorn, 10.12.1816.

[5] The Ambassador's memorandum (P.R.O. Works 6: 356/4, 1–11) presents a garbled version of the whole story.

[6] Joseph Farington noted in his diary that the Baker family had "large dealings in land and timber . . . and the principal management of the Borough of Rochester for the Government." (Farington Diary, B.M. Print Room, typescript pp. 6680, 17,6.1815, and 6944, 3.6.1816.) The partnership "Baker & Son" consisted of Samuel Baker of Rochester and George Baker of Montagu Place, London. For a while Samuel Baker was in partnership with Nicholson of Rochester. Together they contracted for the Ordance Barracks at Chatham. Nicholson later joined with Joliffe and Banks in contracting for the Sheerness Dock Yard. (Pasley, *Practical Architecture*, p. 185.) The architect, Sir Herbert Baker (1892–1946) was the grandson of Sarah's brother-in-law Thomas Baker.

difficulties, such as the County Courts, Lincoln, or the Temple Church, London.[1] Soon after the rescue of Millbank he was involved in the development of the Savoy Precinct. This project followed naturally from the building of Waterloo Bridge. Between 1817 and 1823 Smirke organised the transformation of the Precinct, cluttered with the ruins of a Tudor Hospital, three chapels and an eighteenth century barracks, into the broad vista of Lancaster Place and Wellington Street. Using the system adopted by Nash in Regent Street, Carlton Gardens and the Trafalgar Square district, he considered the general plan and proposed a set of standard designs with which speculative builders (in this case Messrs. Baker and Henry Peto) or privately commissioned architects, had to conform. In addition, he himself took direct responsibility for the Duchy of Lancaster Offices. This block formed the most important unit in the scheme: one half of Smirke's original proposal for a grandiose entrance to the Strand from Waterloo Bridge.[2]

The start of the Savoy project coincided with Smirke's rescue operation at Millbank. Once again, the proximity of the Thames to a valuable building site presented him with an opportunity of displaying his ability in matters of construction. But he was not yet prepared to use concrete except in an emergency. His construction of the Savoy embankment in 1819 followed a well-established practice: the use of a timber grid similar to Fig. 5. One of Baker's labourers who worked on the foundations "all hours of the day and night, up to my middle frequently in water and mud," recorded the following description.[3] Smirke "determined upon a novel expedient to obtain a sound foundation . . . neither piling, or using a cofferdam . . . His plan was this. The excavators took away the mud from the river bank and then excavated the soil down to a depth of about 4 ft. below the common level of the shore, having to pump out the water every tide. When they had got a certain position ready where one of the new walls was intended to be erected, I had to put down a row of fir sleepers at regular intervals; each sleeper was a foot or two longer than the intended foundations were wide. The sleepers were 7 by 7 or what is called quarter timbers. These were merely laid in the mud and slush left behind after the water was pumped out. Upon and across these for their whole length were laid very long *fir* half timbers 14 by 7 so as to form a sort of monster grating, the joists being kept 2 in. or 3 in. apart. These were spiked down upon the sleepers with very long and strong spikes; this all had to be done as rapidly as possible, so as to enable the bricklayers to get a sufficient quantity of their wall built upon it before the tide came up again to float the wooden platform away . . . The brickwork was then carried up the height intended for the level of the new wharf, and then dry rubbish was filled in between the walls up the level of the ground, and in this state it remained for two or three years; and then the building was proceeded with to completion. We all prophesied settlements, fissures and cracks, and some the downfall of the building; but Sir Robert Smirke was right and we were wrong, for there is not a building in London more free from the defects named than that is." Almost 10 years later, in response to a request from his former master, Smirke provided Soane with a diagram of the riverside foundations and the information that "the expense of building the wall, excavating, piling, pumping and everything, was about £50 per yard."[4]

But as the redevelopment of the Savoy Precinct progressed, so did Smirke's confidence in the value of concrete foundations. By 1825 he could claim that he "would not scruple to use a grouted foundation for a Naval Dock, or other massy work of equal importance, in the softest soil."[5] The buildings in Lancaster Place, adjoining the Duchy of Lancaster Offices, were put up between 1820 and

[1] J. Mordaunt Crook, "The Restoration of the Temple Church: Ecclesiology and Recrimination," *Architectural History*, 1965, VIII, 39–46.

[2] P.R.O. D.L.41/76, draft pencil letter to Treasury, July 1815. For the history of the Savoy development, with its attendant wrangles between Exchequer, Duchy, tenants and builders, cf. P.R.O. D.L. 41/76 (including street plan), D.L. 41/79 and D. L. 41/45.

[3] Newnham MS. II, ff. 121–2. For the date of the embankment, cf. R. Somerville; *The Savoy* (1960), p. 113.

[4] Sir John Soane's Museum, Soane Correspondence, Division II, Cupboard I.

[5] Pasley, *Practical Architecture*, p. 41.

SIR ROBERT SMIRKE: A PIONEER OF CONCRETE CONSTRUCTION

1823, and this time concrete was used. The whole block was contracted for and rented out by Samuel Baker. Pasley lists Lancaster Place as the second instance of the use of mixed concrete foundations in this country, and provides us with a full description and diagram.[1] The foundations (Fig. 6) on average 8 ft. wide and 8 ft. deep, consisted of three sections: firstly rammed rubble, secondly mixed concrete and thirdly brick footings enclosing sleepers and chain timbers. Smirke was responsible for "the general design," Baker for "all the details of execution." The building was "founded on very bad soil" consisting of "soft mud . . . intersected in all directions by the remains of the foundations of old houses and wharfs. . . . Fragments of the old walls were laid [in trenches], in the same manner as large flat stones might have been used. These were well sluiced with water and mixed with grouted gravel . . . and rammed down flat . . . Then the rough foundation was continued to the proposed footing of the brickwork, with grouted gravel alone; and at that level transverse sleepers were used at intervals of about 4 ft. apart (but all under piers none under the position of Doors or Windows), upon which were laid two longitudinal chain timbers 9 in. square, completely surrounded by brickwork, . . . and a few feet higher, opposite to the top or uppermost offset of the brick footing, a third chain timber of the same scantling was bedded in the middle of the wall." As for the brick footings, they had "a base of 5½ bricks, which by four successive offsets on each side, at intervals of three courses apart, diminished at the height of 3 ft. to 3½ bricks, which is the thickness of the front and back walls of the lowest storeys of the buildings, which are used for stables or stores. Over this the walls of the basement floor of the inhabited part . . . are 3 bricks thick, . . . gradually diminishing by successive offsets to 1½ thick in the uppermost storey of the house." All in all "the height, from the top of the brick footing to the coping of the parapets . . . is 88 ft.; to which adding the footings and foundations, it appears that there is a mass of brickwork and rough masonry and grouting nearly 100 ft. in height, that is on the side nearest to the river. If no failure should hereafter occur in this great building," Pasley concluded, "it will afford a most convincing proof of the efficacy of [concrete foundations]." The building survived until the 1920s.

IV. Later Works in Concrete

Smirke's use of concrete at Millbank had been an emergency operation. The new technique was slow to gain acceptance for general purposes. But in 1822–23 he began work on a town house for Sir Robert Peel, built upon a foundation of concrete. The house in question was one of a row of terraced houses, No. 4 Whitehall Gardens. The site of Whitehall Gardens was a strip of Crown property, formerly known as Privy Gardens, which lay between Whitehall and the Thames. In the days before the Victoria Embankment, Peel's property ran right down to the river, and was frequently flooded at high tide. Artificial foundations of some sort were therefore essential. The adjoining houses had all been raised on a foundation of piles. No. 4 was unique in resting on concrete. Writing in 1826, the year of the building's completion, Pasley noted that "Mr. Peel's house . . . affords another instance of a grouted foundation in very bad soil, and as all other buildings near it are piled, time will show which of these two expedients is the most effectual."[2] Baker contracted for the foundations

[1] *Ibid.*, pp. 43–5. Pasley also lists "the new Parliamentary Mews . . . about to be built" as another example of Smirke's use of concrete (p. 43). But a marginal entry in the R.I.B.A. copy of this rare volume, presented by J. W. Hiort in 1835, attributes the mews (built 1826, demolished 1854) to Decimus Burton, an attribution supported by J. Elmes, *Metropolitan Improvements* (1831 ed.), p. 139, and plate, The Government Mews, Princes St., Storey's Gate, Westminster. For details *cf.* "Report from the Select Committee on the Office of Works, 1828," *Parliamentary Papers*, 1828, IV, 446–7; and Accounts, *Parlimentary Papers*, 1828, XXI, 617. For Burton congratulating Pasley on the publication of *The Natural Water Cements of England* (1830), cf. B.M. ADD. MS. 41964, f. 62.

[2] Pasley, *Practical Architecture*, p. 43. For Baker's contract for the foundations, cf. Newnham MS. II, f. 117. For full details of Peel's house, cf. letters between Peel and Smirke, B.M. ADD. MSS. 40605–7 *passim*; P.R.O. T55/28, f. 243, and LRRO. 60/925, 28.7.1824; *Private Letters of Sit Robert Peel*, ed. G. Peel (1920), pp. 46, 52, 53; *Survey of London*, XIII (1930), pp. 204–13.

and footings; Westcott and Browning for the remainder of the building. Peel's house cost £14,000, twice as much as its next door neighbours, and contained at least one more novelty. Gas Lighting was supplied by the Gaslight Co., and although by now well-established, it was still sufficiently strange for Smirke to have to reassure his client "that where there is no smell, there can be no escape of gas and therefore no risk of explosion."[1]

Smirke's methods of preparing concrete are illustrated by the later foundations of the British Museum, 1833 onwards (Plate VIII (*a*)) and the foundation of the Oxford and Cambridge Club laid in the Winter of 1835.[2] At the British Museum the thickness of the concrete varied from 2 ft. 6 in. under load-bearing walls to 6 ft. under the great colonnade. At the Oxford and Cambridge Club in Pall Mall the specification was very similar: "the whole area of the building and to an extent of 2 ft. beyond the line of the lowest footing, was covered to a depth of 2½ ft. the depth being increased to 4 ft. under all the walls that rise to the roof." Smirke instructed the contractors that for this "grouted stratum, clean river gravel is to be . . . mixed with lime ground or pounded to a fine powder, twice turned over before it is wheeled to the excavation, and thrown from a height of not less than 6 ft. . . . A man to be kept treading down and puddling the mass as it is thrown down, the proportion of the materials to be six parts of gravel to one of Dorking, Merstham or Haling stone lime."

On the subject of foundations Smirke's considerable experience made him a respected adviser. In 1830 his advice was unofficially requested by the officious Croker with regard to the concrete foundations prepared by G. L. Taylor for the Admiral's House at Sheerness,[3] and in 1835 he supervised the arbitration by Joseph Kay and Thomas Allason which confirmed the negligence of Lewis William Wyatt in preparing foundations for Sherborne House, Gloucestershire.[4] In 1845 the Harbour Commission sought his advice on the foundations to be laid in seven fathoms of water for a new harbour at Dover.[5] His advice was more direct in connection with the construction of the London Bridge Wharves in 1832–34. The wharves on both the City and Southwark sides were the joint responsibility of Sir John Rennie and Sir Robert Smirke, as respective architects of London Bridge and the London Bridge Approaches. In June 1832 Smirke submitted plans, elevations and specifications for the guidance of the contractors[6]: "the ground is to be excavated down to a sufficiently firm stratum . . . a bed of gravel grouted with stone lime, ground or well pounded, and mixed in the proportion of one of lime to six of gravel, is to be laid . . . under the walls at least 4 ft. in thickness, and projecting on each side 2 ft. beyond the footings . . . . This surface should be covered with 4 ins. York landings, and the row of sheeting piles previously driven in front should have their heads raised a few inches above the surface of the stone platform. The face of the wharf walls upon their sectional line are to be of curvilinear form; the footings at the lowest level not less than 7 ft. thick and the thickness of the walls diminishing to not less than 3½ ft. at the top, exclusive of the counterforts . . . at the back of the walls. A grouting of gravel, sand and lime should be laid at the back of the walls as they are raised, and the surface of the wharf at the coping should be about 4 ft. above the Trinity

[1] B.M. ADD. MSS. 40605, f. 313; 40606, f. 251; 40607, f. 30. J. Mordaunt Crook, "Sir Robert Peel: Patron of the Arts," *History Today*, Jan. 1966.

[2] For the Oxford and Cambridge Club specification, cf. *Civil Engineer and Architect's Journal*, 1843, VI, 46. For Smirke's plans of the British Museum foundations, cf. P.R.O. Works 33/3–6, 150, 204, 325, 495, 786.

[3] G. L. Taylor, *Autobiography of an Octogenarian Architect* (1870), I, p. 171. Like Smirke, Taylor was a Greek Revivalist and an enthusiast for concrete: his Proprietary School at Lee, Kent (1836), was built of concrete blocks and copied from the Propylaeum on the Acropolis. Taylor also used concrete foundations at the Clarence Victualling Yard, Gosport, 1833.

[4] H. M. Colvin, *Dictionary of English Architects* (1954), p. 734. Ironically, Smirke had himself provided designs, which were rejected, for the re-building of the house in 1820–21. (Gloucestershire County Records Office, Sherbourne MSS.)

[5] Letter (28.11.1845) in the possession of Smirke's descendant, Sir Edward Malet, Bt.; "Report of the Harbour of Refuge Commission . . . on the several Modes of Construction referred to the Commission," *Parliamentary Papers*, 1847, LXI, 42.

[6] Guildhall Record Office, Bridge Committee Journal, VII, ff. 495–8, 666.

High Watermark." But the most famous project for which Smirke's advice was required was the construction of Nelson's Column. Together with James Walker, the President of the Institution of Civil Engineers, Smirke recommended in 1839 that the column be lowered by 20 ft. As Sir Robert Peel assured the House of Commons "it would be extremely inconvenient should the monument fall in that crowded part of the metropolis."[1]

The repair of Millbank Penitentiary first established Smirke as an expert in the laying of foundations, but it was his reconstruction of the London Custom House (Plate VIII (*b*)) which confirmed his reputation as "the best constructor of his day."[2] His reports exposing the incompetence of its architect, David Laing, and the corruption of his contractor, Henry Peto, like those he prepared in 1819 and 1841 on the conduct of D. A. Alexander at Maidstone Gaol and James Savage at the Temple Church, made him a recognised arbiter in matters of professional practice. Nine years before its collapse the Custom House had become an object of scandal in professional circles. Work began in 1813, but delays in excavation were followed by dilatory construction. Four years later the building was nearing completion. Enquiries by Treasury, Customs and Office of Works officials disclosed excessive expenditure and glaring irregularities in accounting and construction. Subsidence began in 1820 and collapse followed in 1824–5. The beechwood piles, alternately damp and dry, had rapidly decayed. But both Smirke and Sir John Rennie decided that the work had been so badly executed that the foundation would have failed even if oak had been used. In 1812 the elder Rennie had warned Laing that the subsoil was treacherous and the use of piles essential.[3] But the execution of the piling left both architect and contractor open to accusations of negligence, conspiracy and fraud. Indeed Peto's profiteering, Laing's ineptitude and Smirke's reconstruction produced a total expenditure of £435,000 between 1813 and 1829.

Smirke's new concrete foundations followed his usual practice. Once again the contractor was Samuel Baker. The piers and walls were underbuilt "with a solid substratum of gravel grouted with hot stone lime." While the building was shored up with timber supports, under-pinning proceeded in sections 10 ft. long. The new foundation consisted of three layers. Firstly, there was a platform of concrete mixed in the proportions of 1 part pulverised Dorking quicklime to 7 or 8 parts Thames ballast. Above this was placed a course of Yorkshire landing stones each measuring 5 ft. by 5 ft. by 6 in. thick. The final level consisted of twelve courses of brickwork laid in cement wedged close to the original brickwork with pieces of slate. These three layers served to equalise the pressure placed upon the lime concrete stratum.[4] For the sections begun *de novo*, where no under-pinning was required, Smirke refined upon the system adopted at Lancaster Place. Above the concrete and Yorkshire landings were laid "two chain timbers of scantling 9 in. wide and 12 in. deep . . . firmly bedded in the brick footing of the walls, without resting on sleepers in the usual manner." "This judicious arrangement," Pasley assures us, "possesses the advantages of a continued wooden

[1] P.R.O. Works 1/23, ff. 436, 506; Works 1/24, f. 24; *The Builder*, 1844, II, 369; *Civil Engineer and Architect's Journal*, 1840, III, 286, 305.

[2] Sir John Summerson, *Georgian London* (1945) p. 190. For the whole story, cf. J. Mordaunt Crook, "The Custom House Scandal," *Architectural History* 1963, VI, pp. 91–102; R. C. Jarvis, "Laing's Custom House," *Transactions of the London and Middlesex Archaeological Society* 1961, XXX; P.R.O. Works 1/13, ff. 260, 293; Records of H.M. Customs and Excise, MSS. relating to Rebuilding of Custom House, III, *passim:* R.I.B.A. MSS., Pasley to Smirke, 12.1.1838.

[3] Rennie's report was referred to in 1825 by George and Sir John Rennie during their investigations, and again in 1829 during the legal proceedings which followed collapse. (Rennie MSS., Reports 1st Series, VII, ff. 52–3, 17.2.1812; 2nd Series, II, f. 127, 18.4.1825.)

[4] P.R.O. Works 1/13, ff. 141, 353. Pasley, *Observations on Limes etc.*, pp. 267–9. At Chatham in 1834 G. L. Taylor, and his contractor Ranger, refined upon this method by using a screw press to force the concrete up to the bottom of the old wall, thus saving the cost of the intermediate brickwork. (G. L. Taylor, "Account of the Methods used in Underpinning at Chatham Dockyard in 1834," *Transactions of the (R.)I.B.A.*, 1835–6, I, 40.) However, these foundations were insufficiently broad and later subsided (Pasley, *Observations on Limes etc.* (1847 ed.), p. viii).

platform without its disadvantages." On this occasion Smirke also experimented with a different brand of concrete: he tried and rejected a mixture of gravel and Roman cement. "He grouted two equal parts of gravel, the one with Dorking lime, the other with Roman cement; it was found that the latter would not answer at all; for instead of becoming harder, the mixture fell to pieces."[1]

By 1828 Smirke's use of concrete was 10 years old. But his methods were still sufficiently novel for him to be asked the following questions by a Parliamentary Committee[2]:—

"Is that article you mention a compost, technically called grouting?

It is a sort of grouting.

What is the stone you have used in combination with the lime?

It is a stone lime.

Lime ground?

Limestone burnt and pounded.

Then you use a gravel in combination with that?

Yes.

What gravel is that, is it silicious?

No, but it is sufficient to have gravel free from clay and from every other impurity, whether silicious or not.

Do you grind them together?

No, they are well mixed together.

You throw them into the holes, then; therefore it is a combination of pebbles and lime?

It is nothing more."

APPENDIX

Sir Robert Smirke's library was broken up soon after his death in 1867. Literary works passed to his daughter Laura and were sold by Messrs. Hodgson & Co. Ltd., in 1906 (Sale catalogue, Bodleian Library press mark 2591a 2 (2)). A few architectural works may well have been inherited by his younger brother Sydney Smirke, R.A. and auctioned on the latter's death in 1878 (Bodleian press mark 2591e 59). But the bulk of Smirke's collection was disposed of by Messrs. Sotheby, Wilkinson & Hodge in 1868. The auction catalogue (Bodleian press mark MUS. BIBL. III 8°575 (7)) reveals wide interests in topography, antiquities and engineering as well as an unexpected appetite for theology. Apart from its intrinsic interest as the library of a Regency architect, the catalogue supplies several hints as to Smirke's sources of inspiration in matters of style and construction. The following is a list of works in his possession relating solely to building techniques. It is based on the 1868 catalogue, except for those items marked with an asterisk which come from the 1878 sale.

G. Atwood: *A Dissertation on the Construction and Properties of Arches* (1801).

W. Banks: *A Treatise on Mills* (1795).

P. Barlow: *A Treatise on the Strength of Timber, Cast Iron, Malleable Iron and other materials* (1837).

*B. Forest de Bélidor: *La science des ingénieurs dans la conduite des travaux de fortification et d'architecture civile* (Paris, 1739).

W. Bland: *Construction in Arches, Piers etc.* (1839).

L. A. de Cessart: *Description des travaux hydrauliques*, ed. Du Bois d'Arneuville, 2 vols. (Paris, 1806).

E. M. Gauthey: *Traité de la construction des ponts . . . Publié par M. Navier*, 3 vols. (Paris, 1809–16).

J. Knowles: *An inquiry into the means which have been taken to preserve the British Navy . . . particularly from that species of decay now denominated dry-rot* (1821).

[1] Pasley: *Practical Architecture*, pp. 42–3, 45.

[2] "Report from the Select Committee on the Office of Works, 1828," *Parliamentary Papers*, 1828, IV, 394.

SIR ROBERT SMIRKE: A PIONEER OF CONCRETE CONSTRUCTION

J. C. Krafft: *Plans, Coupes et élévations de diverses productions de l'art de la Charpente, exécutées tant en France que dans les pays étrangers* (Paris, 1805).

*D. Laing: *Plans, Elevations and Sections, of Buildings, public and private, executed in various parts of England, including the new Customs House, London* (1818).

R. McWilliam: *An Essay on the origin and operation of the Dry Rot, with a view to its prevention or cure* (1818).

The Mechanics' Magazine, 1823–38, 29 vols.

Modèles de Menuiserie (Paris, 1825).

C. L. M. H. Navier: *Projet pour l'Etablissement d'une Gare à Choisy* (Paris, 1811).

C. W. Pasley: *Observations on Limes, Calcareous Cements, Mortars, Stuccoes and Concrete* (1838) containing autograph notes and MS. insertions.

J. R. Perronet: *Description des projets et de la construction des Ponts de Neuilly, de Mantes, d'Orléans et autres*, 2 vols. (Paris, 1782–83).

C. J. Richardson: *A Popular Treatise on the warming and ventilation of buildings* (1837), presentation copy from the author with autograph note.

J. Rondelet: *Traité Théorique et Pratique de l'Art de Bâtir*, 4 vols. (Paris, 1805–10).

G. Semple: *A Treatise on Building in Water* (Dublin, 1776).

J. Smeaton: *A Narrative of the building, and a description of the Eddystone Lighthouse* (1791).

J. Smeaton: *Technical References to the Plates illustrative of the descriptive Account of the Eddystone Lighthouse*, with MS. insertions.

J. Smeaton: *Reports of the late John Smeaton made on various occasions in . . . his employment as a civil engineer*, 3 vols. (1812).

C. J. Toussaint: *Traité de Geometrie et d'Architecture*, livraisons 1–17 (Paris, 1811).

Transactions of the Institute of British Architects, Vol. I (1836–42).

Transactions of the Institution of Civil Engineers, Vols. I–III (1836–40).

T. Tredgold: *Elementary Principles of Carpentry* (1820).

T. Tredgold: *Principles of Warming and Ventilating Public Buildings, Dwelling Houses, etc.* (1824).

T. Tredgold: *A Practical Essay on the Strength of Cast Iron etc.* (1824 ed.).

L. J. Vicat: *Recherches expérimentales sur les chaux de construction, les bétons et les mortiers ordinaires* (Paris, 1818).

C. F. Viel (de Saint Maux): *Dissertations sur les Projets de Coupoles de la Halle au Blé de Paris* (Paris, 1809).

C. F. von Wiebeking: *Traité contenant une partie essentielle de la science de construire les ponts . . . pour servir de Complément aux Oevres de Perronet* (Munich, 1810).

J. Williams: *An historical account of sub-ways in the British Metropolis, for the flow of pure Water and Gas into the houses of the Inhabitants, without disturbing the pavements* (1828).

PLATE V

Portrait of (Sir) Robert Smirke, by George Dance, R.A. (1809)

PLATE VI

(*a*) From a drawing by T. H. Shepherd (*Metropolitan Improvements*, 1829)
Courtesy of R.I.B.A.

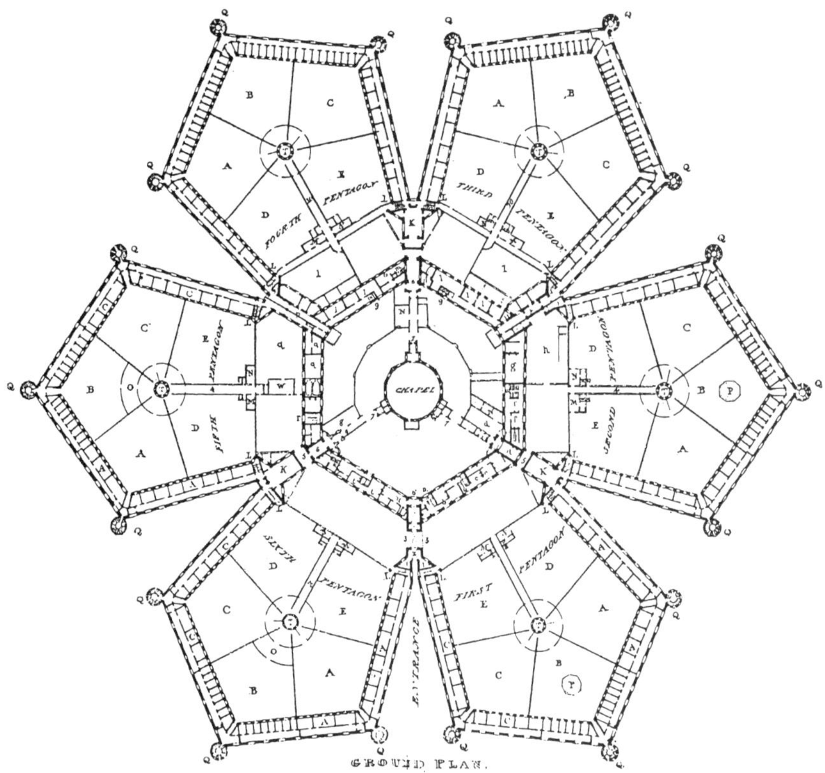

(*b*) From G. Holford, *An Account of the Penitentiary* at Millbank (1828)
Millbank Penitentiary, or the Panopticon
Courtesy of R.I.B.A.

PLATE VII

Capt. C. W. Pasley, R.E.
(painted 1810)

B.A. Add. MS. 41766 Crown Copyright

PLATE VIII

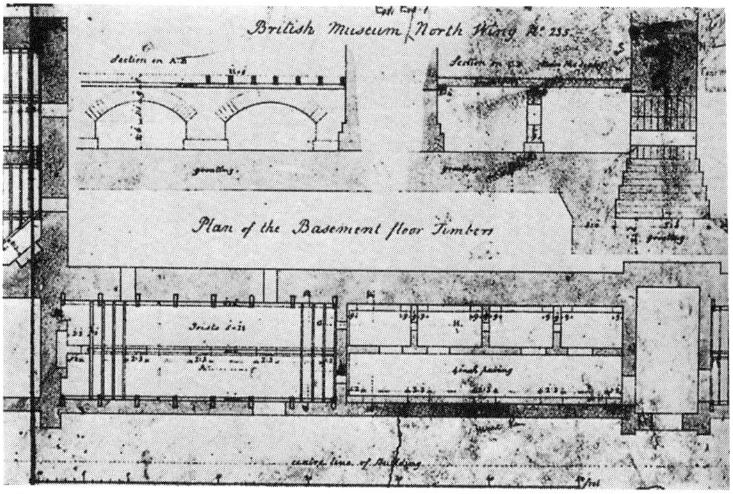

(*a*) Concrete Foundations; British Museum; North Wing. Smirke (1833)

Crown Copyright

(*b*) London Custom House, rebuilt by Smirke (1825–29). From a drawing by T. H. Shepherd, *Metropolitan Improvements*, 1829

Courtesy R.I.B.A.

2

Concrete and the structural use of cements in England before 1890

B.L. Hurst

The years between the invention of Roman cement in 1796 and 1890, by when Portland cement had effectively replaced all its competitors as the binder for mortar and concrete, also saw the development of concrete and its increasing use for foundations and for fireproof flooring. This paper reviews that development and the use of cements in buildings by reference particularly to the *Transactions of the Royal Institute of British Architects*, which was the forum for discussion because building design was the preserve of architects, generally unassisted by engineers. The next paper[44] takes over the story with the advent of reinforced concrete building frames by the specialists who had crossed the Channel and the Atlantic to take out English patents.

Introduction

During the nineteenth century most buildings were designed and drawn in an architect's or surveyor's office, and their drawings included all necessary details of the structure. The 1909 London County Council (General Powers) Act[1] was the first statutory document to lay down floor loadings, permissible stresses and a basis of design and hence it was not until then that consulting engineers started to be concerned with building structures. So during the period covered by this paper, the Institution of Civil Engineers was not the place to seek guidance on common practice, but rather the Royal Institute of British Architects. In those days the Institution of Civil Engineers was concerned with engineering works, leaving all aspects of buildings to architects.

2. Certain types of building were designed by engineers, such as railway stations, but in those cases they generally acted as architects, and parts of buildings were designed by specialist engineers—R. M. Ordish and J. W. Grover designed the dome of the Royal Albert Hall in London for Major General Scott, himself an engineer, but acting in that case as an architect. The remainder of the ironwork, the masonry, the foundations and the fireproof flooring were, however, designed and detailed in Scott's own office. Architects undoubtedly had assistance with their structural designs but that assistance is seldom identified, unless it was obtained from the supplier of the iron beams or the fireproof flooring or the reinforced concrete components. Foundations, masonry, unreinforced concrete and similar non-specialist components were detailed by the architect, any design necessary being either based on experience of satisfactory behaviour or by reference to published formulae, which were generally derived empirically from tests.

3. It is notable that the first paper in Volume 1 of the *Transactions of the Institute of British Architects* is George Godwin's 'Prize essay upon the nature and properties of concrete, and its application to construction up to the present period',[2] which is followed by notes on the concrete foundations at Westminster New Bridewell[3] and on concrete underpinning at Chatham Dockyard[4] and later on M. I. Brunel's experiments on reinforced brickwork.[5] These papers and others from the *Transactions* are referred to later in this paper.

2 CONCRETE AND THE STRUCTURAL USE OF CEMENTS

Binders and aggregates

Cements

4. Today we know what is meant by the terms 'concrete', 'mortar' and 'cement': unless qualified by additional words, cement means Portland cement, mortar means a mixture of Portland cement and sand, and concrete is made with coarse and fine aggregates and Portland cement. Before and during most of the nineteenth century such certainty did not exist, and until 1790 the distinction between cements which set or did not set under water, termed hydraulic or non-hydraulic, was not known, or if known was not really appreciated.

5. John Smeaton's experiments to discover a suitable mortar for use in the construction of the Eddystone Lighthouse were reported in his *Narrative* first published in 1791,[6] where he set out the properties of 'water cements' and distinguished between cements which set by chemical action, under water, and those which merely hardened, by carbonation and evaporation, and which dissolved when placed in water. He deduced that ground volcanic rock—either puzzolana (now usually spelt pozzolana) from Italy or terras (also known as trass and terrace) from Germany via Holland—were needed as a constituent of a water cement. His recipes never came into general use, because in 1796 James Parker patented his Roman cement.[7] Parker's Roman cement, also known as Frost's, was a quick-setting artificial hydraulic cement, 'artificial' meaning manufactured, not 'natural' as were the mined volcanic hydraulic cements. It was used for waterproof renders and for mortar for laying bricks in wet conditions during the first half of the nineteenth century, until such time as true Portland cement became available in 1845–50.

6. 'Cement mortar', as used by Sir Robert Smirke in the brickwork above the lime concrete underpinning to Custom House in 1826 and as used by Mark Isambard Brunel in the Thames Tunnel, was made with Roman cement.[8] The remarkable watertightness of the Thames Tunnel is a tribute not only to the workmanship but also to the use of Roman cement, without which it would not have been possible. Robert Stephenson made a rule never to use lime mortar in the arches of tunnels, but to build them with (Roman) cement exclusively.[8]

7. Mortar or render made with Roman cement is generally a brownish colour, and is much harder than the white or dirty grey lime mortar of the period. It was used for precast embellishments to be incorporated with stucco and for mortars where its quick setting or strength or hydraulic properties were needed. It was, however, seldom mixed with fine and coarse aggregates to make concrete, because it set too quickly. References to Roman cement are to be found up to 1880, when it was still significantly cheaper than Portland cement, but not much evidence of its use is to be found in those later years.

8. Portland cement was the term used by Joseph Aspdin in his patent of 1824, but it was not until about 1845 that I. C. Johnson, the manager of the cement works at Swanscombe, Kent, burnt the new materials at a high enough temperature to produce what we now know as Portland cement,[9] which fairly quickly replaced Roman cement in mortars and renders but, with a few exceptions, was not mixed with aggregates to make concrete for use in buildings until about 1865.

9. In 1850 Dobson, in his *Rudimentary Treatise on Foundations and Concrete Works*,[10] which is basically on engineering works, not buildings, is quite clear: 'concrete is made of gravel, sand and ground lime, mixed together with water' and 'beton' is concrete made with hydraulic lime. Neither Parker nor Frost nor Roman cement appear in the index. Others define concrete as a combination of aggregates, cement and water mixed at the same time, but 'beton' meant mixing the sand, cement and water together first before adding it to the coarse aggregates.

10. Concrete figures largely in the discussion following a paper read to the RIBA in 1875 on 'New materials and recent inventions connected with building',[11] but in many instances it is not apparent if Portland cement or lime was used as a binder, such was the confusion of terminology in general use at that time. It is therefore necessary to be particularly cautious in making deductions from descriptions of 'concrete' or 'cement' in nineteenth-century references.

11. In some particular uses gypsum was used as the binder for mortars and concretes—for flooring in the Nottingham area and for concrete in Dennett's floors.

Aggregates

12. Whilst sand was initially mixed with lime or cement, Roman or Portland, for mortar, a number of materials were used for coarse aggregate to make concrete.

13. Lime concrete for footings and underpinning was generally made with gravel or ballast, but the fireproof flooring systems variously specified coke breeze (obtained from gas works), cinders, broken brick, burnt clay, limestone and even clean sieved rubbish. 'Rubbish' was the term used to describe the arisings from taking down old buildings and was therefore a mixture of broken brick, plaster and mortar.

14. There was a theory that concrete made with aggregates that had gone through fire performed better in a fire than that made with those which had not, hence the use of breeze, brick and burnt clay as aggregate for concrete for fireproof flooring. Crushed brick is a good aggregate for concrete but breeze is much more commonly found. Breeze is liable to be incompletely burnt and to contain a proportion of coal, giving a combustible content to 'fireproof' floors. Breeze also tends to be acidic, because of the sulphates and sulphides originating in coal, and hence can actively promote rusting of embedded iron or steel in damp conditions.

A solution to foundation problems

15. The history of concrete foundations is covered elsewhere, in Chrimes's paper,[45] but it is necessary to mention its early use here, if only because of the use of cement mortar in the brickwork. There is little doubt that Smirke's was the first effective use since Roman times of concrete in foundations, when he used lime concrete in his remedial works for the sinking foundations at Millbank Penitentiary (1817–22) and for the underpinning following the partial collapse of the London Custom House (1825–27).[8]

16. At Custom House he excavated 12 ft wide and 12–15 ft deep, down to a bed of gravel, and filled the trenches with lime concrete, on top of which he bedded large York stone slabs or landings, and then pinned up to the old footings with slate above 12 courses of hard stock bricks laid in (Roman) cement mortar (Fig. 2). The lack of any sign of subsequent movement in the superstructure is a testament to the success of his underpinning work.

17. In the following years, Smirke continued to use lime concrete as an artificial foundation when building on bad ground, sometimes as a raft, as in the rebuilt centre of Custom House, and sometimes as a footing, doubtless in conjunction with brick footings laid in Roman cement mortar.

18. Occasionally a lime concrete footing is found in buildings on a smaller scale, but that must be regarded as an exception and would only have been considered in particularly soft ground; normally the brick footing courses were laid directly on the earth.

19. Portland cement concrete footings started to be use for large and important buildings in the 1860s but did not come into general use until the 1880s and even later for domestic buildings.

The early use of mortar and concrete in flooring

20. Plaster floors for upstairs rooms were used in the east Midlands from Elizabethan times until the nineteenth century, with straw or more commonly reed laid over the joists and then covered with 2 in or more of plaster or mortar made with gypsum or occasionally lime, mixed with burnt clay or broken brick or ashes[12,13] (Fig. 3). Plaster floors were also to be found in the Cotswolds, but survivals are more to be expected in the Nottingham area, where they continue to serve their purposes, but are known to be susceptible to point loads from bed legs, baths and pianos.

21. Recipes for plaster floors vary from neat coarse gypsum, to the use of the waste mixture of lime and ashes from the bottom of a lime kiln, to one-third lime, one-third coal ashes with one-third loamy clay and horse dung.

22. Lime ash concrete was shown by John Foulston as a finish to the stone slabs carried on cast iron inverted tees for the suspended floors of Bodmin Asylum (1818),[14] and it

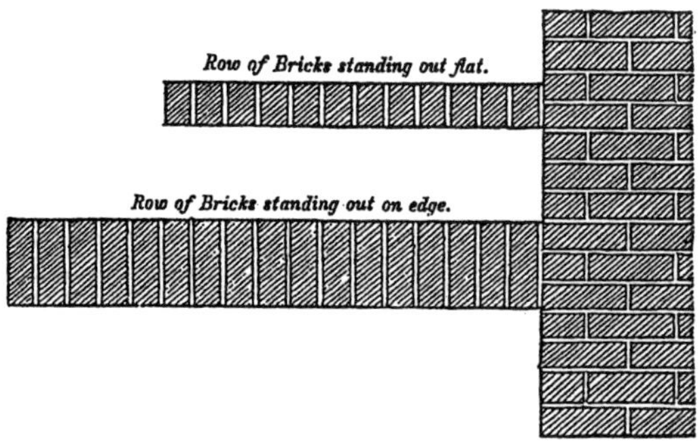

Fig. 1. A method of testing and comparing cements used by Pasley and others. Neat (Roman) cement was used and each brick held up by pressing the point of a trowel firming against it for 2 or 3 minutes. The next brick was similarly set 6 or 7 minutes later. Francis White & Co., the cement manufacturers, repeatedly struck out 29 bricks in this way before the joints gave way

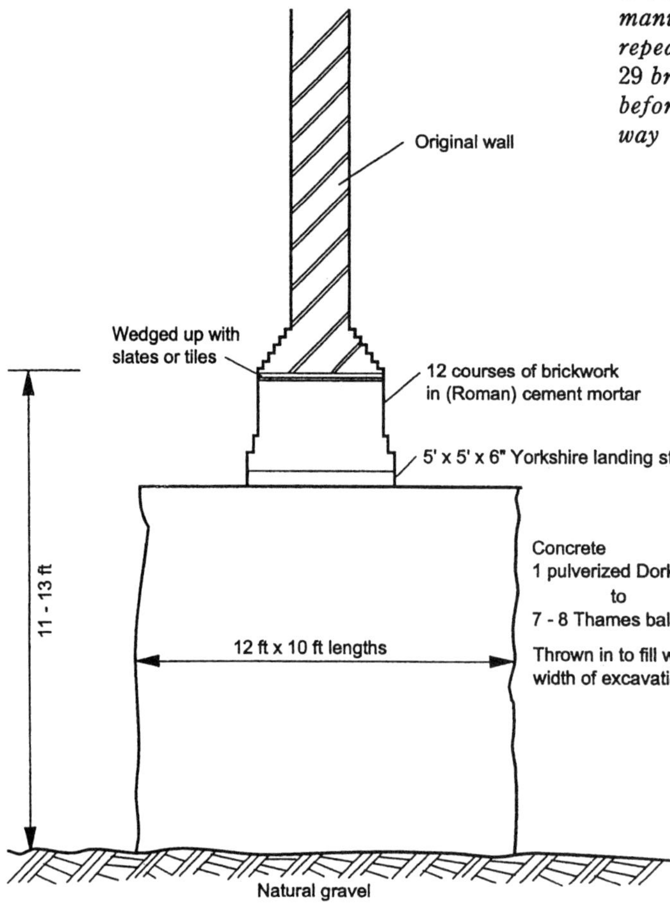

Fig. 2. Section through Robert Smirke's underpinning of Custom House in 1825, as described by Pasley[8]

Fig. 3. Lime ash on reed flooring (by courtesy of Philip Hartley)

appears probable that this is the germ of the idea developed by Henry Hawes Fox for the fireproof flooring he invented for the private asylum he built at Northwoods, near Bristol, in 1833.

23. Fox's patent of 1844[15] for the floor, which was developed and marketed by James Barrett, and is found in very many buildings up to 1885 or even 1890, consists of cast iron inverted tee joists at about 1 ft 6 in centres with stout timber laths spanning between the outstanding flanges to provide formwork for 8 : 1 : 1 (sieved rubbish or road grit : coal ashes : lime) concrete or pugging (Fig. 4). The surviving floors of Northwoods show that this is a far stronger material than would be expected from the weak mix specified. The concrete was laid on ¾ in of 1 : 2 lime mortar pushed through the gaps between the stout laths to provide a key for the plaster on the soffit which completed the encasing to the cast iron tees and provided the first truly fire-resisting floor, since previously the iron joists supporting brick arches or stone slabs had been exposed on the soffit and thus vulnerable to fire. Barrett was not only one of the first to appreciate the need to provide protection against fire to the undersides of the iron joists, but also the first to appreciate the value of composite action, when he wrote in 1853: 'the force of compression acts upon the joists, only through the medium of the concrete, and this material is well known to be of the best for resisting that force'. He also appreciated the 'considerable accession of strength' by building the ends of the joists firmly into the walls.[16]

24. After the patent had expired in 1859, the Fox and Barrett system came into general use because its components—iron joists, timber laths and concrete—were readily available to anyone, and the above principles, having not been clearly enunciated, were not perpetuated. An ordinary lath and plaster ceiling was used, with battens interposed between the ordinary laths and the stout laths to accommodate the bottom flanges of the wrought iron joists, and the finish was timber boards on small joists resting on the top flanges, left exposed above the concrete, which was then little more than pugging. It is in this form that Fox and Barrett flooring is more likely to be found (Figs 5 and 6).

25. A builder from Nottingham, Charles Colton Dennett, also developed a concrete fireproof flooring system using tried material when in 1857 he patented 'arches composed of sulphate of lime and an artificial puzzolana of burnt clay or porous cinders' effectively gypsum concrete, as was already in use for plaster floors in his home town.[17] His arches, however, sprung off cast or wrought iron inverted tees with bottom flanges directly exposed to fire and hence were no more fireproof than a similar floor with brick arches, but they were marketed as being less costly (Fig. 7).

26. As the nineteenth century progressed, losses of buildings in fires produced greater and greater pressures to make buildings fireproof, and indeed a series of Acts of Parliament sought to enforce the requirement, initially for parts of buildings needed as means of escape and latterly for certain types of building. However, there was considerable discussion and confusion about what constituted fireproof or fire-resisting construction, with a school of thought which regarded incombustible as synonymous with 'fireproof', generally citing satisfactory behaviour probably in small fires which were insufficiently hot to cause failure of the exposed structural metal. This led to the fire brigades being unwilling to enter burning 'fireproof' buildings because of the risk of abrupt and catastrophic collapse when the exposed cast or wrought iron columns and beams succumbed to the heat.[18]

27. There was nevertheless an increasing market for fireproof flooring, as the number of patents for novel ways of solving the problem economically displays. Almost all of the patents used Portland cement for screeds and bedding and many incorporated Portland cement concrete. Further details of many of the systems can be found in books by Sutcliffe[19] and Potter,[20] and in a paper by Webster.[21]

Hoop iron in brickwork and concrete

28. Hoop iron—a thin strip of wrought iron from 1½ × 15 gauge (1.8 mm) to ¾ in × 20 gauge (1.0 mm)[8] termed 'hoop' because of its use by coopers for binding barrels—was in general use as bond in brickwork, from about 1830 to

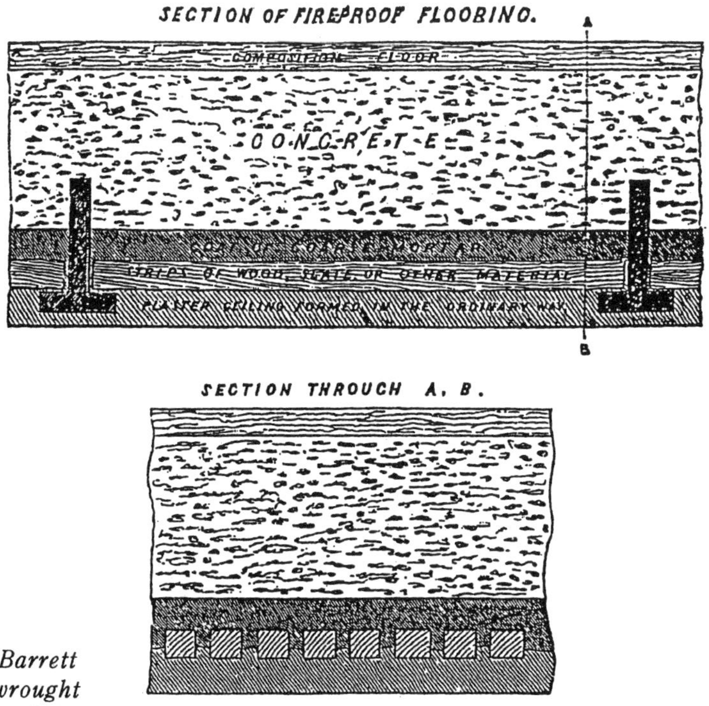

Fig. 5. Fox & Barrett flooring with wrought iron joists, after 1851 or 1852[16]

Fig. 4. Fox & Barrett flooring with cast iron joists, before 1851 or 1852

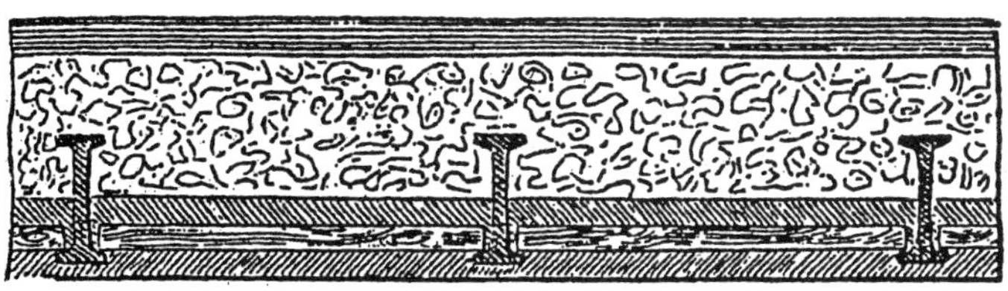

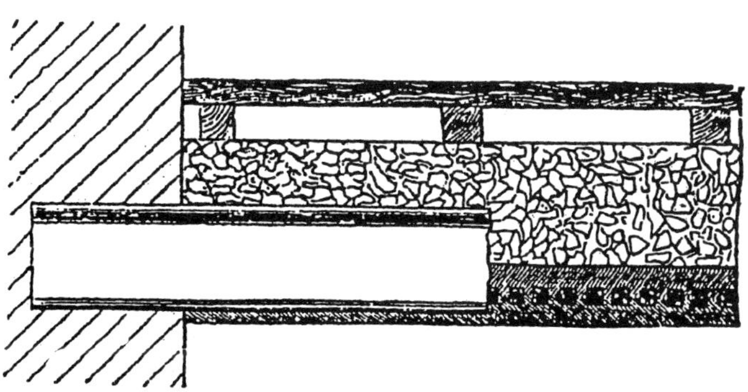

Fig. 6. Fox & Barrett flooring with battens and plaster on ordinary laths

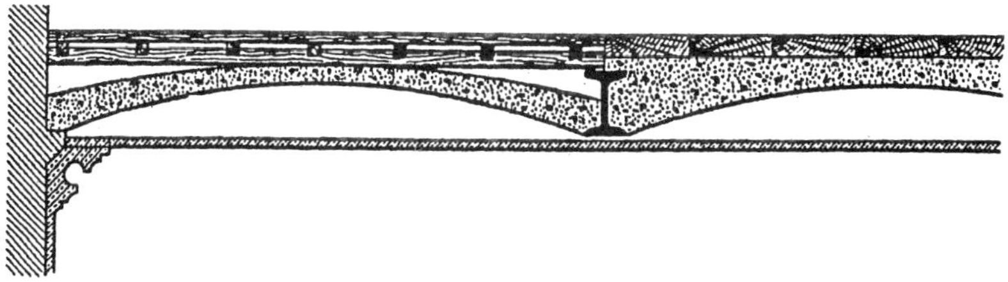

Fig. 7. Dennett's arch floor

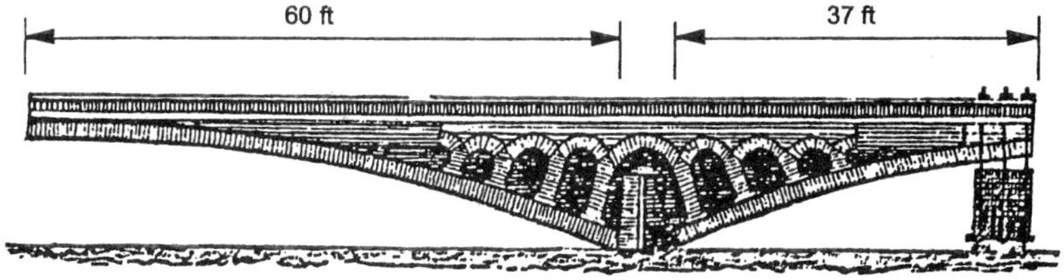

Fig. 9. M. I. Brunel's reinforced brick cantilevers[27]

Fig. 10. Hoop iron bond in party wall at Mansion House Square site

the end of the nineteenth century, as a replacement for bond timbers, particularly chain bond timbers, in the centre of walls, which suffered from rot whatever timber was used.[8] Advocates of in situ concrete walling adopted it to control cracking, and one recommended 'a prodigal expenditure of hoop-iron bond'.[22]

29. It was that material which Brunel and Francis adopted for their experimental brick beams in 1835–8. Francis and Sons, at their Vauxhall (Roman) cement works, and Brunel, at the Thames Tunnel site, both built beams of brickwork laid in neat Roman cement with hoop iron bond laid in the bed joints, and tested them to failure. Pasley made his own tests at Chatham and also tested an unreinforced beam and a reinforced beam built with lime mortar for comparison. When his four-course high brick and cement beam, incorporating five strips of hoop iron, was examined after gradual collapse the centre strip was found to be bent, the two lower strips had fractured and the two upper strips were buckled and the bricks crushed. The unreinforced beam failed abruptly at a small load (498 lb) and the iron strips in the lime mortar beam were 'drawn inwards', that is to say, the bond with the mortar failed (Fig. 8).

30. This clear understanding of the behaviour of reinforced masonry, as reported by Pasley in his *Observations* in 1838,[8] was not adopted and applied practically by the designers of buildings and does not seem even to be mentioned in the *Transactions of the RIBA*, the forum for discussion about building construction.

31. This is perhaps, surprising considering the publicity that surrounded Brunel's works at the Thames Tunnel, where he had also built 'two extraordinary semi-arches for experiment' of brickwork in neat (Roman) cement with hoop iron laid in the courses cantilevering either side of a central pier and rising 10 ft, one 60 ft and the other over 37 ft long and loaded with 62 700 lb (Fig. 9).[5]

32. The only hoop iron that engineers of today are likely to encounter will be as bond in brickwork, as could be seen recently in quantity on the Mansion House Square site at No. 1 Poultry in the City of London, hanging down from the back of the Cheapside elevation where the party walls had been demolished (Fig. 10). In Thomas and Frank Verity's specification for the new building to be erected at 96 Piccadilly in 1890, the contractor was required to allow for providing 5 tons of hoop iron in the brickwork and, incidentally, he was required to use lime concrete footings for the walls but cement concrete foundations for the cast iron stanchions.[23]

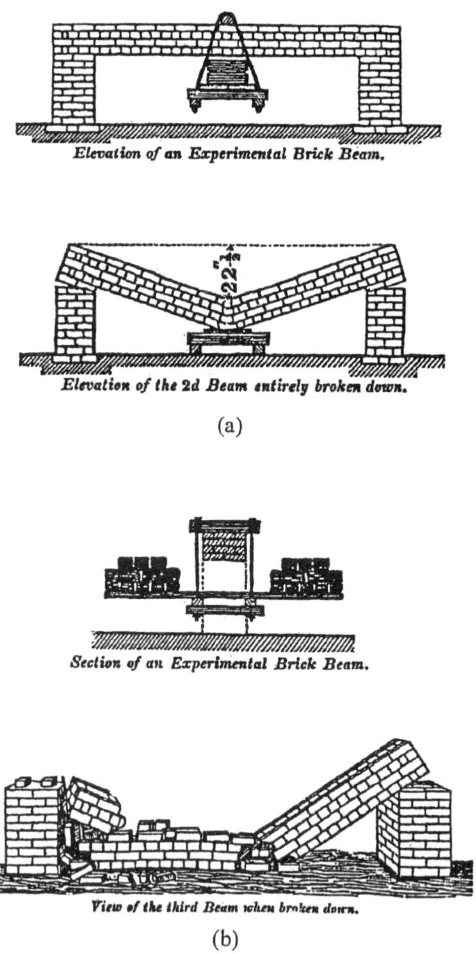

Fig. 8. (a) Reinforced Roman cement beam failed with 4532 lb. (b) Reinforced lime mortar beam failed with 742 lb[8]

Flat floors without iron: 'Bold; be not too bold'

33. The above quotation is taken from the heading of the chapter in George Sutcliffe's book on *Concrete its nature and uses*,[19] published in 1893, on flat concrete floors without iron, of the sort that engineers today would not contemplate building and, if they know of them, would dread encountering in a survey or appraisal of a Victorian building, for fear of knowing what to say. Is over 100 years of not unsatisfactory use a reason to recommend or permit retention of a floor which cannot be shown by calculation to be structurally sound and which would collapse abruptly and catastrophically if anything occurred to disturb its abutments?

34. There were at that time a number of architects and engineers who held that steel or iron joists or reinforcement in floors constituted unnecessary expense, and indeed some who held that they were a positive disadvantage.[24] There are reports of plain concrete floors 10 ft × 10 ft × 4 in thick and even 26 ft × 20 ft × 7 in thick which behaved satisfactorily and stood up to tests. Others built shallow unreinforced concrete arches and even cantilevers. Seddon's reported tests included an unreinforced slab 14 ft 6 in × 13 ft 6 in clear span × 6 in thick which broke suddenly under a weight of 10 tons or 120 lb/sq. ft.[25] Most of these slabs had good bearings on thick solid walls and many were of concrete made with broken brick aggregate, thus ensuring a tenacious material. Sutcliffe[19] goes on to say:

> They are, however, economical, and there is no reason why they should not be used in houses and other buildings where they will not be subjected to intense heat or heavy impingent loads, up to spans of 10, 12 or even (with care) 14 feet.

It is probable that some of these unreinforced floors are still extant, serving their purpose without revealing that they cannot be justified to modern standards—to engineers appraising Victorian buildings Sutcliffe's adage could be changed to 'Beware'! (examples are shown in Table 1).

35. Another form of unreinforced construction which certainly does exist in some quantity is tile creasing—flat floors and arches formed with up to three layers of ordinary flat clay roof tiles bedded in neat cement or strong cement mortar and laid to break bond (Figs 11 and 12).

36. Tile creasing is generally held to have been invented by Charles Fowler in about 1835; he used it for the floors and roofs of Hungerford Market, spanning between iron beams $4\frac{1}{2}$ ft apart to form the terrace.[26] Pasley, however, says it was first proposed by 'the late ingenious Mr. Smart'.[8] Brunel used it for supporting the garden above his drawing office, where cast iron beams at 5 ft centres satisfactorily supported tile creasing spanning 5 ft.[27]

37. That form of construction, flat and arched, can be found today in entrance hall floors, flat roofs, porch roofs, balconies and vaults of early and mid-Victorian houses built in London. It was also used and still exists as an arched non-combustible ceiling to timber floors on cast iron beams in the Palace of Westminster and the Tower of London, and no doubt elsewhere.

Practical application of reinforcement to floors

38. William Boutland Wilkinson's patent of 1854[28] is generally agreed to be the beginning of reinforced concrete in England, and the coffered concrete floor reinforced with old colliery ropes in the ribs in the reinforced concrete house he built in Newcastle in about 1865, as described in detail in the careful records made during its demolition in 1954,[29] is an example of his work. Wilkinson's patent also cites the use of hoop iron as reinforcement. The papers and discussion at RIBA meetings in the 1860s and 1870s, when the use of concrete for building was discussed at length,[11,22,30–35] indicate, however, that neither Wilkinson's patent, nor Lambot's nor Coignet's patents of the following year, were adopted by the building industry.

39. Reinforced concrete floors of the 1860s and 1870s were not distinguished from the other forms of fireproof flooring which were being developed to satisfy the demand arising out of public concern about the increasing cost of damage and the loss of life caused by building fires and to enable compliance with the resulting legislation. The majority of fireproof flooring systems incorporated rolled or fabri-

Table 1. Examples of concrete floors without reinforcement[14]

No	Length (ft)	Breadth (ft)	Thickness (in)	Where used	Authority	Remarks
Flat slabs						
1	20	20	14	Atrium Vestae, Rome	Prof. Middleton	Ancient Roman work
2	6	6	3	Footway of bridge	F. Caws	
3	14·5	7	6	Brigade depots	Col. Seddon	1 P.c. to 4 breeze, slag, brick, or burnt clay, to pass $\frac{3}{4}$ in mesh
4	10	10	4	—	F. Caws	—
5	12·5	11·5	6	—	C. A. Adams	—
6	21	12·5	13	Warehouse	F. Caws	1 P.c. to 4 brick
7	26·5	19·5	7	—	F. Caws	Thicker for 9 in around margin, 1 P.c. to 4 brick
Arched slabs						
8	—		9	12 to 5	Stables	W. B. Wilkinson Granite-concrete
9	—	9·5	11 to 4	Corridor	W. B. Wilkinson	Granite-concrete
10	50	12	11 to 3	Warehouse	Broughton	Carrying 'immense weight of machinery and men'
11	—	12	15 to 6	Pantechnicon	Lockwood	—
12	19	13	11 to $4\frac{1}{2}$	House	J. Tall	—
13	40	16	11 to $4\frac{1}{2}$	Drawing room	J. Tall	—
14	70	16·5	7 to $3\frac{1}{2}$	Roof of barn	Trench	Intrados rising 1 in per ft of span—1 P.c. to 5 gravel
15	90	20	15	Roof of warehouse	W. C. Street	Segmental arch with 3 ft rise. —1 P.c. to 5 Thames ballast
Cantilevers						
16	—	4	8 to 3	Balcony	Potter	1 P.c. to 5 brick
17	50	4	11 to 3	Balcony	Broughton	1 P.c. to 4 ballast (part, 1 to 6).

cated iron joists at 1 ft 6 in – 3 ft centres supporting various forms of cellular clay tile or concrete usually made with breeze aggregate, sometimes containing a particular form of reinforcement.

40. These floors led in the 1890s to the ordinary filler joist floor—with iron or later steel joists at about 3 ft centres embedded in a slab of breeze concrete—which still survives in quantity and will be frequently encountered by engineers working on buildings up to the First World War (Fig. 13).

41. The other type of fireproof floor, which continued in use during the 1920s and can indeed be found up to the Second World War, is formed with hollow clay tubes spanning between steel joists, with the ends notched to embrace the bottom flange and finished with concrete, usually with breeze aggregate, or with screed; examples of this were by J. A. King (Fig. 14). Homan & Rogers (Fig. 15) and Fawcett (Fig. 16).

42. Thaddeus Hyatt, who crossed the Atlantic to England in the early 1870s and was responsible for over 40 patents concerned with pavement lights and concrete flooring, had a series of tests carried out by Kirkaldy and published in 1877 a pamphlet setting out why reinforcement was only needed in the bottom of the Portland cement concrete slab and why all iron needed to be protected by concrete cover from fire.[18] The need to provide cover to protect iron was stated as early as 1838 by Pasley, who suggested a coating to tiles laid in cement,[8] and in 1853 by Barrett, who referred to the plaster to protect the iron joists from the action of fire from below,[16] but it was not generally understood until the 1870s and 1880s that it was not sufficient just to provide a noncombustible structure to achieve fire resistance. Hyatt and the tests he had carried out certainly helped this understanding.

43. Other patents were taken out for what was effectively reinforced concrete, by Scott[36] of Royal Albert Hall fame; Matthew Allen,[37] who in 1862 was one of the first to advocate Portland cement and breeze concrete; F. G. Edwards,[38] for I section joists with a breeze concrete nailing strip in the top (Fig. 17); and others, but none of them came into general use or can be expected to have survived in any quantity.

44. However, it would not be unexpected to find examples of the two floors mentioned earlier—Fox & Barrett and Dennett (or later Dennett & Ingles).

In situ walls

45. The other type of patent in the lists for 1860, 1870 and 1880 for concrete work is for apparatus for forming concrete walls, taken out by Tall (7), Drake (3), Payne (2) and others. These were for formwork systems and sometimes included loose descriptions of reinforcement and recipes for concrete—there is even one system of climbing formwork! Most of them were more ingenious than useful and do not appear to have benefited their inventors or the industry, and their products do not appear to have survived in any quantity. They were generally advocated as a less costly way of providing 'workers' dwellings', a philanthropic tendency of that time, and hence it would not be totally unexpected to find mid- and late Victorian small houses with *in situ* concrete walls. Anerley New Church, London, a remarkable neo-Gothic in situ concrete building by W. J. E. Henley, manager of the Concrete Building Company (1883), now converted to housing, shows that it is always wise to be alert for the unexpected.

46. Discussions at the RIBA were a dialogue between the disciples of concrete and the sceptics, who thought that even if it remained standing when the formwork was struck it would crumble away within a year or two. There was probably right on both sides, because they were generally talking of different materials—the sceptics had in mind examples of concrete made with lime or defective cement; and the disciples carried out tests and supervised the work closely to ensure a satisfactory product. It is probable that some examples of the latter survive, such as 63 Lincoln's Inn Fields, with in situ mass concrete walls to Simmon's patent and with Hyatt's floors[39] (Figs 18 and 19).

Precast walls and precast dressings to brickwork

47. It is not generally realized that a substantial proportion of the bay columns with

14 CONCRETE AND THE STRUCTURAL USE OF CEMENTS

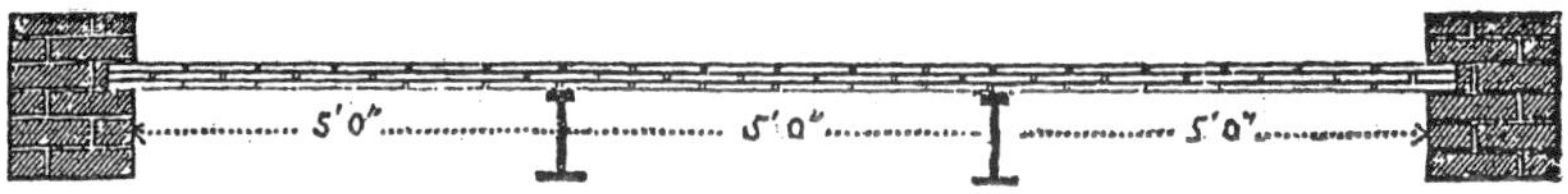

Section of a Flat Roof of plain Tiles and Cement.

Fig. 11. Flat roof of plain tiles and cement (tile creasing)[27]

Fig. 12. Underside of roof formed in tile creasing

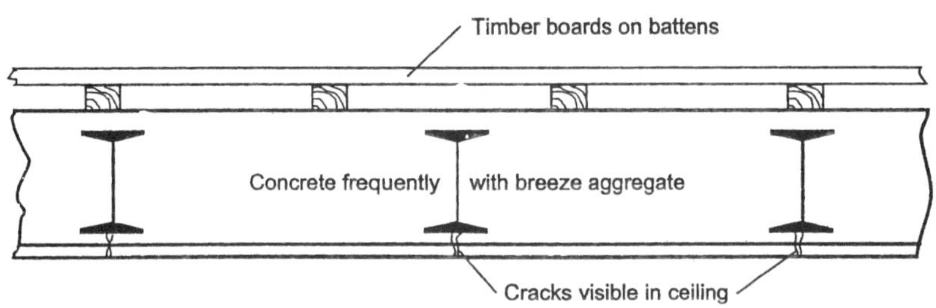

Fig. 13. Ordinary filler joist floor, with iron (later steel) joists at 2 ft 6 in to 3 ft 6 in centres

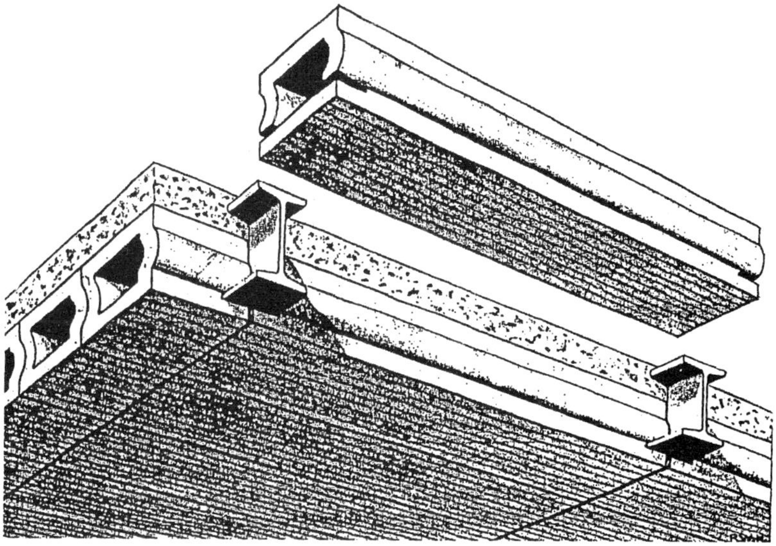

Fig. 14. King floor: fireclay tubes usually 3 ft long × 6 in wide × 4 to 8 in deep

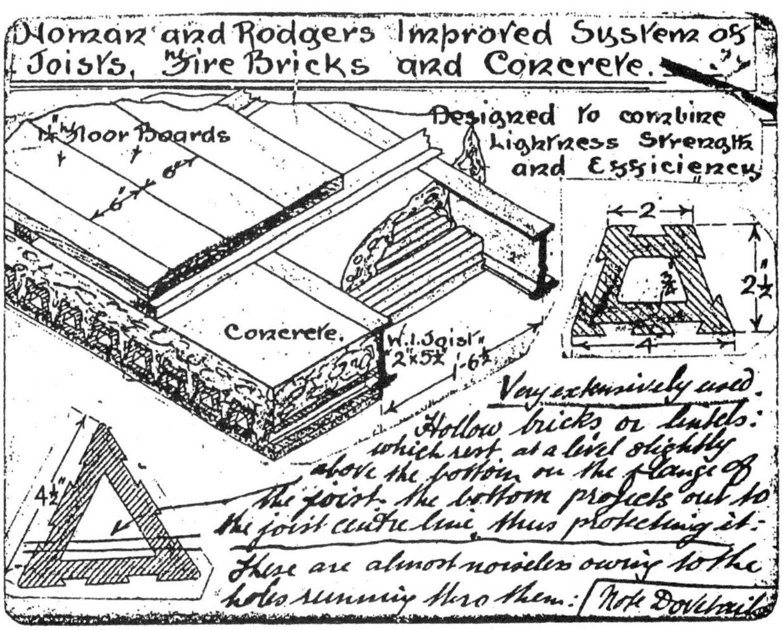

Fig. 15. Homan & Rogers floor

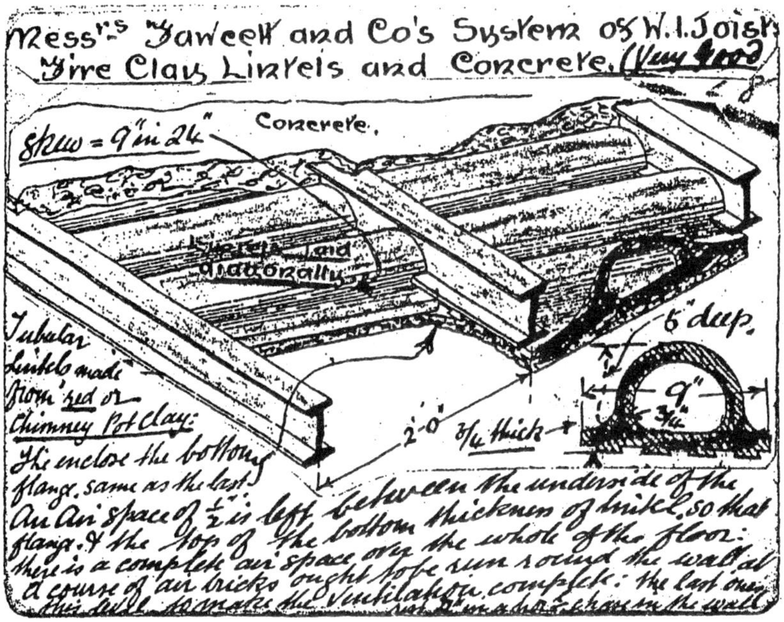

Fig. 16. Fawcett's floor

Fig. 17. F. G. Edwards reinforced concrete joists with nailing strips[38]

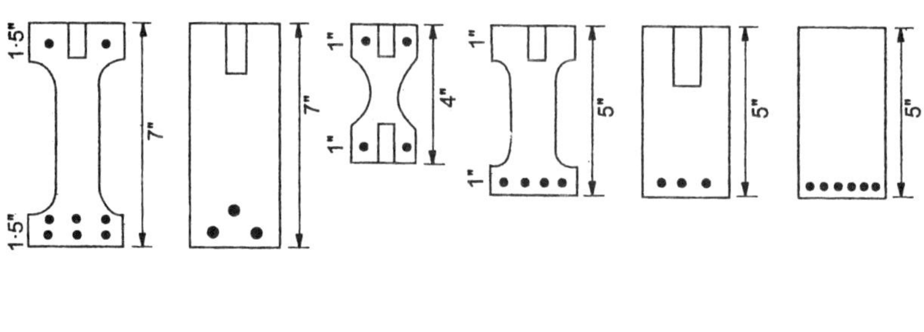

Fig. 18. 63 Lincoln's Inn Fields, with in situ mass concrete walls and Hyatt's floors

ornate capitals and other painted decorative features on the elevations of late Victorian and Edwardian speculative housing are of precast concrete—an economical and more durable substitute for the soft, easily carved stone alternative—nor is it realized that the use of precast components in buildings pre-dates most other uses of concrete.

48. From the early years of the nineteenth century, stucco decoration was precast, using Roman cement, rather than being formed in situ, and, in 1835, William Ranger used precast concrete blocks, made with lime slaked with boiling water, for building dock and sea walls and a number of buildings in London. Precast concrete, or, as it was generally termed, 'artificial stone', continued to be used, as whole blocks, or probably more commonly facings to mass concrete walls, usually with dovetail strips or ribs to bond them to the in situ work.

49. Recently there has been an opportunity to inspect an ingenious precast system, patented by William Lascelles in 1875, in the form of the Central Buffet and Dock Manager's Office at the Royal Albert Dock in London's docklands, whose elevations have been conserved and restored[40,41] (Fig. 20).

50. Lascelles's system[42] employed 3 ft × 2 ft breeze concrete slabs, screwed to timber studs. They were $1-1\frac{1}{2}$ in thick, reinforced with two diagonal wrought iron rods. Similar panels could be used on the inside in lieu of plaster and on the floors in lieu of boards, but were not at Royal Albert Dock. The wall slabs at Royal Albert Dock had a rough cast finish but they could be cast to resemble brickwork or tile hanging and coloured accordingly. The Central Buffet also incorporates a further layer of precast components outside the slabs, and fixed through them to the timber framing, of columns, with bases and ornamental capitals, friezes, and decorative panels, and a precast concrete cornice incorporating the gutter.

51. Lascelles's concrete could be self-coloured—red, buff or grey. Red, obtained by lining the mould with a grout of 'Spanish brown', was used at the Royal Albert Dock buildings, and the finish on surviving unweathered areas is good enough to be mistaken for terracotta. Even the fire surrounds and overmantels in the Central Buffet are Lascelles's precast concrete.

52. Lascelles's brochure lists the range of architectural dressings of all kinds supplied by the company, many no doubt from stock, and it would have been from workshops such as his that the speculative builder would purchase the bay columns and other decorative work for his speculative housing.

53. Lascelles's system is likely to have been chosen for the buildings at Royal Albert Dock on the grounds of speed and economy. Moulds for precast work could be made quickly and cheaply in Lascelles's joinery works so that the precast concrete components could be delivered to site within weeks of being ordered. In his contribution to the discussion at the RIBA in 1876,[42,43] Lascelles said the slabs were removed from the mould in three or four days and were then ready for fixing to the walls.

54. To produce such decoration the only alternatives would be terracotta or stone; the former would take months for pattern and mould making, drying and firing, while stone would have to be sourced as well as carved, so precast concrete was undoubtedly both quicker and cheaper.

55. It is interesting to note that one of the latest concrete repair techniques—realkalization—has been used to conserve a substantial proportion of Lascelles's units at the Royal Albert Dock buildings[41].

56. One of the leading Victorian architects prepared to use this new material was Norman Shaw, who seems to have been retained by Lascelles to produce typical house designs, and it was he who designed the elaborate façade of Lascelles's award winning pavillion at the Paris Exhibition of 1878.

Summary

57. The primary purpose of this paper has been to introduce engineers of today to the development of the structural uses of cements in buildings during the nineteenth century and hence to describe forms of construction they are likely to encounter when working on Victorian buildings.

58. Until 1890 concrete was not used in Britain for building frames, only for slabs,

lintels and occasionally walls. Most forms of fireproof flooring which had been developed since 1850 incorporated concrete or screed and by 1890 this would have been made with Portland cement, which had by then replaced the lime, and occasionally gypsum, used as cement in the 1850s and 1860s.

59. Fireproof flooring frequently finished with timber boards on battens was used in commercial and institutional buildings, but not exclusively, except for the stairs, landings and corridors forming the escape routes. It was also used in the higher class of domestic buildings, but only in the lower floors, or sometimes at basement and second-floor levels, to separate the kitchens from the family living rooms and to separate the bedrooms from the rooms below.

60. Reinforced concrete beams and columns would be unusual, except for lintels, although concrete casing to rolled wrought iron beams (steel would be unlikely before 1890) and occasionally to columns is not uncommon. The concrete would more than likely be made with coke breeze aggregate. Concrete in floors had either breeze or occasionally broken brick aggregate. Gravel was uncommon, but some patent systems used crushed limestone.

61. Two other forms of construction have been mentioned: hoop iron, which can be expected to be encountered in the brickwork of substantial buildings, sometimes in quantity, sometimes only in a couple of courses in each storey height; and tile creasing, which was used for fireproof flooring in houses and also in external construction where timber would be liable to rot.

62. During the second half of the nineteenth century, a number of people had published and patented the principles of reinforced concrete, but the forms of construction in general use indicated that no real understanding of the behaviour of reinforcement in conjunction with concrete existed in general practice. The ingredients were all there waiting for Hennebique and others to compile the recipes.

References

1. London County Council. *London County Council* (General Powers) Act. HMSO, London, 1909.
2. Godwin G. Prize essay upon the nature and properties of concrete, and its application to construction up to the present period. *Trans. Inst. Br. Archit.*, 1836, **1**, 1–37.
3. Abraham R. Concrete used at Westminster new Bridewell. *Trans. Inst. Br. Archit.*, 1836, **1**, 38–39.
4. Taylor G. L. An account of the methods used in underpinning the long storehouse at His Majesty's Dock Yard, Chatham, in the year 1834. *Trans. Int. Br. Archit.*, 1836, **1**, 40–43.
5. Brunel M. I. Particulars of some experiments on the mode of binding brick construction. *Trans. Inst. Br. Archit.*, 1836, **1**, 61–64.
6. Smeaton J. *Narrative of the building of the Eddystone Lighthouse*. London, 1791.
7. Parker J. *A certain cement or terras to be used in aquatic and other buildings and stucco work*. Pat. 2120, 1796.
8. Pasley C. W. *Observations on limes, calcareous cements, mortars, stuccos and concrete, etc.* London, 1838.
9. Skempton A. W. Portland cements 1843–1887. *Trans. Newcomen Soc.*, 1967, **35**, 117–152.
10. Dobson E. *A rudimentary treatise on foundations and concrete works*. London, 1850. (Reprinted Bath, 1970.)
11. Smith T. R. On new materials and recent inventions connected with building. *Trans. R. Inst. Br. Archit.*, 1874–5, 199–216, 221–230.
12. Barley M. W. *The English cottage and farmhouse*. Routledge & Kegan Paul, London, 1961.
13. Stevens H. J. Plaster floors in Derby district. *Trans. R. Inst. Br. Archit.*, 1854, 63.
14. Foulston J. *The public buildings erected in the west of England*. London, 1838.
15. Fox H. H. *Fire-proof roofs, floors and ceilings*. Pat. 10047, Patent Office, 1844.
16. Barrett J. On the construction of fire-proof buildings. *Proc. Instn Civ. Engrs*, 1852–3, **12**, 244–272.
17. Dennett C. C. *Floors and ceilings*. Pat. 685, Patent Office, 1857.
18. Hyatt T. *Portland-cement-concrete combined with iron, as a building material*. London, 1877. (Reprinted by ACI, Detroit, 1976.)
19. Sutcliffe G. L. *Concrete its nature and uses*. Crosby Lockwood & Son, London, 1893.
20. Potter T. *Concrete: its use in building and the construction of concrete walls, floors etc.* Winchester, 1891, 2nd edn; 1908, 3rd edn.
21. Webster J. J. Fire-proof construction. *Proc. Instn Civ. Engrs*, 1891, **105**, 249–286.
22. Wonnacott T. H. On the use of Portland cement concrete as a building material. *Trans. R. Inst. Br. Archit.*, 1870–1, 175–180.
23. Verity T. & F. T. Specification of Work required to be done in the erection and completion of Club Premises at Nos. 96 & 97 Piccadilly, 1890.
24. Tall J. Discussion of paper by Payne. *Trans. R.*

Inst. Br. Archit., 1875–6, 232.

25. SEDDON H.C. *Experiments with concrete slabs.* School of Military Engineering, Chatham, 1880.
26. FOWLER C. On terrace roofs. *Trans. Inst. Br. Archit.*, 1836, **1**, 47–51.
27. PASLEY C. W. *Observations on limes, calcareous cements, mortars, stuccos and concrete, etc.*, Part 1. London 1847, 2nd edn.
28. WILKINSON W. B. *Construction of fireproof buildings etc.* Pat. 2293, Patent Office, 1854.
29. CASSIE W. F. Early reinforced concrete in Newcastle-upon-Tyne. *Mag. Concr. Res.*, 1955, 25–30.
30. BURNELL H. H. Description of the French method of constructing iron floors, and discussion. *Trans. R. Inst. Br. Archit.* 1853–4, 36–74.
31. LEWIS T. H. Fire-proof materials and construction. *Trans. R. Inst. Br. Archit.*, 1864–5, 109–126.
32. BLOMFIELD A. W. Remarks on concrete building. *Trans. R. Inst. Br. Archit.*, 1870–1, 181–187.
33. SEDDON H. C. Our present knowledge of building materials and how to improve it. *Trans. R. Inst. Br. Archit.*, 1871–2, 143–157, 177–184.
34. PAYNE A. Concrete as a building material. *Trans. R. Inst. Br. Archit.*, 1875–6, 179–192, 225—254.
35. CATES A. Concrete and fire-resisting constructions. *Trans. R. Inst. Br. Archit.*, 1877–8, 296–312.
36. SCOTT H. Y. D. *Floors and roofs.* Pat. 452, Patent Office, 1867.
37. ALLEN M. *Stairs, etc.* Pat. 244, Patent Office, 1862.
38. EDWARDS F. G. *Concrete building etc.*, Pat. 2941, Patent Office, 1891.
39. SIMMONS W. Concrete Floors, *The Builder*, 1888, 421–2.
40. RUSSELL L. Panel Preservation, *New Builder*, 1995, 32–34.
41. MCFARLAND B., WOODHOUSE J. and JEANES C. Royal Albert Dock, 'the gentle touch' *Constr. Repair*, 1995, **5**, No. 3, 20–23.
42. LASCELLES W. H. *Constructing buildings.* Pat. 2151, Patent Office, 1875.
43. LASCELLES W. H. Discussion of paper by Payne. *Trans. R. Inst. Br. Archit.*, 1875–6, 185.
44. CHRIMES M. M. The era of the proprietary reinforcing systems. *Proc. Instn Civ. Engrs Structs & Build.*, 1996, **116**, 296–316.
45. CHRIMES M. M. Concrete foundations and substructures: a historical review. *Proc. Instn Civ. Engrs Structs & Build.*, 1996, **116**, 344–372.

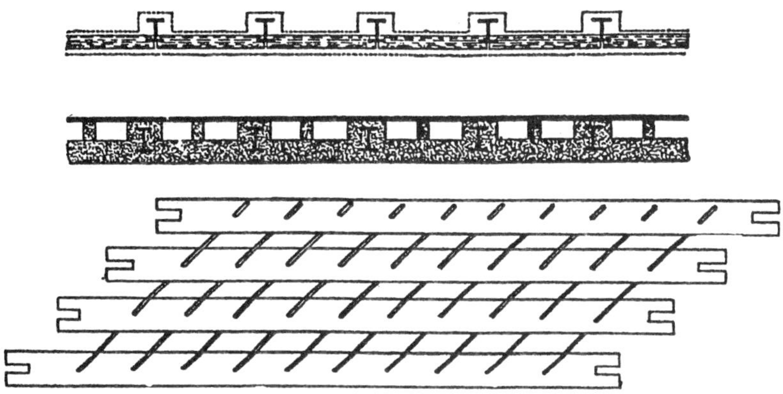

Fig. 19. Hyatt's 'gridiron' floor[18]

Fig. 20. Lascelles's system at Central Buffet

3

W.B. Wilkinson (1819–1902) and his place in the history of reinforced concrete

Joyce M. Brown

HIS EARLY CAREER

Reinforced concrete is an indispensable modern building material, fundamental to the construction of a wide variety of engineering structures. The principle which lies behind it is that of embedding metal in concrete in such a way that the two materials share the load in an efficient manner. For example, in a beam subjected to bending, the metal is placed in a position to take most of the tensile forces, leaving the concrete to resist the compressive forces. Various, somewhat primitive attempts to combine concrete and iron were made in the first half of the nineteenth century,[1] but the earliest evidence of what can be considered reinforced concrete in the modern sense is in a British Patent dated 1854, and the credit for this achievement rests with a hitherto obscure figure in engineering history, W. B. Wilkinson, a plasterer and concrete manufacturer of Newcastle-upon-Tyne.

William Boutland Wilkinson (for portrait see Plate XXVI (*a*) was born on 2 January 1819 in St. Peters, a district on the quayside of Newcastle-upon-Tyne, the son of Joseph Wilkinson, a potter (1772–1837), and Sarah (née Boutland) (1786–1856). There were three children, Joseph (born in 1817), who became a grocer, William and Anne.

At the age of 11, Wilkinson was sent for three years to Dr. Bruce's Academy in Newcastle, a reputable school at which Robert Stephenson, the engineer, had also received some education prior to Wilkinson's time there. A half-yearly letter[2] to his parents in 1833 describes in the formal style of the day that he has "in this half-year studied Mensuration of Solids, Cylindric Rings, Bricklayers Work, Tilers Work, Plasterers Work, Glaziers Work, Paviors Work, Plumbers Work, Timber Measure" as well as Chemistry, Geography, Scripture and English Literature. He was evidently a good student for he was awarded a medal, inscribed "Industry Rewarded". This technical education lasted until he was 14 when he became apprenticed to Robert Robson, a plasterer of Clayton Street, Newcastle. The indenture[2] is dated 18 October 1833, and by its terms he received wages of 3 shillings per week in the first year, increasing by a shilling per week every year until he received 9 shillings per week in the seventh year. We have one glimpse of him in this period in a letter[3] to his mother in May 1838 from Howick Hall, where he was doing a plastering job. After his apprenticeship ended he set up in business on his own in Newcastle in 1841 at 3 New Bridge Street. He is described in the local directories[3] as "manufacturer of

---

[1]Some of these are considered briefly later in the present paper.

[2]In the possession of Wilkinson's grand-daughter, Mrs. M. Wasteneys Smith of Riding Hill, daughter of Edith (née Wilkinson) and the Rev. F. E. Powell.

[3]e.g. Robson's *Commercial Directory*, 1841.

W. B. WILKINSON (1819–1902)

vases, plasterer and modeller" and as "plaster of Paris and Roman Cement manufacturer". His business had moved by 1847 to 43 Prudhoe Street and he continued to live at home in St. Peters until the death of his mother in 1856, when he took up residence at 26 Lovaine Place, Newcastle.

It appears from the evidence of street directories and advertisements in local papers that Wilkinson began work as a plasterer and gradually expanded his business so as to offer an increasingly large range of cement and artificial stone goods as an important side line. These goods included flags, hearths, chimney tops, troughs, spouting, chimney pieces, drain pipes as well as ornamental work in the form of plaster figures, busts, brackets, pendent baskets and "vases and pedestals suitable for the lawn or greenhouse" (Plate XXVII).[1] Gradually, however, Wilkinson became more and more interested in concrete manufacture and the advertisements of 1854 state that "CONSUMERS OF PORTLAND CEMENT are respectfully informed that they can be supplied with that Article, of such Quality as can be depended on, at a Reasonable Price".[2] The advertisements from then on concentrate more and more on flooring: "Improved Artificial Stone-indented Surface FLAGS . . . at 2s. 3d. per Superficial Yard"[3] and "Wilkinson's Improved Concrete Flooring", which "is not affected by Frost, Damp or Heat; has no Smell, and its Durability is guaranteed".[4] Wilkinson also recommended his granite concrete paving, for constructing acid tanks and stated that "our material is extensively used for tanks in the chemical works on the Tyne and Wear".[5] In 1854, he applied for a patent for "construction of footpaths, roadways and other floorings or pavings" but although the Provisional Specification was accepted, the application was either abandoned or considered void and the patent was not granted. Wilkinson's firm also carried out work in connection with coating the iron ribs of ships with cement,[6] and a letter to *The Builder* in 1872 shows him describing the manner of laying cement on brick-work to line rain-water tanks.[7]

Wilkinson is thought to have had some business connection with William Aspdin, who opened his cement works in Gateshead in 1852, and it is feasible that he might have obtained his Portland Cement from him. Another manufacturing interest was his widely advertised fibrous slabs "for nailing to joists or stoothings, instead of laths and two coats of plaster".[8] The second half of his 1854 patent describes how to make hollow tubular fireproof partitions. Apart from the 1854 Patent, there is another curious patent applied for in 1889 (No. 19,482) for "Improvements in 'Fish' plates for rail joints on railways" describing a cast or pressed plate of steel or iron to be used at the joints in rails to take some of the weight of the vehicle and carry it smoothly over the joint. Wilkinson was then 70, but had evidently lost none of his vigour or interest in new inventions.

Wilkinson was obviously a shrewd business man and it is interesting to see his method of maintaining his firm's reputation in the teeth of a lost tender for a plastering job. On 13 February 1857 he published in the *Newcastle Journal*, in the form of an advertisement, a letter his firm had written to Mr. Robert Robson, who apparently gave the contract over Wilkinson's head, in spite of a verbal promise, to Mr. Alderman Dodds.[9] The job in question was the plastering of the

[1]Ward's *North of England Directory*, 1853, 34–35.
[2]*Newcastle Chronicle*, 17 Nov. 1854, and *Newcastle Courant*, 27 Oct., 3 Nov., 10 Nov., and 17 Nov. 1854.
[3]*Newcastle Chronicle*, 15 Aug., 29 Aug., 12 Sept. 1856.
[4]e.g. *Newcastle Courant*, 20 Jan. 1865.
[5]Letter to *The Builder*, 14 Oct. 1882.
[6]Letter to *The Builder*, 2 Dec. 1871.
[7]Letter to *The Builder*, 27 Jan. 1872
[8]e.g. Ward's *North of England Directory*, 1883–4, 84.
[9]Mr. Robert Robson was presumably the man with whom Wilkinson had served his apprenticeship. Mr. Alderman Dodds was Ralph Dodds, whose plastering business Wilkinson is said to have taken over, though the evidence for this is not known.

W. B. WILKINSON (1819–1902)

Corporate Buildings in St. Nicholas' Square—quite an important commission—and Wilkinson's letter states: "We regret exceedingly that you should have thought it consistent with honour and fair dealing to make a tool of us in the way you seem to have done, as, although no written engagement was entered into (other than a tender in the usual way, which among tradesmen is always considered binding on honour, if asked for, used, and accepted) yet it was perfectly understood, both in the Council and out of it, that we were, with you, the successful competitors for this department of the buildings, our tender being 1,249 *l.*, while that of Mr. Alderman Dodds was, we believe, several hundred pounds more, and, if we mistake not, our name was mentioned at the letting as the party who would be taken by you for the plaster work . . . . You can, we presume, scarcely plead incapacity, as we employ the largest number of men, and have the largest establishment of the kind in the North of England, a connection which any firm may be proud of, and which we have no doubt this incident will strengthen . . . . ."

Another anecdote which illustrates Wilkinson's business acumen is recounted in a letter to the editor of *The Illustrated Carpenter and Builder* of 30 March 1923, by Hubert Todd, formerly Wilkinson's London manager. Having executed quite a big contract for a royal person, Wilkinson experienced some difficulty in having his account settled, although the statement of account was rendered several times to the agent. Mr. Todd writes: "He (Wilkinson) was much too upright a man to enter into a contract and then forgo payment just because his client was of Royal blood, and he had the same respect for the client's character (i.e. good name) as he had for his own. After some thought he wrote the agent one very short letter. It was as follows: 'When the enclosed account was passed for payment I think it must have slipped out of the bundle.' The cheque came by return of post."

Wilkinson evidently did a lot of business and could claim in 1883[1] that his concrete pavement was "used at about 8,000 places in all parts of the kingdom"; and in 1887[2] he supplied an impressive list of patrons which included the names of many members of local aristocratic families. A contributor to *The Illustrated Carpenter and Builder*,[3] Mr. Frank Jay, claims that Wilkinson laid a granite concrete tennis court at Sandringham in 1883 for King Edward VII, then Prince of Wales, and that his concrete paving was adopted by the War Office and used in practically every barracks and troop stable, as well as by Royalty, noblemen and landed gentlemen in their coach and hunting stables, farm buildings, kennels, etc. An obituary notice[4] says that the paving was laid even as far away as Lisbon. Wilkinson also had an office at 13A Great George Street, Westminster and one also in Sunderland. From all this it is obvious that Wilkinson had a good eye to the current market and not only realised the value of advertising, but concentrated on selling the goods most likely to be in demand.

Wilkinson's Reinforced Concrete Invention

Although in 1854 in his Patent Wilkinson had shown that he knew how to construct concrete floors reinforced with iron, there is no evidence of this in the advertisements. In 1856, two years after the Patent, a series of advertisements appeared[5] headed FIREPROOF CONSTRUCTION: W. B. WILKINSON'S PATENT, but these refer to "an INCOMBUSTIBLE SUBSTITUTE for ordinary Lath-and-plaster Partitions on wooden or other Floors where Bricks or other heavy Materials cannot be used"—that is, the second rather than the first part of the patent. In 1887,

[1]Kelly's *Directory for Durham, Northumberland, Cumberland & Westmorland*, 1883, 7.
[2]Advertisement in *Catalogue of Royal Mining Engineering & Industrial Exhibition, Newcastle-upon-Tyne*, 1887, 355.
[3]*The Illustrated Carpenter and Builder*, 16 March 1923, 355–56.
[4]*The Builder*, **83**, 25 Oct. 1902.
[5]*Newcastle Chronicle*, 23 May, 30 May, 6 June, 13 June, 27 June 1856.

W. B. WILKINSON (1819–1902)

however, W. B. Wilkinson & Co. was one of four firms chosen to represent concrete manufacture at the Newcastle-upon-Tyne Royal Mining Engineering & Industrial Exhibition, and among his 19 exhibits there were eight types of concrete floor including "a fire-proof flat Concrete Floor; with Iron joists, and girders, to carry a safe load of 4 cwt. per superficial foot." Another exhibit was a "Concrete Bar; with Iron inserted, and carrying weights, to show the transverse strain it is capable of bearing"[1] and a "Machine for testing the tensile strain of Cements".[2] None of these exhibits is quite in the line of the Patent, but they do show Wilkinson's interest in the behaviour of concrete under stress. It is also perhaps necessary to say that, even if he did not advertise it, Wilkinson was making use of his invention. although we have today knowledge of only one example, to be described later in this paper.

The Patent itself, No. 2293, had been applied for by Wilkinson as early as 27 October 1854, when he left his Provisional Specification, with his Petition, at the Patent Office. He described his invention as "Improvements in the Construction of Fireproof Dwellings, Warehouses and other Buildings, parts of the same" and claimed that his was much cheaper than existing methods. The Specification was accompanied by two sheets of drawings to show longitudinal and transverse sections of a warehouse building with typical details of alternative forms of fire-resisting construction. One of these sections is reproduced in Fig. 1. To construct ceilings of arch form, Wilkinson describes how the walls (A) are to be built up to the requisite height for the first floor (B); a centring is then fixed over the whole area to be floored, composed of narrow strips with a small space between each and lightly nailed to keep them in position. A concrete mixture composed of crushed bricks, metallic scoria and Portland cement containing some fairly coarse material is thrown on to the centring and beaten in solid. Wilkinson says "a number of strips (C) of hoop iron laid on edge are imbedded in the mass across the crown of the arch, at distances of about 2 ft. asunder, or at greater or according less distances, to the desired strength of the floor, and reaching the full length across the floor or floors, the under edge of the same being set at or near the crown of the centring, so that in this low position the strips may act with more power as tension rods to the floor". The concrete is then filled up to the desired level and may be trowelled smooth and finished in cement, or laid with boards, as shown at D. The centring should not be removed for at least a month. This is one method. If ceilings of arch form are not required and when speed is needed in the construction, a flat platform of wood is to be erected to the ceiling line and the floor is to be composed of plaster, air-slacked lime, hay, ashes and breeze, and wire rope (Wilkinson probably had in mind second-hand colliery rope) "or iron in other forms of tension". A floor of this construction is shown at G. The wire rope H, H, is secured at its extremities by imbedding it in the mixture of concrete while in a soft state, and forming the ends into loops, or opening out the strands and twirling them in various directions, so that the rope cannot be drawn out under any force short of the breaking weight. In ordinary dwelling houses Wilkinson suggests the ropes should be placed about 9 in. apart and the depth of the floor should be a sixteenth of the span, viz. a floor 1 ft. in depth for a 16 ft. span, the breaking weight of which Wilkinson says is 2½ tons per superficial yard. The floor can be lessened in weight by using hollow bricks, field-drain pipes, and coke in lumps in the concrete, as shown at G. Where a rigid or unyielding floor would be unsuitable Wilkinson proposed to form the surface of the flooring of wood in the usual way as shown at I, I.

The Specification goes on to describe how to make partitions which can be set on the reinforced floor in positions where a wooden floor would not allow; these are light partitions made of pre-

[1]Details of its dimensions and strength are given in a letter by Wilkinson to *The Builder*, 9th July 1887.
[2]Catalogue of the Exhibition.

W. B. WILKINSON (1819–1902)

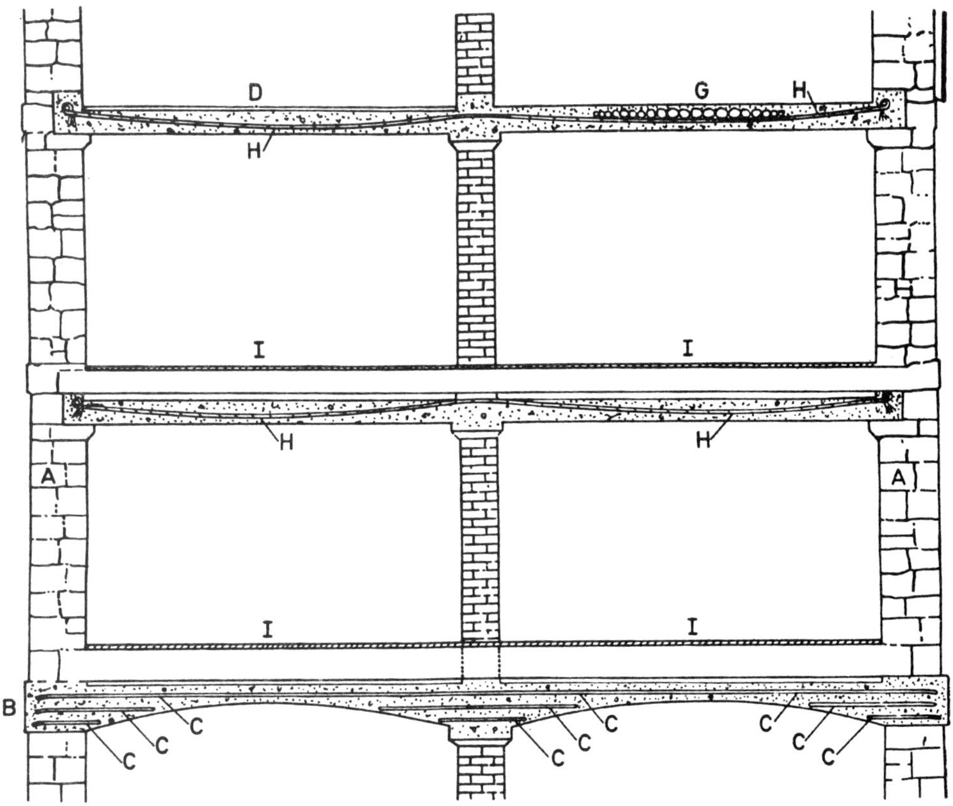

Fig. 1. W. B. Wilkinson's Patent Specification No. 2293 of 1854.
Part of his Fig. 2.

fabricated cellular blocks dovetailed together, composed of Paris plaster or plaster of Paris and air-slacked lime; the cavities run horizontally.

It will be clear to the engineer that this Specification describes nothing less than modern reinforced concrete. Interesting points in it are Wilkinson's understanding of the need to provide a bond between iron and concrete, shown by his flaying out the ends of the iron rope, or forming the ends into loops; his suggestion of protecting the concrete with thin deals laid on a stratum of sifted sand so as to keep the concrete moist and thus to prolong its setting time, "slow drying adding greatly to its strength and durability"; the economy of the system and the relative lightness of the floors compared with timber, brick arch, or concrete and iron girder floors. The key words are obviously his reference to the wire rope "or iron in other forms of tension", and an important point to notice is the way in which the reinforcement follows the line of tension: rising to the upper part of the beam to withstand negative bending moments over supports and falling to the lower part of the beam in mid-span.

If there is any doubt as to Wilkinson's understanding of the structural principles he employed in the Patent, this is completely dispelled by a letter he wrote to *The Builder*, published on 5

W. B. WILKINSON (1819–1902)

January 1884. In this he described experiments he had made prior to his Patent to test resistance to load, degree of deflection and ultimate breaking-point of a concrete beam reinforced with wire rope in the manner of the Patent. It is worth quoting this letter almost in full because the earliness of the experiments makes it an important document in the history of reinforced concrete.

"I differ from your correspondent.... as to the value of iron in a state of tension in combination with concrete for the formation of beams, floors, etc. A beam of concrete resting on its ends without a firm and unyielding abutment is a thing not to be depended on to carry much more than its own weight; but with a firm abutment and suitable depth of section it will carry a good load, with a wire rope or ropes, or with iron in any cheap form, such as the old tyres of wheels, firmly embedded in the concrete and forming a curve (with the ends close up to the upper surface of the beam at the ends, and dipping down in the centre to near the underside), it will carry safely a very heavy load, and will not yield until the iron begins to draw. If wire rope be used there will be considerable elasticity without damage, as the beam, if cracked, either by impact or dead weight, will spring up when the load is lessened. After the beam has cracked, the upper line is in a state of compression, and the wider the beam, of course the greater the power of resistance. A thick concrete floor so constructed may be considered as a number of beams united together, each receiving support from its neighbours, and exceedingly strong. The iron, however, should be curved, not straight; a straight rod of iron has not much supporting power, even if held firmly at the ends; it has a great deal when bent and embedded in a rigid material like concrete. There is sufficient adhesion of the concrete to the iron to prevent its drawing, but a precaution may be taken by beating up the ends a little, or, if rope be used, by twirling the ends.

"In 1855 I took out a patent for fireproof construction on this principle, and included concrete arching. Previously to doing so I made some experiments to test the bearing-power of old wire rope, which is plentiful and cheap. I made a beam, 21 ft. long, 16 in. wide, 1 ft. 1$\frac{3}{4}$ in. deep, composed of plaster of Paris, a little lime, chopped straw, and coke, with a wire rope, about 1 in. diameter, embedded as above described. I first tried its strength at a bearing distance of 18 ft. 6 in. It deflected with its own weight about $\frac{1}{8}$ in., and with two men on the centre there was a slight increase; with 19 men standing as near the centre as possible it went down 1$\frac{1}{2}$ in. The span was increased to 21 ft., and a load of 3 tons put on, causing an additional deflection of $\frac{3}{32}$ in. In 5 days this had increased to $\frac{3}{4}$ in. On removing this load it ascended $\frac{1}{2}$ in. Four tons were then put on the centre and allowed to remain a month, when the deflection had increased to 1$\frac{1}{2}$ in.; on taking this off it ascended 1 in. The beam was at last entirely broken down, the rope being torn asunder with 5$\frac{1}{4}$ tons. This was conclusive as to the bearing-power of the rope. I also experimented with some small beams of plaster of Paris. One, 3 ft. long, 4 in. deep, 1$\frac{3}{4}$ in. wide, with a piece of iron wire, bent to shape, and slightly turned at the ends, distance between points of support 2 ft. 7$\frac{1}{2}$ in.; it bore 4 cwt. in the middle with a deflection of $\frac{3}{16}$ in. A piece of Memel timber, of exactly the same size, made the same deflection with the same weight. This little plaster beam, without iron, would probably not have borne 14 lb. Good concrete would, of course, after a month be much stronger than the beam of plaster, and an additional rope would have given much better results."

CONTEMPORARY "FIREPROOF" CONSTRUCTION

Wilkinson's patent belongs to a period in which there was a growing interest in "fireproof" construction.[1] From the late eighteenth century onwards serious attempts had been made to

[1]See J. J. Webster, "Fire-proof construction", *Min. Proc. Inst. Civ. Eng.*, **105** (1891) 249–288 and S. B. Hamilton, "A short history of the structural fire protection of buildings", *Nat. Bldg. Studies Special Report* No. 27, H.M.S.O. 1958.

construct fireproof floors, the urgency of doing this in mills and warehouses lending an impetus to inventors. For a long time the brick-arch floor led the field, a system in which brick arches sprang from iron beams. Hollow bricks or pots were shortly introduced to make the floors lighter and at the same time less combustible. Sir William Fairbairn, for example, devised a floor in which arches were formed of plate-iron in 3 ft. widths, the space between them and the floor-line being filled with concrete, but the iron was not encased in the concrete. This was partially achieved in a floor construction devised by Dr. Henry Fox (Patent No. 10,047, 1844) and improved by J. Barrett and known as the "Fox & Barrett system" (Fig. 2).

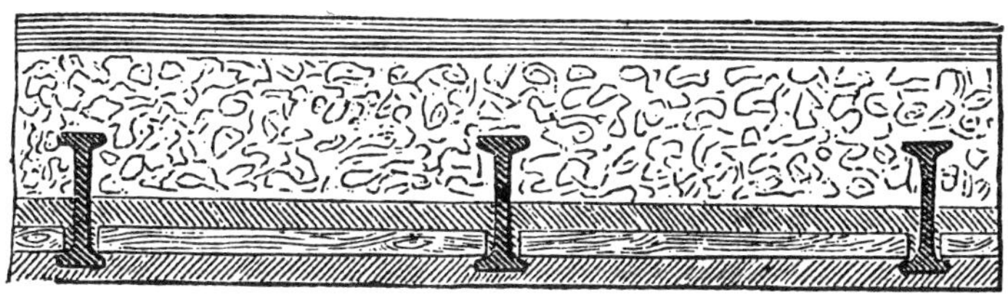

(a)

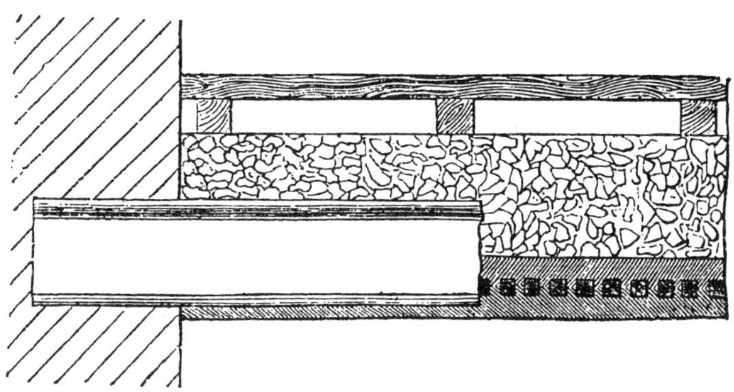

(b)

Fig. 2. Fox and Barrett Floor (1853).
(*a*) Section through wrought iron joists.
(b) Section with joist shown broken to reveal ceiling construction.

Fox and Barrett floors were widely advertised and much employed from the late 1840s. As early as 1833–34 Fox had constructed a floor at Northwoods asylum, Gloucestershire.[1] In its

[1]Described in *Notes on a Visit to Northwoods, Gloucestershire* by J. C. Christopher, London, Oct. 1848 and in *The Builder,* **6** (1848) 541–542.

W. B. WILKINSON (1819–1902)

original form, the I beams in Fox's system were made of cast-iron and were set in a pugging composed of $\frac{8}{10}$ of road grit, $\frac{1}{10}$ of coal ashes and $\frac{1}{10}$ of lime, mixed with water. However, between 1848 and 1849 concrete was used,[1] and by 1851 or 1852 the I beams were made of wrought iron. Barrett described the system in a paper read at the Institution of Civil Engineers in January 1853[2]: small rolled joists about 20 in. apart rested upon the main girders and strips of wood about 1¼ in. square were laid on the bottom flanges about 1½ in. apart. The concrete was laid upon these strips and between the joists, the floor being finished in wood or tiles as desired. To sum up, in England at the time of Wilkinson's patent, the main system of flooring used was the brick arch and iron beam, rapidly being superseded by the I beam partially set in concrete. There was a growing interest in other fireproofing methods, some of them employing iron and concrete in combination (see, for example, George Nasmyth's Patent No. 12,260, 1848; James Frost's Patent No. 4,710, 1822), but there is no evidence of their use.

The French, however, had from the end of the eighteenth century employed metallic floors and it is important to trace their development there, since the credit for the invention of reinforced concrete is usually attributed to the French. The most commonly employed system was devised by an architect, Ango, in about 1782[3] and was frequently used in Paris until about 1850. The floors were formed by means of beams which spanned from wall to wall, each beam composed of two wrought iron girders rectangular in section, the lower one straight and the upper one curved or arched, but placed on edge and tied to the lower at each end; these beams were set in a filling of hollow pots and plaster[4] (Plate XXVIII (*a*)).

There were many variants of the Ango system,[5] but the next important advance came when Vaux replaced the trussed girder by a flat wrought beam (Plate XXVIII (*b*)). The iron beams, their extremities resting on the walls, were connected by short inter-ties of bar iron running in a transverse direction with their ends bent to lap over the joists; these were embedded in plaster. This system was widely used in Paris in the 1850s and was certainly in existence already by 1845 when tests were carried out on it[6] by government officials.

Vaux used flat beams without flanges, but those who developed his system used I beams. Several such systems were exhibited in the Paris Exhibition of 1849, of which that of Baudrit is a typical example (Plate XXVIII (*c*)). By 1852,[7] these I beam floors were in general use in Paris. At first, the beams were set in plaster or in hollow pots, but there was no conception of an interaction between the iron and plaster beyond the fact that plaster may have helped to keep the metal in place. Concrete gradually replaced plaster, for it was in the middle decades of the century that rapid advances were made in the development of stronger and quicker-setting cements.

A Frenchman who saw the potential of concrete as a building material was the contractor, François Coignet. He was chiefly interested in the use of concrete as a monolithic material, but also used it to enclose iron members for floor construction. This is briefly described in an English patent of 1855 (No. 2659) and more fully in his book published 6 years later.[8] From the latter it

[1]See *On the Construction of Public Buildings and Private Dwelling-Houses on a Fire-Proof Principle*, by Fox & Barrett, London 1849.
[2]*Min. Proc. Instn. Civ. Eng.*, **12** (1853) 244–265.
[3]A manuscript by Ango entitled *Art de faire les planchers incombustibles, approuvé par l'Académie royale d'architecture en* 1782 was sold in Paris in 1855.
[4]Typical dimensions are given in *La Revue Générale de l'Architecture et des Travaux Publics*, IX, 1851, col. 192.
[5]See C. L. G. Eck, *Traité de Construction en Poteries et Fer*, Paris 1836.
[6]Tests described in *La Revue Générale de l'Architecture et des Travaux Publics*, VII, 1847, col. 479; the system described col. 478–482, Pl –46.
[7]When H. H. Burnell visited Paris. See his paper "Description of the French method of constructing iron floors", *Trans. R.I.B.A.*, 1854, 36–38 and discussion 41–52.
[8]*Bétons Agglomérés Appliqués à l'Art de Construire*, Paris 1861, 97–101.

is clear that small wrought iron I joists were placed in a concrete slab, with 5 to 6 cms. of concrete below the lower flange and the same thickness above the upper flange. The joists were spaced at 1 m. intervals, linked with tie rods, and the size of the joists varied in proportion to the span, e.g. 12 cm. deep for a 5 m. span, 16 cm. for a 6 m. span and so on (Fig. 3). This system was first used by Coignet for the flat roof of his own house built in 1853 in St. Denis (Rue Charles-Michel). Here the joists appear to have been a little smaller than those described in the 1861 book, being 12 cm. deep for a 6 m. span, although the concrete had the same thickness, namely 27 cm.[1] The architect's drawings of this house (without constructional details) are reproduced by Collins.[2] As an advance on the floors of Fox and Barrett, the iron beams are completely encased in concrete, giving greater protection and stability, but the position of the iron in the centre of the slab suggests that Coignet had little conception of the main principle of reinforced concrete.

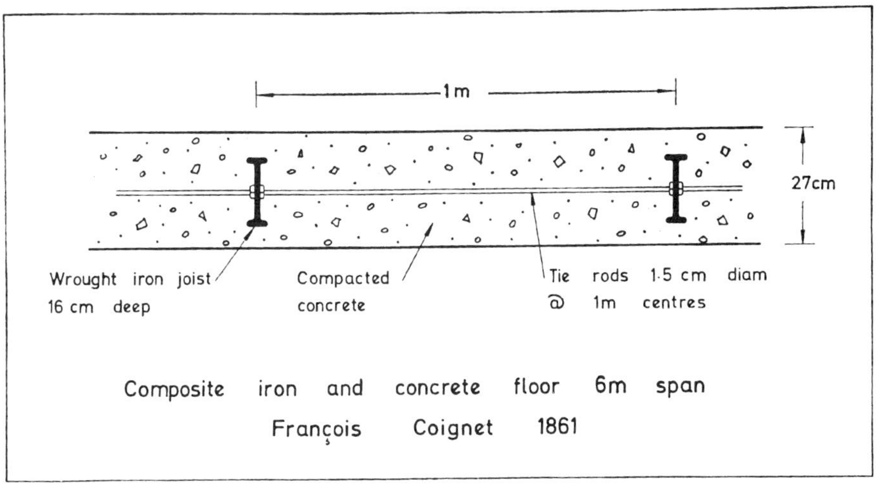

Fig. 3. Coignet floor, drawn by A. W. Skempton.

It may be remarked in passing that Coignet's concrete was undoubtedly better than Wilkinson's. The concrete in the roof of the house in the Rue Charles-Michel was formed of gravel, lime and cement in the proportions of 5 to 1 to 1½, with no stone larger than 1 cm. and the whole mixture very dense. Wilkinson's by contrast must have been very coarse, composed, the Patent advises, of three parts of crushed hard-burnt bricks and scoria, passed through a riddle of 2 in. mesh,

[1]Report by M. Verrine for the Société des Ingénieurs Civils, summarised in *Nouvelles Annales de la Construction,* **8**, 1862, col. 149–51. Apparently a few centimetres of concrete were later added as an extra protection from the weather. This presumably accounts for the total depth of 13 cm. of concrete over the top flange of the I beam noted by a committee of the Concrete Institute who visited the house about 55 years after its construction (*Trans. Conc. Inst.,* **2** (1910) 38–39). The address of this house is incorrectly given in this reference as Rue des Poissoniers also in St. Denis. Coignet did build a house in the Rue des Poissoniers (in 1856–57) but it had a vaulted concrete roof (Coignet, *The Builder,* **17** (1859) 436–37).

[2]Peter Collins, *Concrete,* Faber & Faber 1959, 28, Pl.2.

and two parts of hammer-broken stones or bricks, half broken to 2 in. in diameter, and half to the size of a quarter brick, all mixed with Portland or other cement of good quality.[1]

Mention must also be made of another Frenchman, a contractor, Joseph Louis Lambot, who left a Provisional Specification at the Patent Office on 13 February 1855, for "An Improved Building Material to be used as a substitute for wood" where he claims "I make this substitute of a network, or a parallel set of wires, or metallic bars or rods, imbedded or cemented together with hydraulic or other cementing matter, so as to form beams or planks of any suitable size".[2] Lambot demonstrated his mesh system by making a boat using a skeleton of wire embedded in concrete. Such a boat was exhibited by him at the Paris Exhibition of 1855, but the French in celebrating the Centenary of Reinforced Concrete in 1948 in Paris, dated it back to 1848 when Lambot is said to have replaced a wooden boat on his lake with one of hydraulic lime cement, with walls only 2 in. thick and reinforced with mesh. Similarly, Joseph Monier made tubs of reinforced concrete for orange trees in 1849, and took out a number of patents from 1867 onwards for these and other small structures. Although these mesh systems have some place in the history of reinforced concrete, Wilkinson's use of reinforcement in a structural context seems more important.

In fact, after Wilkinson's patent, the next clear expression of principle seems to be by H. Y. D. Scott in 1867 (Patent 452), who describes a floor made of concrete reinforced with iron rods placed in the lower part of the slab, and speaks of "the iron in combination with the concrete to take the tensile strain and the concrete to take the compressive action". The interest in and activity concerned with fireproofing methods in these middle decades of the century was astonishing, and yet between Wilkinson and Scott, at least as far as the study of the British patents of the period reveals, there was no invention which could be described as reinforced concrete.

Wilkinson's Concrete Cottage

There is very little evidence remaining to show to what extent Wilkinson made use of his Patent. One example which has been recorded was a little cottage built on his concrete works site on a plot of ground behind Ellison Place in Newcastle (Fig. 4). The cottage, occupied at one time by his coachman, was pulled down in 1954, but a careful record of it was made at the time by Professor W. Fisher Cassie.[3] Since this may have been the earliest reinforced concrete house in the world, some attempts to date it accurately have been made, but its actual date of construction cannot be verified by the usual processes, since it stood on a piece of waste ground not technically in any street. Newcastle was surveyed in 1855–64, no cottage appearing on the map then, but it is shown on a map in the City Engineer's Office, dated 1885. To narrow the range of date is not easy. A draft lease still exists,[4] of 23 July 1866, in which Wilkinson leased the land on which it stands from the Master and Brethren of St. Mary Magdalene Hospital, the date

[1]Professor Cassie's tests on Wilkinson's concrete as used in the cottage (*The Structural Engineer*, **33** (April 1955) 137) gave on 2 in. cubes;

aver. strength in compression 3,800 lb./sq. in.
max. ,, ,, ,, 4,500 ,,
density 109 lb./cu. ft.
aggregate a hard clinker; organic matter 2 per cent.; loss on ignition 37 per cent.;
absorption of water after 24 hours immersion 27.3 per cent.

[2]Letters Patent were never granted on Lambot's Provisional Specification as application was not made within the time prescribed.

[3]"Early reinforced concrete in Newcastle upon Tyne", *The Structural Engineer*, **33**, (4) April 1955, 134–37 and "The first structural reinforced concrete", *Magazine of Concrete Research*, March 1955, 445–49, the latter article translated into French, Italian and German in *Bulletin du Ciment*, 1957.

[4]In Newcastle City Archives.

W. B. WILKINSON (1819–1902)

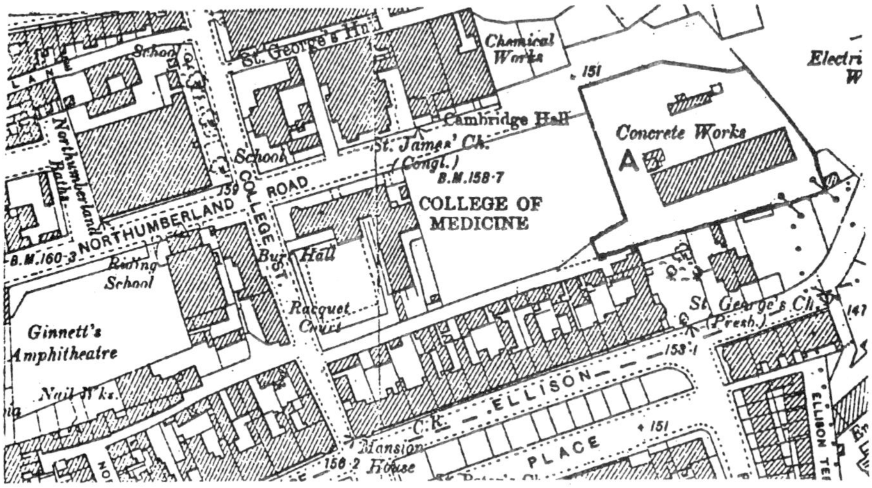

Fig. 4. Map of central Newcastle-upon-Tyne showing the concrete cottage **A** and the access lane leading to No. 5 Ellison Place where Wilkinson lived from 1866.

of tenure effective from 1 August 1865. A date of 1865 for the cottage is given by a correspondent, W. H. Thorpe,[1] in a letter to *Engineering* dated 15 May 1925,[2] and Hubert Todd, when he first saw the cottage in 1886 thought it was upwards of 20 years old then.[3] Wilkinson's move to 5 Ellison Place in 1866 and the lease of the land behind it for his concrete works at the same date lend a certain amount of substance to a date around 1865 for the cottage. Wilkinson's firm was bought a few years after his death by W. Ferguson & Son, Ltd., but this firm no longer has any of the old records.

The cottage had one large room on the ground floor, 16 ft. by 14 ft. with a bay window (Plate XXVI (*b*) and (*c*). A concrete staircase, cast *in situ*, rose to a floor above. According to Professor Fisher Cassie's measurements, the walls were 9 in. thick in the first storey, 12 in. on the ground floor and 18 in. in the foundations. The ceiling of the ground floor was panelled and had deep ribs. Professor Fisher Cassie describes the probable method of construction in his article in *The Structural Engineer*:[4]

> "The most interesting part of the building was the first floor which consisted of precast plaster moulds which had quite obviously been supported in position on timbering while Wilkinson's old colliery ropes were laid between the moulds in a rectangular pattern. The spaces between the moulds had then been filled with coarser plaster *in situ* and with concrete and covered by a a concrete slab having a granolithic surface. Beams in this floor were approximately 26 in. apart and were about $6\frac{1}{2}$ in. deep. The reinforcement of these beams was $\frac{5}{16}$ to $\frac{3}{8}$ in. twisted wire rope placed in the bottom but sloping up approximately 30° at the ends before being anchored in the larger beams or in the wall. The plaster *in situ* filling between the plaster

[1] Probably William Henry Thorpe, elected A.M.I.C.E. 1890
[2] *Engineering*, **119** (1925) 678.
[3] *The Illustrated Carpenter and Builder*, **92** (1923) 438.
[4] *The Structural Engineer*, **33** (1955) 134–37.

W. B. WILKINSON (1819–1902)

moulds had effectively protected the reinforcement for 90 years. The concrete slab which overlay the hollow tile ceiling and beams was $1\frac{1}{2}$ in. thick and contained $\frac{3}{16}$ by $\frac{3}{8}$ in. thick steel flats in the bottom running in both directions. Here again the material was unrusted and the span was approximately 12 ft. in each direction."

The method employed here will be seen to be exactly that of the Patent (Plate XXIX and Fig. 1).

This cottage was no doubt not the only example of Wilkinson's reinforced concrete method. W. H. Thorpe in the same letter speaks of "other work . . . . carried out by Wilkinson, at about the date indicated (i.e. 1865), or a little later, including house construction, stable with flat roof, reinforced paving . . . and a floor of 14 ft. span to carry a baker's oven", and Wilkinson's former London manager, Hubert Todd, says "He (Wilkinson) built several all-concrete houses (walls, floors, stairs and roof) at a very early date". But if any of these structures remain, they have so far not been identified as Wilkinson's work.

Wilkinson's Later Years

Wilkinson seems very rapidly to have established a flourishing business and as he extended it moved into bigger premises. By 1857 his firm had become a registered company, W. B. Wilkinson & Co. In 1860 or so he had his cement works at Stockbridge; in 1879 in Elswick Court off Northumberland Street; by 1897 his office was in 13 St. Mary's Place and his works on the piece of land he leased from 1865 behind Ellison Place. On his marriage in 1866 he set up house at 5 Ellison Place; by 1877 he had acquired another house at Whitley Bay. He became a City Magistrate and was also a Chairman of the Newcastle and Gateshead Gas Company, Vice-Chairman of the Newcastle and Gateshead Water Company and a Director of the Redheugh Bridge Company.

Wilkinson retired a few years before his death. There is no doubt that he was a clever craftsman and a good businessman, combining his technical skill and originality with foresight, acumen and shrewd judgment. From a humble start in life, he became a very wealthy man. He had a kindly disposition and gave generously to charity. A typical Victorian head of family, he was respected and held in awe by his wife and children. He liked things done in style, drank wine with his meals, was formal in dress and manner and rode in a carriage and pair with his own footman. He married Elizabeth Wilkinson, a cousin's child 13 years younger than himself, whom he had carried as a child on his shoulder; she cried so much after their marriage for the sister from whom she had never been parted that he sent for the sister and she lived with them thereafter and survived both Wilkinson and his wife. He had six children: Elizabeth (1867–1951); Beatrice (1868–1945); Isabella (1869–1950); Edith (1870–1927); William (1871–1883)[1]; Charles (1872–1937).

Wilkinson died on 13 October 1902 in his 84th year, and is buried at Jesmond Old Cemetery.[2] He left about £170,000, bequeathing over £3,000 of it to charity, mainly to church societies. Details of the will are given in the Appendix. The Vicar of Newcastle, in his sermon on 19 October, in St. Nicholas' Cathedral, asked the congregation to remember Wilkinson, "one of the benefactors of this diocese and cathedral church, and the quiet unobtrusive helper of many causes and persons about which the world knew nothing". There is a stained glass window to his memory in St. Nicholas' Cathedral, Newcastle.

Wilkinson's contribution to the history of reinforced concrete is undoubtedly original and early, and this paper has tried to show that so far as remaining evidence reveals he was the first person to demonstrate the principles of modern reinforced concrete. While he is not usually totally ignored in the literature, he is often passed over in a brief reference, perhaps because so

[1]Wilkinson put a window to his memory in St. Paul's Church, Whitley Bay.
[2]A full account of the funeral appeared in the *Newcastle Daily Chronicle* on 16 October 1902.

little has been known about him. Professor Cassie, Dr. Hamilton,[1] and W. R. Howard[2] have all recognised him as the first exponent of structural reinforced concrete, and Huberti in a more recent book has given him a substantial treatment.[3] Lambot's boat and Coignet's floors, however, have tended to steal the day in most discussions of the subject.

In the end, however, Wilkinson cannot be shown to be the pioneer from whom stems directly the further development of reinforced concrete in the nineteenth century. His work was isolated and probably unknown to William E. Ward[4] and Thaddeus Hyatt, who in the 1870s began to analyse and calculate the properties of concrete and iron when combined, or to Wayss, Hennebique and others who in the 1890s brought about the rapid development of reinforced concrete as a structural material. Had Wilkinson's firm been comparable in size with Coignet's contracting business and the structures undertaken of comparable magnitude, his name might have been associated with reinforced concrete in a more telling way than it is. Perhaps like most innovators, Wilkinson was a little before his time, and unable to grasp the full significance of his own discovery. But that is not to minimise his achievement.

ACKNOWLEDGMENTS

I am grateful to Professor A. W. Skempton of Imperial College, London, who suggested and guided the research and has criticised the manuscript; in particular, I am grateful for his help in assessing the work of François Coignet, notably in footnote 1 on p. 137 where the confusion about Coignet's house is, one hopes, finally dispersed.

I also wish to make acknowledgment to Professor W. Fisher Cassie of the University of Newcastle-upon-Tyne for his interest and loan of the plates reproduced in Plate XXIX; to Wilkinson's grand-daughter, Mrs. M. Wasteneys Smith of Riding Mill, for information about Wilkinson's family; to the City Engineer's Office, Newcastle, and to Mr. Michael Cook, the City Archivist, for help in trying to date the cottage; and to Miss Joyce Gurr for the photographic work.

APPENDIX

WILKINSON'S WILL

Excerpt from *The Newcastle Daily Chronicle*, 24 November 1902.

"Probate has been granted in the estate, valued at £171,494 19s. 10d., including personalty of the net value of £161,269 12s. 10d., of Mr. W. B. Wilkinson of 5 Ellison Place, Newcastle and Belvedere House, Whitley, who died on 13 October. The will and codicil have been proved by Charles Joseph Wilkinson, of Sebergham Castle, Cumberland, the son; Edward Armorer Hedley, of Windsor Crescent, Newcastle; and Fred Emley of Gateshead.

The testator bequeaths £500, and during her widowhood an annuity of £1,200, and the use of his two residences, to his wife,[5] and £500 to her for his old work people or for any charitable object she may select; £10,000 each in trust for his four daughters,[6] Elizabeth, Beatrice, Isabella

[1] *A History of Technology*, ed. C. Singer *et al.*, V, 1958, 488–9.
[2] Centenary Presidential Address 1954. *Trans. Soc. of Engineers*, March 1954, 16.
[3] G. Huberti, "Die erneuerte Bauweise" in *Vom Caementum zum Spannbeton* by G. Haegermann, G. Huberti & H. Möll. Bauverlag GmbH, Wiesbaden-Berlin 1964. 1, Teil B, 24–29.
[4] The earliest example of reinforced construction in America is said to be Ward's house on Comly Avenue in Port Chester, N.Y., 1871–76.
[5] His wife had predeceased him, in 1897.
[6] Only two of his children married: Isabella, who married the Rev. J. W. Walton, and Edith, who married the Rev. F. E. Powell.

W. B. WILKINSON (1819–1902)

and Edith, and a further sum of £10,000 each on the death or re-marriage of their mother; £22,000 to his son, Charles Joseph.

He also bequeaths £500 each to the Newcastle Infirmary, the Prudhoe Memorial Convalescent Home, and the Church Institute; £250 each to the Cathedral Nurse and Loan Society, the Curates' Augmentation Fund, the Society for the Propagation of the Gospel, the Society for Educating the Poor in the Principles of the Church of England, the Society for Promoting Christian Knowledge, and the Diocesan Fund, Newcastle; £200 each to Edward A. Hedley and Fred Emley, and annuities of £50 each to his brother Joseph, his sister-in-law Elizabeth, and to Sarah Wilkinson, Margaret Wilkinson and Mary Craven. The residue of his property he leaves to his son."

PLATE XXVI

(*a*) W. B. Wilkinson (1819–1902) in old age.

(*b*) Wilkinson's reinforced concrete cottage. (c. 1865)

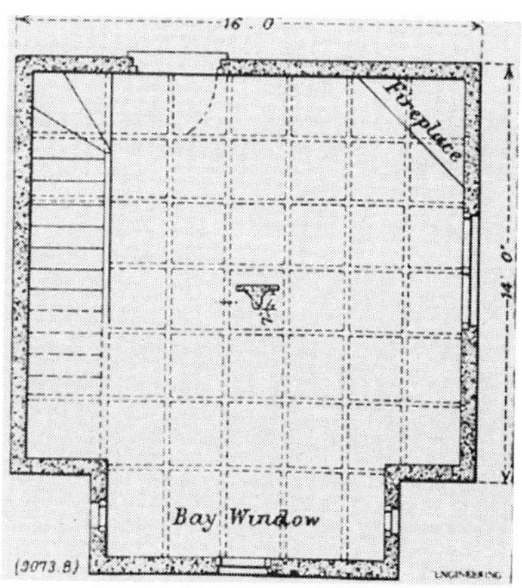

(c) Plan of Wilkinson's concrete cottage. (*Engineering* **119** (1925) 678)

PLATE XXVII

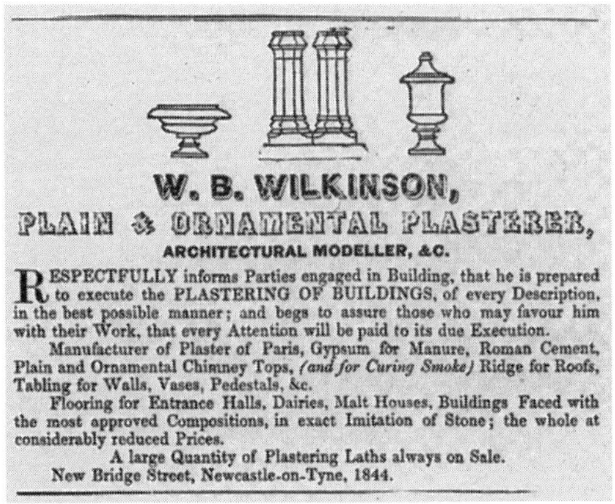

Wilkinson's advertisements of 1844 and 1853.

PLATE XXVIII

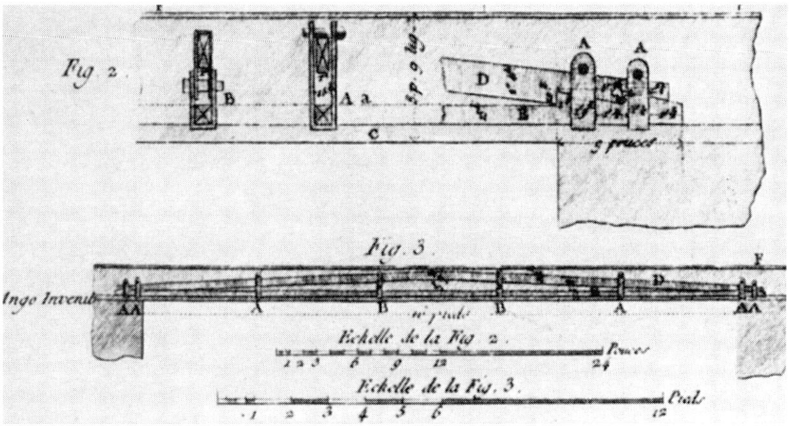

(*a*) Ango's Floor from Encyc. Méth. V (1787) Pl. 52. Wrot. iron bars of rectangular section set in concrete.

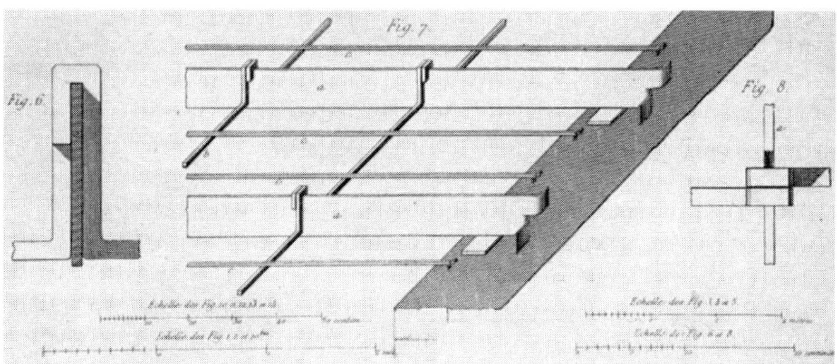

(*b*) Vaux's Metallic Ceiling from *Revue Gén. de l'Arch. et des Trav. Pub* 7 (1847) Pl. 46.

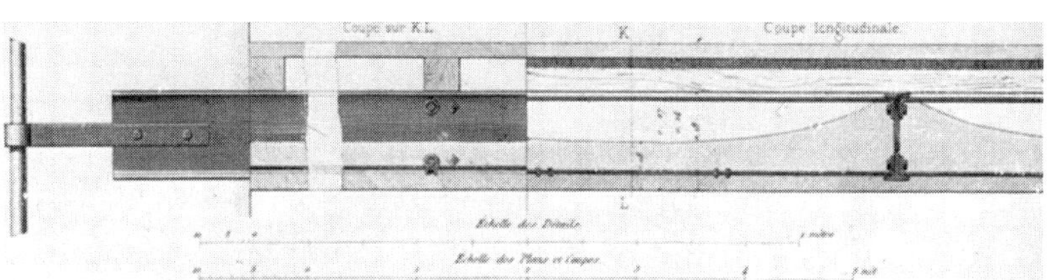

(*c*) Baudrit's Metallic Ceiling. *Rev. Gén. etc.*, 9 (1851) Pl. 12.

Contemporary "Fireproof" Construction.

PLATE XXIX

(a) Coffered ceiling in reinforced concrete and plaster.

(b) Reinforcement exposed in the soffit of a secondary beam.

(c) A broken beam showing wire-rope reinforcement.

(d) Reinforcement exposed in a secondary beam and slab.

Ceiling of Wilkinson's reinforced concrete cottage.

(W. Fisher Cassie)

4

Portland cements, 1843–1887

A. W. Skempton

Introduction

Portland cement has been for many years of immense importance in civil engineering and building work, and its manufacture has brought a great industry into being. The history of Portland cement has therefore attracted attention from several writers, among whom may be mentioned Michaelis (1869), Redgrave (1880), Quietmeyer (1912), Davis (1924), and Spackman (1929); whilst Haegermann (1927) has given an account of the development of testing methods, and Gooding and Halstead (1952) and Halstead (1961) have contributed material on the English pioneers of cement manufacture. It is still, nevertheless, an open question as to whether Joseph Aspdin actually produced a Portland cement, or whether he can merely be credited with inventing the name, which occurs in his patent of 1824. Indeed, it is usual nowadays to attribute the first production of a Portland cement to I. C. Johnson in 1845, when he was employed as works manager for the firm of J. B. White & Sons at Swanscombe. But, even then, there seems to be no clear idea as to how the cements of that period compare with the material of more recent times, nor as to the differences between the early Portland cements and the "Roman" cements which, in the 1840s, were far better known and more widely used. Neither has any explanation been offered for the remarkable delay before Portland cement concrete was combined with iron or steel reinforcement: for reinforced concrete can scarcely be said to have existed, as a practical proposition, until the 1880s.

In the present Paper an attempt is made to answer these and other questions by means of a quantitative examination of the available data on Portland cements of various periods, with special attention to measurements of strength.

Part I. The Early Portland Cements, 1843–71

John Grant's Tests, 1860–67

The first systematic tests on Portland cement were published by John Grant (1819–88) in 1866 and 1871 whilst he was working on the Main Drainage of London for the Metropolitan Board of Works. These tests necessarily constitute the foundation of our enquiry. They fall into three principal groups:—

(i) tension tests on Portland cements mixed with sand in proportions varying from 1 : 0 to 1 : 5 (by volume), at ages ranging from 7 days to 1 year;

(ii) a more restricted set of tension tests on Roman cements;

(iii) a comprehensive series of compression tests on Portland cements; also on mixes from 1 : 0 to 1:5 and, at the age of 1 year, on mixes down to 1 : 10.

The tension test was introduced by French engineers about 1848 using the briquette shown in Fig. 1 with a sectional area 5 × 7 cm. = 35 sq. cm. This was later reduced in size to 4 × 4 cm. and in this form was adopted by Grant in 1859 as his "Mould 1" (see Fig. 2 (*b*)). He also introduced a lever testing machine, made for him by Adie (Fig. 3), which so improved the loading system as to give results about 15 per cent. greater than those obtained by the earlier rather crude system of simply adding weights to a scale pan suspended from the briquette.[1] In October 1866 Grant modified the briquette ("Mould 2") to avoid stress concentrations, by rounding the corners of the neck (Fig. 2 (*c*)), and this resulted in a further improvement of 25 per cent.; as proved by numerous comparative tests. Ten years later a modification was introduced in Germany (Fig. 2 (*d*)), and this was widely adopted after its appearance in the first German Standard Specification of 1878. It was little better than Grant's "Mould 2," although more convenient. At the same time, however, Wilhelm Michaelis devised a double lever testing machine to give a more consistently axial load (Fig. 4) and this increased the test results by 15 per cent.

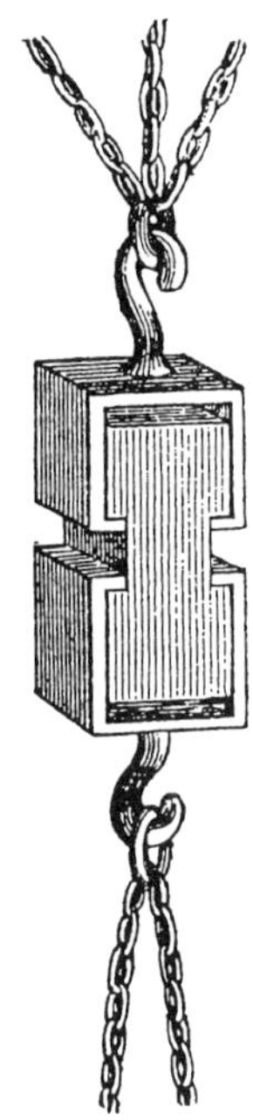

FIG. 1. BRIQUETTE USED 1848–66 FOR TESTING PORTLAND CEMENT
(G. F. White 1852)

Thus, to compare tension strengths on a uniform basis the following correction factors should be applied:—

Mould 1, direct loading	1·15 × 1·25 × 1·15	= 1·65
Mould 1, single lever	1·25 × 1·15	= 1·43
Mould 2, single lever	1·15	= 1·15
German Mould, double lever	1·0	= 1·0

The factor for Mould 1, single lever, is in reasonable agreement with independent tests by Candlot (1886).

[1] For the comparison of results obtained by various loading machines, including the old method of direct loading, see Grant (1880).

PORTLAND CEMENTS, 1843–1887

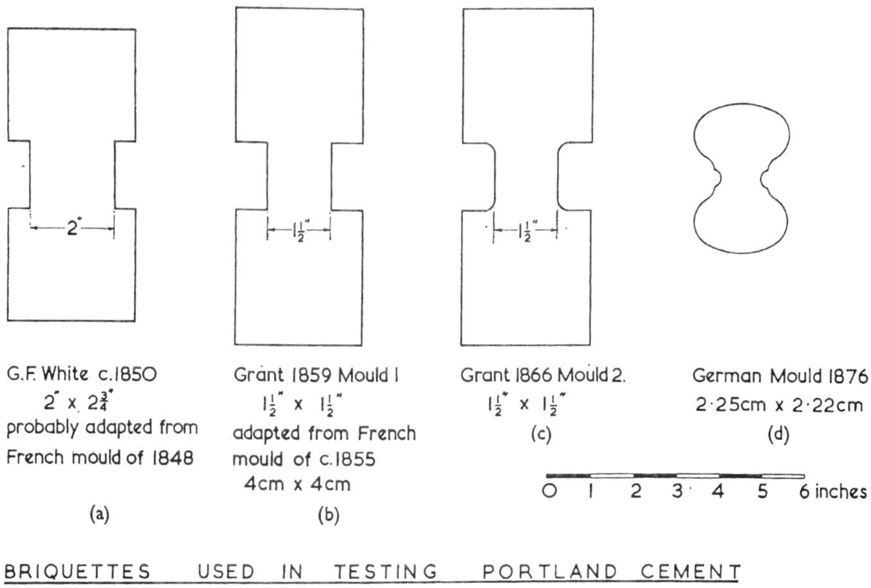

Fig. 2.

Grant's first series of tension tests were carried out in 1861–62 on two cements.[1] Individual results show considerable scatter but by double interpolation (plotting strength against age of specimen, and strengths at various ages against the proportion of sand) sufficiently reliable data can be obtained. The density of both cements was 1·40 and since the density of the sand would have been about 1·5, the proportion of 1 : 3 by weight, which later became standardised, is equivalent to 1 : 2·8 by volume. The strength-age graph for this mix is plotted in Fig. 5 and the general results are given in Table 1.

TABLE 1

Grant's Tension Tests (Converted to the German Standard)

(Specimens made in 1861 and 1862 with two Portland cements using clean pit sand. Strengths in lb./sq. in.)

Mix (by vol.)	7 days	28 days	3 months	6 months	12 months
1:0	320	465	570	630	660
1:1	130	245	360	430	510
1:2	75	160	255	320	375
1:3	45	110	175	220	270
1:4	30	70	120	155	200
1:5	—	40	80	115	145
1:3 (by weight)	50	115	185	235	290

[1] See Tables XVI and XVII in Grant's paper of 1866. After 1 day in air the specimens were cured under water until tested.

PORTLAND CEMENTS, 1843–1887

Routine tests were constantly being made, and these show a decided improvement around 1863 which was maintained, but not extended, during the next 5 years, as can be seen from Table 2. From 1860 each figure in this Table is the average of at least 50 tests, and the grand (weighted) averages are based on not less than 2,000 tests. The 1859 result is calculated from 130 tests. There can be no doubt that the manufacturers were making considerable efforts to improve their products in response to Grant's strict control of cement quality and, as just mentioned, this improvement came into full effect by 1863.

Fig. 3. Grant's Briquette Testing Machine, 1859
(Grant, 1866, Pl. 4)

In 1868 Grant carried out a second set of tension tests with various proportions of sand, and at ages from 1 month to 1 year (Grant 1871, Table 9). The interpolated results for 1 : 3 (by weight) are plotted in Fig. 5. It will be seen that the 28-day strength is now 160 lb./sq. in., as compared with 115 lb./sq. in. in 1861–62; but the strengths at 12 months are almost the same in both cases.

PORTLAND CEMENTS, 1843–1887

Grant carried out compression tests in 1860 on nearly 180 specimens, with the size and shape of an ordinary building brick, using six different mixes ranging from 1 : 0 to 1 : 5, each being tested at 3, 6 and 9 months (Grant 1866, Table XXX). Equivalent "cube" strengths can be deduced by applying a factor of 0·8 to the "brick" strengths.[1] Then, in December 1865 he cast 12 in. cubes with mixes from

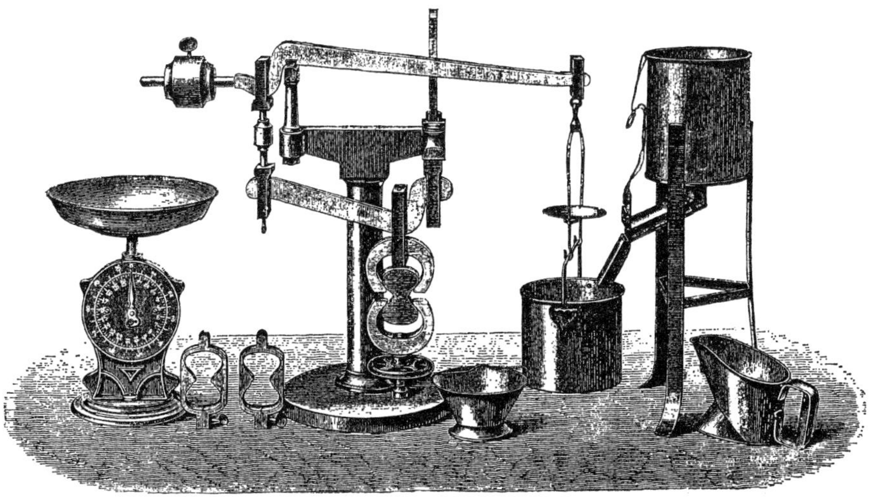

Fig. 4. FRUHLING AND MICHAELIS DOUBLE-LEVER BRIQUETTE TESTING MACHINE, 1876
(*Nouvelles Annales de la Construction*, Jan., 1882)

TABLE 2

TENSION STRENGTH OF PORTLAND CEMENTS (1 : 0) AT 7 DAYS

(Average of Routine Tests for Metropolitan Board of Works (lb./sq. in.))

Manufacturer	1859	1860–62	1862–63	1863–65	1867–71
Burham	cements	360	400	430	420
Hilton	not	310	410		400
Knight	specified	320			410
Robins	by	350	440		400
White	name	330	360		400
Grand average	270	340	420	430	410

1 : 1 to 1 : 5 and tested them at about 4 weeks (Grant 1866, Table XLII). Owing to the limitation of his compression machine the specimens were not taken to their ultimate crushing strength, and only the stresses causing the "first symptoms of failure" were given. However, from the 1860 tests, where "first signs" and the ultimate strength were recorded it is possible to say that the ultimate strength

[1] The specimens were 9 in. long × $4\frac{1}{4}$ in. broad × $2\frac{3}{4}$ in. high, loaded vertically. With a height/breadth ratio of 0·65 the factor to convert to cube strengths would be about 0·85 for a square base, but with a length/breadth ratio of about two a further reduction of 5 per cent. is necessary. The former figure is deduced from tests by Gillmore (1888), Gonnerman (1925) and Bradford (1945), whilst the effect of length is found from tests by Mather (1945).

PORTLAND CEMENTS, 1843–1887

is roughly 50 per cent. greater than the stress when failure is first noted. On the other hand the strength of the cements available in 1865 was approximately 33 per cent. greater (at 28 days) than those of 1860. Hence the 1865 results can be converted to ultimate cube strengths in 1860 by applying a factor of 1·5/1·33 = 1·12. Finally, in 1867, about 40 cubes (6 in. and 12 in.) were cast and cured in air or water for one year, mixed in proportions from 1 : 2 to 1 : 10 of cement to sandy gravel (Grant 1871, Series D, Tables I and II). The tension tests show that little improvement had taken place during the 1860s in the strengths at one year, and the 1867 cube tests can therefore be taken as applying with little error to any period within this decade.

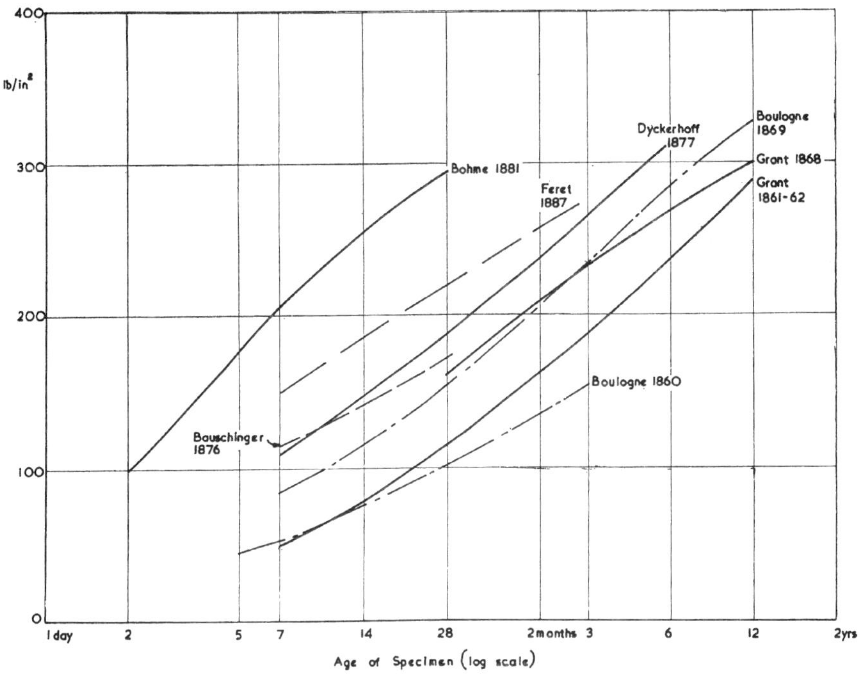

Fig. 5. TENSION STRENGTH OF PORTLAND CEMENT (1 : 3 by weight)

In spite of the assumptions that have had to be made, the foregoing data are found to be consistent when plotted in graphs relating compression strength and age for a given mix, and strength and mix for a given age. The results obtained from such double interpolation are given in Table 3, and the gain in strength with time for the 1 : 3 mortar[1] is shown in Fig. 6.

From the 12 in. cube tests of December 1865, the 28 day strength for a mix of 1 : 3 by weight at this date is found to be about 1,200 lb./sq. in.

Boulogne Cement, 1860–70

It is an interesting fact that the strengths of the Portland cement manufactured at Boulogne are very similar to those of the contemporary English cements. The Boulogne factory, the first in France, was

[1] According to accepted nomenclature a cement mixed with sand is a "mortar." This does not imply, of course, that it is necessarily to be used as a mortar for brickwork or masonry.

PORTLAND CEMENTS, 1843–1887

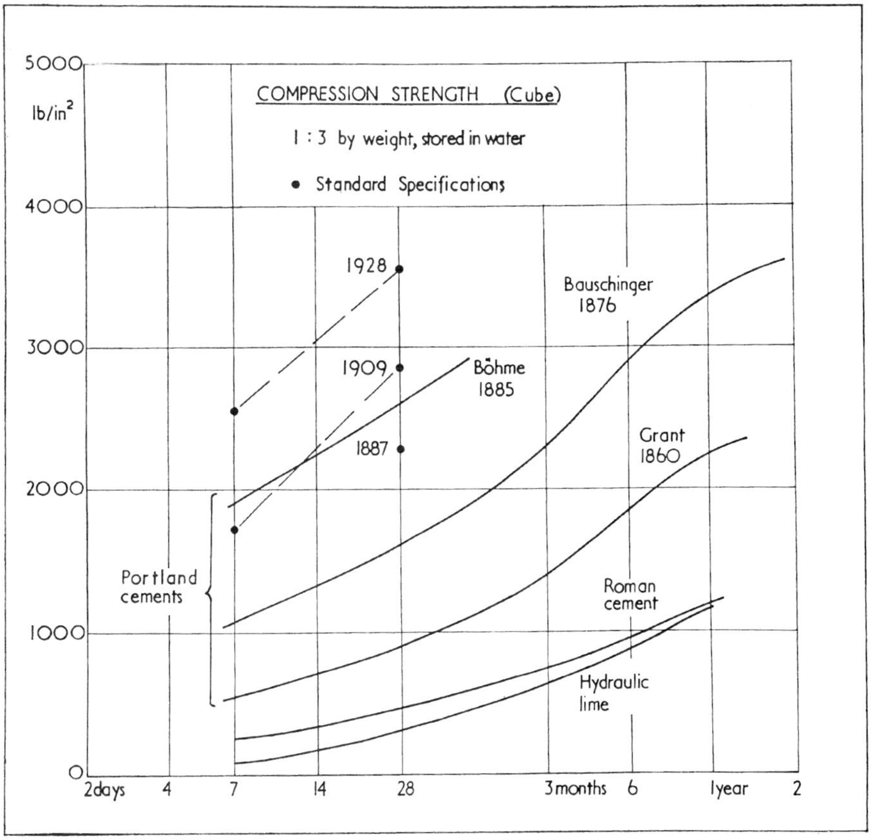

Fig. 6.

TABLE 3

GRANT'S COMPRESSION TESTS ON PORTLAND CEMENT WITH SAND AND SANDY GRAVEL

(All expressed in 1860 strengths, lb./sq. in.)

Mix (by vol.)	28 days	3 months	6 months	12 months
1:0	2,500	3,300	4,100	4,900
1:1	1,600	2,150	2,700	3,350
1:2	1,100	1,600	2,100	2,600
1:3	800	1,250	1,700	2,150
1:4	600	1,000	1,450	1,800
1:5	450	800	1,200	1,550
1:3 (by weight)	900	1,400	1,850	2,250

started by Dupont and Demarle in 1853. Tests were carried out on this cement soon after 1860 by Leblanc (1865) and about 1869 (Gillmore 1871). The 7 and 28-day tensile strengths are given in Table 10 and it will be seen that they are much the same as Grant's results for 1860 and 1868 respectively (see also Fig. 5). In the 1869 tests on Boulogne cement Gillmore also determined the 7-day compression strength; perhaps the earliest example of a test which has now been widely adopted. The importance of using a comparatively dry mix, and compacting the concrete as fully as possible, had long been known. However, Leblanc was probably the first to demonstrate the existence of an *optimum* water-cement ratio. Tests made in 1864 on a 1:4 mix, at 1 month, gave the following results:

Water/cement	0·66	0·78	0·94
Tensile strength (lb./sq. in.)	70	85	60

Summary of the Situation, 1860–70

Taking as a standard the 28-day compression strength of a 1:3 mix (proportions by weight) we have found the following results from Grant's tests:

1860	900 lb./sq. in.
1865	1,200 lb./sq. in.

At the time of the 1862 Exhibition there were about a dozen Portland cement firms in England, with a total annual output of 200,000 tons. One of the largest companies was J. B. White & Sons of Swanscombe. They employed 750 men; the factory was powered by eight steam engines and the production amounted to 30,000 tons of cement. During this decade the manufacture of Portland cement developed substantially in France and rapidly in Germany. In England the cost was stabilised at around 2s. to 2s. 3d. per bushel and the cements, in loose packing, weighed between 105 and 115 lb. per bushel (1 bushel = 1·28 cu. ft.). The raw material contained about 25 per cent. "clay" ($SiO_2 + Al_2O_3 + Fe_2O_3$) and a typical chemical analysis of the cement[1] would be as follows:

SiO_2	Al_2O_3	Fe_2O_3	CaO	MgO
23%	7%	4%	60%	1%

This might be expected to yield roughly 55 per cent. dicalcium silicate and 10 per cent. tricalcium silicate. These proportions indicate a poor Portland cement, by modern standards, with a low early strength: but quite definitely a Portland cement, and nothing else. The composition was somewhat variable, as might be expected at this stage in the development of cement technology, yet it is doubtful if the better manufacturers allowed the CaO content, for instance, to wander outside the limits 56 to 64 per cent. Having established the characteristics (especially the strength properties) of Portland cement for the period around 1860, when comprehensive data first became available, we can now turn to a study of the earlier cements.

Tests on White's Portland Cement, 1847–52

The earliest experiments on the strength of Portland cement, from which wholly reliable information can be obtained, were carried out under the supervision of James Rendel (1799–1856) at Henry Grissell's Regents Canal Ironworks in London on 10 and 31 December 1847. The specimens were 9 in. square in section and 18 in. high, and they were tested in a 75 ton hydraulic compression machine (Fig. 7). The cement was that of J. B. White & Sons. Details of the tests were given in *The Builder* of 15 July 1848 and in a valuable Paper read to the Institution of Civil Engineers in May 1852 by G. F. White (from which the illustrations in Figs. 1 and 7 are taken). George Frederick White (1816–98)

[1] Average of analyses given by Lipowitz (1868) and Michaelis (1869) of six different cements.

was the son of John Bazley White and he effectively directed the firm for many years. He frequently visited Cherbourg, Havre and other harbours where Portland cement was used and later in 1852 wrote a pamphlet giving the first substantial set of tension tests, to which reference will be made subsequently.

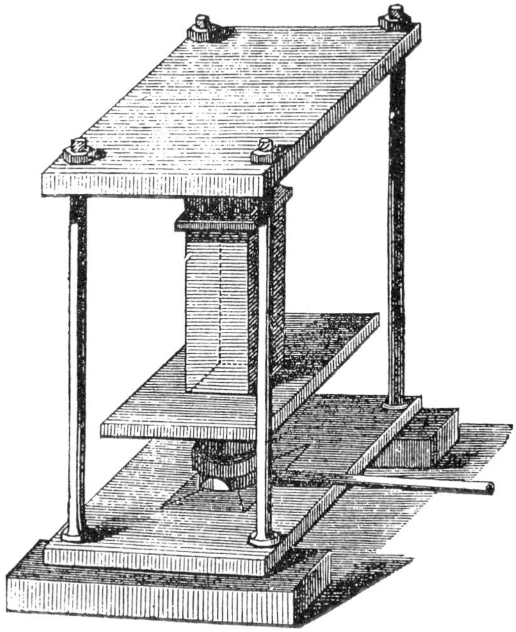

Fig. 7. 75-TON HYDRAULIC COMPRESSION MACHINE USED IN 1847 FOR TESTS ON PORTLAND CEMENT CONCRETE BY J. B. WHITE & SONS (G. F. White, 1852)

Of the 1847 series, three tests were made on specimens cut from large blocks of rubble masonry set in Portland cement mortar (one part cement to two or three parts sand). These blocks were probably made in connection with trials on various cements to be used at Cherbourg where work was just starting on the final stages of the great breakwater, and it may be noted that about 6,000 tons of White's cement was exported to Cherbourg during the period 1848–53; the first major use of Portland cement (Bonnin 1857). The strength of small brick piers set in Portland and in Roman cement was also determined, but the important tests from our point of view were those made on Portland cement simply mixed with sand or sandy gravel. The results are given in Table 4 and they are seen to be similar to the strengths obtained by Grant in 1860. The comparison may not be quite so close as would appear, since White's specimens were placed under water immediately after preparation and after 7 days' immersion they were cured in air until tested. These conditions differed from Grant's but the consequential effects on the strength were probably not important.[1]

Compression and tension tests on neat cement were made in 1851. These also give strengths very

[1] A small increase (about 5 per cent.) might be added to the compression strengths in Table 4, to convert to "cube" strength, but such a refinement is not justified in view of the variations in curing conditions.

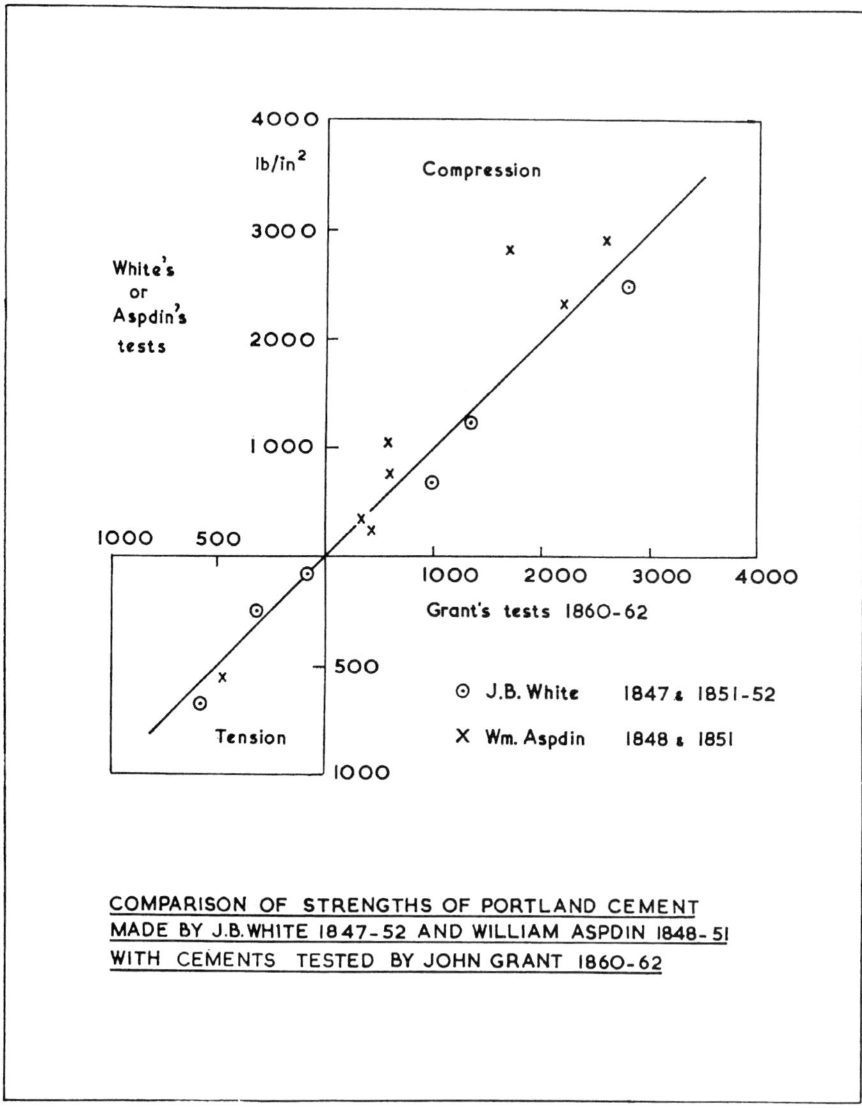

Fig. 8.

similar to those found by Grant. Of greater interest is a series of no less than 60 tension tests in 1852. These were carried out for routine control, during a 6-month period, on neat Portland cement at ages of 2 and 7 days and also on a 1:2 mix (by volume) at 7 days. The briquettes were of the original French form (Figs. 1 and 2 (*a*)) with dimensions anglicised to 2 in. × 2¾ in., and they were cured under water. The average results (see Table 4) are, again, closely similar to the strengths found by

Grant in his 1861–62 series. It is worth mentioning that in White's 1852 tests the standard deviations of the tensile strengths at 7 days were ±10 per cent. and ±22 per cent. for the 1 : 0 and 1 : 2 mixes respectively. From a general view of Table 4 it may be said that J. B. White's cement showed little variation during the period 1847–52 and had strengths equal to about 90 per cent. of the strengths measured by Grant in 1860–62 (see Fig. 8). Thus the 28-day cube strength of White's cement (1 : 3 by weight) in this period can be taken as 800 lb./sq. in. and from information given by G. F. White (1852) the density in loose packing was around 105 lb. per bushel, whilst the cost was 2s. 6d. per bushel.[1]

TABLE 4

STRENGTH OF PORTLAND CEMENT (J. B. WHITE & SONS) 1847–52
(compared with Grant's Tests 1860–62)

	Date of test	Mix	Age of specimen	Strength (lb./sq. in.)	Grant's tests (lb./sq. in.)	White/ Grant
Compression	10.12.1847	1 : 0	30 days	>2,100	2,500	—
	8.3.1851	1 : 0	40 days	2,490	2,700	0·9
	31.12.1847	1 : 2	52 days	1,250	1,350	0·9
	31·12·1847	1 : 3	52 days	690	1,000	0·7
	31.12.1847	1 : 10	52 days	280	—	—
Tension..	20.9.1851	1 : 0	6 months	680	630	1·1
	1852	1 : 0	2 days	110	—	—
	1852	1 : 0	7 days	250	320	0·8
	1852	1 : 2	7 days	70	75	0·9

It may be added that G. F. White (1866) gave the average results of a large number of tension tests made on cement supplied to the Metropolitan Board of Works, and also sent out to France, during 1860–65. These are rather higher than Grant's tests in 1861–62 and conform to the general improvement which, as already mentioned, took place around 1863 (see Tables 2 and 10).

Tests on Robins & Aspdin's Portland Cement, 1848–51

The publication in July 1848 of the tests on White's cement stimulated William Aspdin (1815–64) to have similar tests carried out on the cement which he was manufacturing at Northfleet, where he had entered a partnership with Robins in January of that year. In a letter to *The Builder* of 9 September 1848 he gave results of two compression tests on 1 : 0 and 1 : 3 mortars at 14 days. On 18 September further tests were made at the Robins & Aspdin wharf at Great Scotland Yard, and these were continued, with a more powerful machine, at Messrs. Robinson's works (previously Bramah's) in Pimlico on 26 September. The specimens were the same size and shape as those tested a year earlier by White. The two latter series (i.e. the tests at Great Scotland Yard and at Pimlico) were reported by *The Builder* on 30 September 1848 and in an advertisement by Robins & Aspdin on 7 October. The results of all the tests are given in Table 5, together with a briquette test on neat

[1] It is of interest to record typical prices of various materials for the period around 1848 when Portland cement first became of importance:—

Sand and gravel	2d. to 3d. per bushel
Hydraulic lime	8d. to 10d. per bushel
Roman cement	1/2 to 1/6 per bushel
Portland cement	2/- to 2/6 per bushel

cement carried out at the Great Exhibition in September 1851 (*The Builder*, 27 September 1851).[1] Table 5 shows that the Robins & Aspdin cement during the period 1848–51 was, on the average, about 20 per cent. stronger than the cements tested by Grant in 1860–62. But the specimens were probably cured in air, all the time, and this might lead to somewhat higher strengths. A figure of 1,000 lb./sq. in. for the 28-day cube strength (1 : 3 by weight) is perhaps the best estimate that can be made.

From an advertisement by Robins & Aspdin, announcing that they had purchased the Northfleet cement works of Maude & Aspdin (*The Builder*, 22 January 1848), we know that their cement weighed about 100 lb. per bushel and cost 2s. per bushel. Another advertisement (*The Builder*, 11 March 1848) reports that the cement "is manufactured to set in from five to sixty minutes." This is not necessarily the "final set," as later defined, but it certainly indicates a quick setting cement not very different, in this respect, from the Roman cements. Robins & Aspdin's cement seems on the whole to have been rather stronger than White's. But it was more variable. The strength ratios in Table 5, for example, show a standard deviation of ±30 per cent. as compared with only ±14 per cent. for White's cement. The latter was preferred for engineering works.

TABLE 5

STRENGTHS OF PORTLAND CEMENT MANUFACTURED BY ROBINS, ASPDIN & CO. 1848–51

(compared with Grant's Tests 1860–62)

	Date of test	Mix	Age of specimen (days)	strength (lb./sq. in.)	Grant's tests (lb./sq. in.)	Aspdin/ Grant
Compression	4.9.48	1 : 0	14	2,330	2,200	1·1
	26.9.48	1 : 0	35	2,920	2,600	1·1
	26.9.48	1 : 1	35	2,830	1,700	1·7
	4.9.48	1 : 3	14	780	600	1·3
	18.9.48	1 : 4	28	1,060	600	1·8
	18.9.48	1 : 5	28	240	450	0·5
	18.9.48	1 : 6	28	350	*c.* 350	1·0
	18.9.48	1 : 7	28	470	—	—
	26.9.48	1 : 9	35	120	—	—
Tension..	23.9.51	1 : 0	1 month	620 or 460	470	1·3 or 1·0

I had hoped, at this point, to introduce a set of tests carried out on Aspdin's cement in February 1853, but the published results cannot be trusted. William Aspdin was a restless person. Having moved from Rotherhithe to Northfleet in 1847 he left there towards the end of 1851 and set up a factory at Gateshead in partnership with Ord. Messrs. Robins & Co. continued to operate the Northfleet works, but not necessarily with an unchanged technique, and their advertisements in 1852 refer to "great improvements during the present year," whilst in the 1860s their cement was consistently among the best of those supplied to the Metropolitan Board of Works (see, for example, Table 2).

In September 1852 advertisements appeared announcing that "Messrs. Aspdin, Ord & Co. have much pleasure in informing customers . . . that their works are now in operation." Between 8 and 12

[1] *The Builder* reports a briquette test on neat cement at 1 month, giving a strength of 620 lb./sq. in. when corrected to German Standard. The Report on Class 27 of the Great Exhibition gives a strength of 460 lb./sq. in. (German Standard) for a test on the same size briquette, also at 1 month, but for a 1 : 3 mix. This latter strength is inconceivably high, and the cement must have been neat. Thus we have two values (see Table 5). The result given by *The Builder* is more probably correct.

January 1853 a dozen or more test specimens were made, and shipped to their yard in Abingdon Street, Westminster, 3 weeks later. The specimens included 7 in. cubes for compression tests as well as prisms 4 in. × 4 in. × 3 in. high, and briquettes. The results were reported in *The Builder*, 26 February 1853, but they are in almost every case fantastically high: at least 300 or 400 per cent. greater than Grant's tests and very substantially above even the best German cements of the 1880s. From a circular issued by Aspdin & Ord in 1853 (quoted by Redgrave and Spackman 1905) it is known that their cement was sold at 2s. per bushel. In 1856 the Gateshead works were taken over by I. C. Johnson, who had left White's a few years earlier and had already set up two factories in Kent. William Aspdin then went to Germany where, after two false starts, he opened a cement factory at Lägerhorf near Itzehoe. This became the nucleus of an important branch of the Alsen Company (Gooding and Halstead, 1952).

Summary of the Situation, 1847–52

Test results published in 1848 on cement manufactured by J. B. White & Sons in the preceding year, and on cement made in 1848 by Robins, Aspdin & Co., show that the materials produced by these firms were similar in strength to the cements on the market 12 years later (and tested by Grant). The latter were unquestionably Portland cements and there is no reason to doubt that the cements of 1847–48 were also Portlands. No detectable change occurred during the next 4 years, and three chemical analyses for the period 1849 to 1852 reveal the following average composition:[1]

SiO_2	Al_2O_3	Fe_2O_3	CaO	MgO
22%	6%	5 %	59%	1%

This is not significantly different from the average analysis quoted earlier for the cements of the 1860s, but probably the variations from the mean were greater. By 1852 the manufacture of Portland cement was still confined to England. Six firms were then in existence:

Joseph Aspdin, Wakefield.
J. B. White & Sons, Swanscombe.
Robins & Co., Northfleet.
Charles Hilton, Faversham.
I. C. Johnson, Rochester.
Aspdin & Ord, Gateshead.

White & Sons, if not the others, were already exporting to the Continent, and Portland cement had become a serious competitor to Roman cement.

Tests on Maude & Aspdin's Cement, 1843; *with Notes on "Roman" Cement*

We must now move back a few years into more controversial territory to examine the data on a cement manufactured in 1843 by the firm of J. M. Maude, Son & Co. of Rotherhithe with whom William Aspdin was then associated. William Aspdin had been employed in his father's cement works at Wakefield until 1841. We next hear of him in a circular issued by Maude & Co. late in 1843.[2] The relevant passages read as follows:—

PATENT PORTLAND CEMENT

The manufacture of this cement has for many years been carried on by Mr. Aspdin at Wakefield . . . (but) owing to the heavy charges attending its conveyance to the London market its consumption there has necessarily been limited, and although its superiority over other cements has never

[1] Analyses by Pettenkofer, Faist and Winkler, quoted by Lipowitz (1868). They refer to English cements, probably White's.

[2] The 1843 circular is quoted by Redgrave (1880).

been contested by those who have been induced to give it a trial, the high price at which alone it could be supplied has hitherto proved a serious impediment to its more general introduction into the metropolis. Messrs. J. M. Maude, Son & Co. have now the satisfaction of announcing to the public that they have made arrangements with the son of the patentee for the purpose of carrying on the manufacture of this valuable cement at their extensive premises at Rotherhithe; and whilst they will thus be enabled to supply it at a considerably reduced price, they have also the satisfaction of stating that in consequence of improvements introduced in the manufacture, it will be found . . .infinitely superior to any cement that has hitherto been offered.

Production must have started not later than the summer of 1843, for in October of that year tests were made by Messrs. Grissell & Peto,[1] contractors for the new Houses of Parliament, on brick beams set in mortars consisting either of Roman cement or the cement already being manufactured at Rotherhithe. The tests were reported in the circular already mentioned and in *The Builder*, 8 June 1844. The object was to compare the strength of the beams made with the two types of cement and thus to obtain a comparison between the cements themselves. Unfortunately this type of test is not very reliable; though far better than no test at all. And as the results provide the only source of information so far discovered on the strength of Portland cement before J. B. White & Sons (with I. C. Johnson) produced their cement in 1845, the tests must be examined with the greatest care. A necessary preliminary step is to understand something of the nature and strength of Roman cement.

The so-called Roman cement was first made by James Parker in 1796 at Lambeth, and after 1799 at Bankside, Southwark, in partnership with Charles Wyatt.[2] Septaria (or "cement stones" as they were later called) were brought from the shores around Sheppey, on the north Kent coast, and burnt in kilns at a temperature just below that causing clinkering, and then ground to a powder. The septaria were calcareous nodules, containing about 30 per cent. clay, which had been weathered out of the London Clay cliffs in the vicinity. A typical chemical analysis[3] of the cement would be:

SiO_2	Al_2O_3	Fe_2O_3	CaO	MgO
25%	9%	8%	54%	2%

The higher content of alumina and iron oxide, as compared with Portland cement, imparted the property of setting very quickly (5 to 20 minutes), but the development of strength was largely due to dicalcium silicate. Owing to natural variations in the stones, and also with the variability of early Portland cements, the compositions of the two types of cement could overlap. A cement with 56 per cent. CaO, for example, might give either a Roman or Portland cement. The vital difference lay in the temperature of firing. From time to time some of the material in Roman cement kilns would become clinkered. The clinkers were not only more difficult to grind than the lightly burnt material, but with fairly high clay contents they would have little value as a cement. Consequently in Roman cement manufacture the clinkers were rejected. On the other hand, with lower clay contents (say around 26 per cent.) the clinkers would include some tricalcium silicate, and, after grinding, a Portland cement would be the result. Probably, when Roman cement stones having an unusually low clay content were accidentally overburned, Portland cement clinker was, in fact, produced. But in general the Roman cement manufacturers avoided over-burning and, indeed, the tendency was to use rather lower temperatures than those originally adopted by Parker, as by this means the cost was reduced. Also, after 1810, when Parker's patent lapsed, septaria from Harwich were increasingly used and they contained rather more clay than the Sheppey stones.

[1] Thomas Grissell (Henry's elder brother) and Morton Peto. Their firm carried out many large contracts during the period 1830–46.
[2] Charles Wyatt (1751–1819); apparently unrelated to the famous architects, the brothers James and Samuel Wyatt.
[3] Michaelis (1869) quotes several analyses of the Sheppey cement.

PORTLAND CEMENTS, 1843–1887

The term Roman cement may be adopted for the whole group of quick-setting cements with clay contents higher than those of the Portlands, but in the trade "Roman cement" came to mean a material made by burning a mixture of the cheaper and more abundant Harwich stones with a small proportion of Sheppey stones. "First quality" Roman cement was made predominantly with Sheppey stones, and another of equal quality was produced from cement stones found in the Upper Lias strata near Whitby. This was variously known as "Atkinson's" (William Atkinson introduced it to the London Market around 1813) or "Mulgrave" (the raw material came from Lord Mulgrave's estates) or "Whitby" or "Yorkshire" cement. The "Medina" cement made from cement stones in the Solent area was also of good quality. Not surprisingly, several attempts were made to manufacture this type of cement "artificially" from a suitable mixture of chalk and clay. Such attempts were stimulated by the successful full-scale synthesis of hydraulic lime by Louis Vicat (1786–1861) in 1818. By that time hydraulic limes, made by burning limestones containing 12 to 20 per cent. clay, were being used for making concrete required for the foundations of bridges, canal locks, etc., in France. But, as yet, limestones of this category were not widely known and Vicat's discovery that hydraulic lime could be manufactured from a mixture of the commonly occurring relatively pure limestones and clay was an event of considerable importance. Within a year a factory had been established at Meudon near Paris and the product proved to be of great benefit for building and civil engineering works. James Frost, a London builder, went to France to study the techniques and, by increasing the clay content from 15 up to about 30 per cent., he succeeded in making an "artificial" Roman cement. This he achieved around 1830 at his works at Swanscombe.

Frost's cement was by no means the first of its type, and various patents can be traced back to that of Edgar Dobbs in 1810 for making cements by somewhat similar processes. One of these patents was obtained in 1824 by Joseph Aspdin (1778–1855) for his "Portland Cement," the manufacture of which he started at Wakefield between 1825 and 1828. But, so far as I have been able to discover, Aspdin and Frost were the only two manufacturers who managed to produce commercially successful cements. On an experimental scale, however, Charles Pasley[1] produced a good Roman cement in 1836, after several years research at Chatham Dockyard. His work is fully described in a most informative account of Roman cements in general, and his own tests in particular, from which we can see that on more than one occasion he would have produced Portland cement if only he had realised the significance of his over-burnt low-clay mixtures. This, I think, is what Joseph Aspdin did realise. Although exactly when he did so, and whether by chance or design, are questions which cannot be answered.

Practically the entire output of Roman cement, however, was made from the naturally occurring cement stones. Francis and White were the most important manufacturers of the "natural" Roman cements from about 1820 to 1850. Originally in partnership, their establishment was situated at Nine Elms, Vauxhall. In 1834 they bought Frost's factory at Swanscombe as well, but in 1837 J. B. White took over these works entirely and Charles Francis continued at Nine Elms only. Parker's firm, known in this period as Wyatt, Parker & Co., lasted until 1846 when it was transferred to J. M. Blashfield. Atkinson's cement continued to flourish and other manufacturers included Turner & Montague, and Samuel Sheppard who had started a factory at Faversham in 1816. By 1845 it was officially estimated that a million tons of cement stones had been taken from the Harwich area, where as many as 300 boats with a crew of three or four men were employed in dredging for the stones which had accumulated on the sea bed near the shore (Thurston 1939).

Roman cements were used chiefly as mortar in brickwork, especially in civil engineering works, and as stucco on the external walls of London houses. It was also used, on a smaller scale, for making

[1] Colonel (later Sir Charles) Pasley (1780–1861). His book on cements was published in 1838. An excellent cement of similar composition was manufactured commercially by Jean Lebrun at Moissac (Tarn) from about 1845 (Reports of the 1855 Exhibition).

PORTLAND CEMENTS, 1843–1887

pre-cast concrete. Large quantities of Roman cement mortar were employed by Sir Marc Brunel (1769–1849) in constructing the Thames Tunnel. The ordinary lightly-burned Roman cement weighed about 80 lb. per bushel. In 1825 it cost 1s. 8d. per bushel. By 1843 the cost had fallen to 1s. 4d. and by 1851 to 1s. 1d. The density of first-quality Roman cements may have been a little greater, but almost certainly not more than 90 lb. per bushel. Parker's original price was around 5s. per bushel. Even in 1826 Atkinson's cement was still sold at 4s. although by 1846 the price had fallen to 2s. 3d., and this, I believe, was the most expensive Roman cement on the market. Typically, at the time of Grissell & Peto's tests in 1843 the cost of Roman cements might be taken as ranging from 1s. 4d. to 2s. and their weight from about 80 to 85 lb. per bushel.

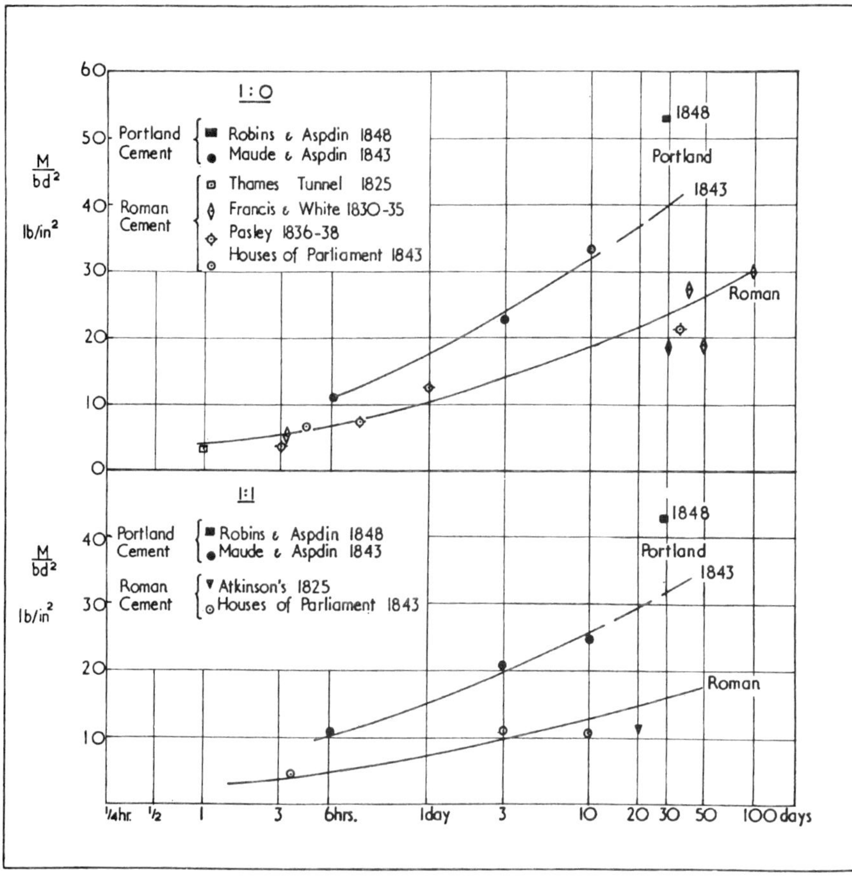

Fig. 9. Tests on Brick Beams with Portland and "Roman" Cement Mortars, 1843–48

It was common practice to assess the strength of Roman cement by building out a number of bricks from a wall, in the form of a horizontal cantilever 9 in. deep; the bricks being mortared in neat cement on the broad face and placed in position after the cement joining the previous brick to its neighbour had set. Practically every cask of cement used in constructing the Thames Tunnel, from

1825 onwards, was tested in this way. The bricks were added every 5 or 6 minutes and for the cement to be satisfactory 12 bricks had to be built out in about 1 hour, before failure occurred.[1] According to Pasley (1838) it was more usual to allow 15 minutes for each brick, and he records that with Francis & White's cement 15 bricks could be stuck out in about 3½ hours. For a longer term test the cantilevers could be built at the rate of a brick an hour or even one brick per day. Pasley built a cantilever of 31 bricks, in as many days, using his "artificial" cement. The best quality Francis & White cement regularly achieved 29 bricks in a month, and 33 bricks was regarded as about an upper limit.

A slightly different type of test was carried out by building a brick beam with cement mortar, using a temporary timber formwork if necessary, and then applying loads until failure took place. White[2] records a brick beam spanning 3 ft., set in a 1 : 1 mortar of Atkinson's cement, which was loaded centrally after 20 days in May 1825 (White 1832). Pasley gives the results of several similar tests, at ages up to 100 days. Now if M is the maximum bending moment at failure, and if b and d are the breadth and depth of the beam or cantilever, then $R = M/bd^2$ is a measure of the strength or adhesion of the cement mortar, independent of the actual dimensions of the beam. Values of R for the various tests just described are plotted in Fig. 9.

TABLE 6

VALUES OF R (LB./SQ. IN.) FROM BRICK BEAM AND CANTILEVER TESTS USING MAUDE & ASPDIN'S PORTLAND CEMENT, OCTOBER 1843

(Figures in brackets are extrapolated)

Mix (vol.)	Cantilevers	Beams	
	c. 6 hours	3 days	10 days
1:0	11	(23)	(33)
1:1	11½	20½	25
1:2	10	16	28
1:3	9	13	21
1:4	—	—	19

The tests by Grissell & Peto in October 1843 included the cantilever type as well as brick beams loaded after 3 and 10 days. The rate of building the cantilevers is not stated, but no serious error will be involved in assuming that the bricks were placed at 15 minute intervals. Roman cement mortars, used neat and with 1 : 1, 1 : 2 and 1 : 3 parts of sand (by volume) were able to support 17, 14, 11 and nine bricks respectively. The beams were built only with 1 : 1 mortar. The values of R calculated from these tests are also plotted in Fig. 9 and are seen to be in good agreement with the previous results. The lines drawn in these graphs represents the relation between R and time for good quality Roman cements.[3] The basis has now been laid for comparing Roman cement with the Portland cement of Maude & Aspdin. The 1843 tests by Grissell & Peto lead to the values of R given in Table 6. The results are plotted in Fig. 9.

[1] Sir Marc Brunel. *Correspondence and Reports to the Thames Tunnel Committee* (MSS. Inst. Civ. Eng. library).
[2] John White, District Surveyor of Marylebone, not to be confused with John Bazley White, the cement manufacturer.
[3] In drawing these lines we can be guided by the fact that 1 : 1 Roman cement has a strength or adhesion equal to about 67 per cent. of neat cement. This is shown by the 1843 cantilevers, by beam tests in 1848, by tension tests made by Grant (1866) and by an important series of adhesion tests by Pasley (1838).

PORTLAND CEMENTS, 1843–1887

The differences between Maude & Aspdin's cement and the Roman cements are striking, and it will be noted from Table 6 and Fig. 9 that even the 1 : 3 mortars made with the Portland cement are stronger than the 1 : 1 Roman cement mortars. This evidence provides convincing proof that Maude & Aspdin were producing a cement in 1843 decidedly better than first quality Roman cement of the period.[1] Moreover, from their circular of that year we learn that the Portland cement weighed about 100 lb. per bushel. This density is roughly 20 per cent. greater than that of Roman cement, and almost certainly indicates a more heavily-burnt cement. The cost was 2s. 3d. to 3s. per bushel. Thus it is difficult to avoid the conclusion that Maude & Aspdin's cement was significantly different from the Roman cements; and in the direction we would expect if it were tending towards proper Portland cement. This point may be investigated further by estimating the strengths of this and other cements, and by examining the historical context.

TABLE 7

TENSILE STRENGTHS OF ORDINARY ROMAN CEMENT AND 1861–62 PORTLAND CEMENT (LB./SQ. IN. CONVERTED TO GERMAN STANDARD)

Proportion by weight	Tensile strength at 28 days		Roman/ Portland
	Roman	Portland	
1 : 0	140	465	0·30
1 : 1·5	65	215	0·30
1 : 3	35	115	0·30
1 : 3 / 1 : 0	0·25	0·25	—

In comparing the strengths of various cements of different density it is necessary to make the comparisons on a weight for weight basis. Thus, if sand weighs 120 lb. per bushel a 1 : 3 mix by weight will correspond to a 1 : 2 mix by volume for a Roman cement (80 lb. per bushel); to a 1 : 2·5 mix by volume for the 1843 Portland (100 lb.), and to a 1 : 2·8 mix by volume for Grant's Portland cement of 1860 (112 lb. per bushel). White's pamphlet of 1852 and Grant (1866) provide data from which the 28-day tensile strength of ordinary quality Roman cement mortars can be found, and the corresponding strengths are also known for Portland cement as tested by Grant in 1861–62. The results, in Table 7, show that Roman cement bears a consistent ratio to the Portland cement, whether the mix is 1 : 0, 1 : 1·5 or 1 : 3 by weight, and that for both types of cement the ratio of the 1 : 3 mix to the neat cement is the same. These conclusions imply that we can calculate the compression strengths of Roman cements for a 28-day test on a 1 : 3 mix (even though no test of this nature was actually carried out in the mid-nineteenth century) from a knowledge of the compression strengths of neat Roman cements. These latter data are given by Charles Francis in a letter to *The Builder* (5 August 1848), where he quotes the 30-day strength of Nine Elms cement first quality and ordinary quality, and by White (1852) who gives the 40-day compression strength of Atkinson's, Sheppey and ordinary Roman as well as Portland cement (average of two tests in each case). The results, converted to 28-day values, are set out in Table 8.

The 1 : 3 strengths can be computed in three ways. In the first place, the 1 : 3 strength of Roman and Portland cements is approximately 35 per cent. of the neat strength. In Table 7 this ratio was 25 per cent., as will be seen, but an examination of many results for both types of cement shows ratios

[1] It is possible that the strength of Parker's original cement may have been higher than that of the best Roman cements made in 1843.

ranging from 0·2 to 0·5, averaging 0·35 and tending strongly to concentrate on this value. The strength of neat Roman cement, ordinary quality, is 860 lb./sq. in. Thus one estimate of the corresponding 1 : 3 strength is 860 × 0·35 = 300 lb./sq. in. Secondly, the ratio of compression to tension strengths, for a given mix, is around seven to nine for Portland cements of the later ninetcenth century and for the somewhat improved Roman cements of the same period. If we take a ratio of eight, and remember that the 1 : 3 tensile strength of ordinary Roman cement is 35 lb./sq. in. (Table 7), then another estimate of the 1 : 3 compression strength is 35 × 8 = 280 lb./sq. in. Thirdly, the average of four tests by Grant shows that, for a given mix by weight, Roman cement is 0·31 times as strong as his Portlands. Hence yet another estimate of the 1 : 3 compression strength of Roman cements is 900 × 0·31 = 280 lb./sq. in. These latter values confirm the first method, which can therefore be used to predict the strength of best quality Roman cement; and the procedure is checked by the value obtained for White's Portland, namely 800 lb./sq. in., exactly the result obtained earlier in this paper from a detailed comparison with Grant's tests. Finally it may be noted that Table 8 shows the strength of ordinary Roman cement to be about 75 per cent. of the first quality product. This is confirmed by a series of tension tests by Grant (1866) which reveal that the average strength of four ordinary Roman cements, at ages from 7 days to 6 months, is 70 per cent. of the strength of Medina cement; perhaps the best available at that time.

TABLE 8

COMPRESSION STRENGTHS OF VARIOUS CEMENTS (LB./SQ. IN.) *c.* 1850

Type of cement	Make	Compression strength at 28 days: 1 : 0	Compression strength at 28 days: 1 : 3 (weight) computed	
Roman, 1st quality	Francis	1,280	440	} 410
	Atkinson's	1,160	400	
	Sheppey	1,125	390	
Roman, ordinary	Francis	955	330	} 300
	White	765	270	
Portland	White	2,300	800	

Returning now to the Grissell & Peto tests, we find from Fig. 9 that the 1843 Portland is 1·7 times stronger than first quality Roman cement (1 : 0), and for the 1 : 1 mortars the ratio is 2·0. On a weight basis this is reduced to 1·9. The average of these two ratios is 1·8. Hence, taking 400 lb./sq. in. as the typical strength of 1 : 3 (weight) first quality Roman cement at 28 days, the corresponding strength of the 1843 Portland is about 400 × 1·8 = 720 lb./sq. in. But the strength of this cement can be estimated by an entirely different route. Cantilever tests, at the age of about 1 month, were made in 1848 on the Portland cement of Robins & Aspdin (*The Builder*, 30 September and 7 October 1848). The values of *R* (see Fig. 9), expressed as ratios of the corresponding values for Roman cement are 2·3 and 2·7 (or 2·55 on weight basis) for neat cement and 1 : 1 respectively. Brick beam tests were also made, using Roman cement in the proportions 1 : 1 and 1 : 2 and Portland cement in the proportions 1 : 3, 1 : 5 and 1 : 6. Not all the dimensions of the beams are given, so *R* cannot be evaluated. But the ratio of loads causing failure are remarkably consistent and show, on a weight for weight basis, that the Portland is 2·3 times stronger than the Roman cement. Thus there is a considerable body of evidence indicating that the 1848 Robins & Aspdin cement was about 2·4 times as strong as Roman cement. Yet we have already found that the 1843 Maude & Aspdin cement was 1·8 times stronger.

PORTLAND CEMENTS, 1843–1887

If the latter result is correct, then by the same token we should calculate a strength of 2·4 × 400 = 960 lb./sq. in. for the 1848 cement. This is almost identical with the result of 1,000 lb./sq. in. derived previously by comparing the 1848 cement with tests by Grant on Portland cements in the 1860s. Thus, taking the 1848 cement as having a strength of 1,000 lb./sq. in., we calculate the 1843 strength as 1,000 × 1·8/2·4 = 750 lb./sq. in. The approximate nature of these calculations should again be emphasised. Nevertheless it would seem justifiable to conclude that the 1 : 3 (weight) compression strength, at 28 days, of Maude & Aspdin's Portland cement of 1843 was not less than 700 lb./sq. in. and, further, that a reasonable estimate for the range of strength would be 700 to 750 lb./sq. in.

Summary of the Situation, 1843–47

The quantitative investigations now completed may be summarised as in Table 9. I have not hesitated to group Maude & Aspdin's cement with the Portlands since the data seem to admit no other interpretation.

TABLE 9

Type of cement	Details of cement	Cost per bushel s. d.	Weight per bushel lb.	Compression strength at 28 days 1 : 3 (weight) lb./sq. in.
Roman cements	Ordinary	1 4	80	300
	First quality	2 0	85	400–450
Portland cements	Maude & Aspdin, 1843	2 3	100	700–750
	J. B. White, 1847	2 6	105	800
	Robins & Aspdin, 1848	2 0	100	1,000±
	Tested by Grant, 1860	2 3	110	900

The historical sequence of events also leads to the same conclusion. The publication of the tests by Grissell & Peto at the Houses of Parliament created a minor sensation, since they revealed a cement very obviously superior in strength to the Roman cements then available. Now at this period I. C. Johnson (1811–1911) was works manager to J. B. White & Sons and he said[1] that

"my employers, attracted by the flourish of trumpets that was then being made about the new cement, desired to be makers of it, and some steps were taken to join Aspdin in the enterprise, but no agreement could be come to."

Thereupon Johnson started a series of trials which led him to find that when he pulverised some material which had been accidentally clinkered in the kiln a more powerful cement was produced than any he had so far achieved. Trial and error led to a mixture of five parts of chalk with two parts of Medway clay, as giving the best results, and with these proportions of the raw materials White & Sons went into full production in 1845. On 19 April 1845 they advertised in *The Builder* as "Manufacturers of Roman and Portland Cement." This form of words appeared repeatedly until October when it was changed to read

"PORTLAND CEMENT of best quality manufactured by J. B. WHITE and SONS . . . ."

[1] A long statement made by Johnson on his part in the development of Portland cement is quoted by Redgrave (1880).

PORTLAND CEMENTS, 1843–1887

The fact that White's adopted the same name for their cement as that associated irrevocably with the Aspdins seems to me to be of decisive significance. If, in fact, as Johnson has subsequently claimed, and as most historians believe, he had invented a genuinely new cement, surely the one name which would not be used was the name already adopted by Joseph Aspdin at Wakefield and by William Aspdin at Rotherhithe.

The truth would seem to be that the "break-through" had been made by 1843 at the latest, and more probably a good deal earlier by the elder Aspdin. Just what Joseph Aspdin did in Wakefield, apart from introducing the name "Portland cement," we may never know with any certainty. His first works were set up before 1828 near the foot of Kirkgate. In 1838 he moved to an adjacent site, having been forced to abandon the earlier site as a result of construction of the Manchester and Leeds Railway. Ten years later he moved again, this time to Ings Road, and the factory remained there until 1892. Aspdin himself retired in 1853 and the business was continued by his eldest son James, who had been in partnership since 1841 (Halstead 1961). Reid (1877) says that the kiln at "the parent cement manufactory at Wakefield . . . is of unusual height . . . affording excellent facilities for burning the cement." This remark must, however, apply to the Ings Road kiln. Nevertheless, the cement produced at the second Kirkgate factory must have been good enough, by 1841, to suggest the feasibility of opening new works at Rotherhithe. We may hazard the suggestion that Joseph Aspdin at some time before this date had made the vital discovery that clinkered material, when ground, and when coming from a low clay-content mixture, actually improved the cement quality. It would seem to be safe to assume that by 1843 William Aspdin was including this "over-burnt" material in his cement in sufficient quantity to raise the cement into the Portland category.

Johnson very probably (and George White also) deserves the credit for introducing more consistent and rational procedures. It can, for example, be shown that the mixture of raw materials adopted by White & Sons in 1845 contained about 26 per cent. clay.[1] This would be expected to yield a cement with around 58 per cent. CaO which is practically equal to the CaO content in the known Portland cements of 1849–52. But White himself did not claim explicitly that his firm had invented the process. In his paper of 1852 he wrote

> "Sir Charles Pasley and Mr. Frost stopped short at the all-important point of calcination, and they rejected that part of the kiln as overburnt, which was now found to yield the most valuable product. The aim of Sir Charles Pasley had been to rival the natural Roman cement, and he had succeeded; but it had been reserved for another operator to produce in Portland Cement a material which should far outstrip in cohesive power any of the natural cements known in this country."

It is reasonably certain that the other operator was Aspdin.

PART II. THE GERMAN PORTLAND CEMENTS, 1871–87

The first important Portland cement factory in Germany was established in 1853 at Züllchow near Stettin by Hermann Bleibtreu (1821–81), who had studied science at Giessen University and for a time in London. The factory came into full production in 1855. This was soon followed by the Quistorp works in the same locality and by Grundmann's at Oppeln in 1857. By 1864 at least 15 factories were in operation, including Alsen's at Itzehoe and Lägerdorf near Hamburg, the Bonn works at Obercassel, the "Stern" factory at Stettin and Dyckerhoff's near Mainz, as well as the Perlmoos works in Austria. The industry expanded rapidly, producing 440,000 tons in 1878 and reaching 1 million tons per annum in 1887.[2]

[1] Pasley (1838) quotes the water content and chemical composition of the Medway clay as determined by Faraday. The density of the wet clay was 87 lb./cu. ft. and that of dry chalk powder 40 lb./cu. ft. The content of $SiO_2 + Fe_2O_3 + Al_2O_3$ can then be calculated.

[2] For the history of Portland cement manufacture in Germany, see *Die Deutsche Zement Industrie* (ed. Riepert), Charlottenburg 1927, especially pp. 97–104, 397–505, 947–65.

PORTLAND CEMENTS, 1843–1887

In the beginning the Germans naturally modelled their factories on the English pattern, and adopted the same name for the product, but by the 1870s great improvements were being introduced and, as we shall see, the German cements of *c.* 1885 had strengths about double those of the English cements made 20 years earlier, and three times higher than those available before 1860. The technical changes cheifly responsible for these improvements may be summarised as follows. In the early period the raw materials were proportioned by "rule of thumb" and the cements were consequently erratic in composition. In 1871, however, systematic chemical analysis of the raw materials was introduced at the Dyckerhoff factory,[1] and soon became standard practice in Germany. The beneficial effects of this control are clearly seen in the analyses of nine different German cements made in 1876 (Bauschinger 1879), for their CaO content varied only from 59 to 62 per cent. The same degree of variation is shown by 11 cements analysed in 1881 (Böhme 1883) and a similar group of modern cements would show scarcely less scatter.

The second innovation was brought about by improvements in kiln design and management, resulting in higher and more uniform temperatures throughout (*c.* 1400° C.). In this way the content of CaO could be increased safely, without incurring dangers from free lime, and the cement would contain a higher proportion of the calcium silicates which are chiefly responsible for the strength after hydration. Already by about 1880 the German cements contained rather more than 62 per cent. CaO. This implies that they would yield about 25 per cent. tricalcium silicate and 45 per cent. dicalcium silicate.[2] No further significant increases were possible until the introduction of the long rotary kilns in the early twentieth century, and indeed a lime content of 62 per cent. was not consistently achieved by English and American manufacturers until about 1900–10. Since then, however, there has been a steady tendency to increase the proportion of lime, and a typical modern cement with 64 per cent. CaO would yield about 45 per cent. tricalcium silicate and 30 per cent. dicalcium silicate. Since the former is mainly responsible for the strength at ages less than 3 or 4 weeks, and the latter for the subsequent gain, it will be realised why there have been such remarkable increases in the strength of Portland cement as normally tested at 7 and 28 days: and the first great step in this direction was made in Germany around 1880.

But no matter how excellent the chemical composition may be, its beneficial effects would remain latent if the cement were not ground to a very fine powder. This follows from the fact that the reactions taking place during hydration and setting are essentially surface phenomena. In the early period when cements were comparatively lightly burnt and generally had low lime contents, the question of fine grinding was not paramount. But with harder clinker, resulting from increasing proportions of lime and higher kiln temperatures, it became necessary to pay serious attention to this aspect of the manufacturing process. By 1875 it was known that particles retained on a sieve with 900 meshes per sq. cm. (or 76 meshes per linear inch) possessed little value,[3] and the leading factories were already producing cement with not more than 15 per cent. (by weight) of particles coarser than this size. By 1880 a representative figure was 8 per cent., falling to 3 or 4 per cent. in 1887.

Such, then, were the technical factors leading to the "second revolution" in Portland cement, the results of which we shall now consider in detail. But it would be misleading to leave the reader with the impression that this was achieved solely by the manufacturers. Much was contributed by independent consultants such as Wilhelm Michaelis (1840–1911), by the university engineering laboratories such as that started at Munich in 1871 under Professor Johann Bauschinger (1833–93) and

---

[1] By Rudolf Dyckerhoff (1842–1917), son of the founder of the firm. He had studied chemistry at Heidelberg University.

[2] The estimations of the silicates, given in this paragraph, follow Lea & Desch (1935).

[3] Frederick Bramwell had drawn attention to the inert nature of the coarser particles, in his discussion on Grant's paper in 1866, but in relation to those retained on a 200 sieve (36 meshes per linear inch). He also very properly criticised the density test as a measure of quality in Portland cements.

by the Royal Technical Testing Station at Berlin where, from 1880, cement testing of the utmost reliability was carried out by Dr. Böhme on samples sent by engineers, local authorities and the manufacturers themselves.

TABLE 10

STRENGTH (LB./IN.^{2}) OF PORTLAND CEMENTS, 1847–90

Date	Authority	Cement	Time to final set hrs.	Residue on 900 sieve %	Tension 1:0 7 days	Tension 1:0 28 days	Tension 1:3 (weight) 7 days	Tension 1:3 (weight) 28 days	Compression 1:3 (weight) 7 days	Compression 1:3 (weight) 28 days
1847	G. F. White	White								800
1848	W. Aspdin	Aspdin								1,000
1852	G. F. White	White			250		45			
1859	Met. Board Works Spec.				255					
1859	Grant	6 cements			260	390				
1860–62	Grant				320	465	50	115		900
1860	Leblanc	Boulogne					55	100		
1864	G. F. White	White			350	480	75	130		
1865	Grant									1,200
1868	Grant				420	570		160		
1869	Gillmore	Boulogne			310	420	85	155	800	
1873–75	Bauschinger	Perlmoos								1,570
1875	Bauschinger	Dyckerhoff								1,680
1875	Bauschinger	Schiffendecker								1,700
1876	Bauschinger	5 cements	4	12	375	500	115	170	1,070	1,620
1877	Dyckerhoff	2 cements	10	14	360	550	110	190		
1878	Barreau	4 cements					105	180		
1879	Guarantee for Stern cement				570	710	213	284		
1878	German Standard		½	20				142		
1878	Austrian Standard			20			114	170		
1880	Grant's Specification		2	10	400	550		170		
1883	Swiss Standard			20				213		2,130
1880–84	Böhme	8 cements	6	8	620	740	220	290	1,600	2,100
1881	Böhme	11 cements	6	7	625	770	205	295		
1884–86	Böhme	47 cements	4	5			230	300	1,900	2,600
1885	French Specification		3	—	284	498	114	213		
1887	German Standard		2	10				228		2,280
1887	Swiss Standard		1	15				228		2,280
1887	Feret	6 cements			560	680	150	220	1,250	1,550
1888	Böhme	36 cements	4	3			230	300	2,000	2,800
1890	Candlot	6 cements							1,500	2,250

Bauschinger's Tests, 1873–79

It is evident from Michaelis (1869) that the testing of cement had not advanced very far in Germany during the period when Grant's earlier work was being undertaken. During the years 1876–78, however, Bauschinger carried out a remarkable series of tests on 10 cements, which opened a new era in this field of enquiry. As well as determining their chemical composition, setting time,[1] particle size, volume change, etc., he measured the strengths of 1 : 0, 1 : 3 and 1 : 5 mortars at times varying from

[1] The test for the time to "final set" followed exactly the test introduced by Vicat (1818), which he used for determining the setting time of hydraulic limes and lime-pozzolana mixtures.

7 days to 2 years (cured in air or water) in compression, tension, bending and shear; whilst briquette tests at 7 and 28 days were also recorded. The results were published in 1879 and attracted much attention, although they have been neglected by modern historians. With a single exception the cements fall into two sharply defined groups: (i) five cements with final sets between $1\frac{1}{2}$ and 7 hours, and with relatively high strengths, (ii) four cements setting in less than 50 minutes with decidedly lower strengths. Since this investigation provides the first complete set of data on Portland cement, the results have been summarised in Table 11, and the rate of increase of compression strength (measured on $4\frac{3}{4}$ in. cubes) is plotted in Fig. 6. This curve and the strengths given in Table 11 relate to specimens cured under water.

TABLE 11

SUMMARY OF BAUSCHINGER'S TESTS ON GERMAN CEMENTS MANUFACTURED IN 1876

Property	Average results	
	Five slow set cements	Four quick set cements
Initial set, mins.	60	5
Final set, mins.	240	25
Residue on 300 mesh/cm.^{2} sieve, %	1	2
Residue on 400 mesh/cm.^{2} sieve, %	7	9
Residue on 900 mesh/cm.^{2} sieve, %	12	14
Density, loose packing, gm./c.c.	1·27	1·29
Expansion of neat cement in 28 days, %	0·02	0·035
SiO_2, %	22·2	22·6
$Al_2O_3 + Fe_2O_3$, %	6·0 + 2·8	7·1 + 2·6
CaO, %	60·6	60·0
Strength of 1 : 0 briquette at 7 days, lb./in.^{2}	375	340
Strength of 1 : 0 briquette at 28 days, lb./in.^{2}	500	410
Strength of 1 : 3 briquette at 7 days, lb./in.^{2}	115	85
Strength of 1 : 3 briquette at 28 days, lb./in.^{2}	170	120
Strength of 1 : 3 cube at 7 days, lb./in.^{2}	1,070	620
Strength of 1 : 3 cube at 28 days, lb./in.^{2}	1,620	1,100

In 1877 Bauschinger published the compression strengths of 1 : 3 and 1 : 5 mortars, made with six different cements (manufactured in 1875), and measured on $4\frac{3}{4}$ in. cubes tested at 60 and 150 days. The variations between individual cements were small, and the average strengths for 1 : 3 fall almost exactly on the 1876 curve in Fig. 6. It is therefore possible to estimate the 28-day strengths for these cements. He also tested one of the cements, from Perlmoos, in 1873 at an age of 3 months (using 6 in. cubes) and here again the strength at 28 days may be estimated with fair accuracy by comparison with the 1876 results. The strengths of three of these cements are given in Table 10, and it will be seen that a value of 1,650 lb./sq. in. is highly representative of the better German cements *c*. 1875. The tensile strengths of two slow setting cements made at the Dyckerhoff factory in 1877 (quoted by Grant 1880) are given in Table 10 and plotted in Fig. 5. They are similar to the corresponding results from Bauschinger's tests. Tests on four cements by Barreau (1882) made at Boulogne in 1878 are also practically identical with the Bauschinger and Dyckerhoff results.

Portland cement of exceptional quality was being produced consistently by the Stern factory of Messrs. Toepffer, Grawitz & Co. which had been founded in 1861. By 1879 the annual production of this factory (Plate XXII) amounted to 40,000 tons of cement.[1] About 2,000 routine control tests were made each year, from which it was possible for the firm to guarantee the minimum strengths given in Table 12 (Reid 1879, p. 314).

TABLE 12

GUARANTEED TENSILE STRENGTHS OF "STERN" PORTLAND CEMENT IN 1879

	7 days		28 days	
Mix	Kg./cm.²	Lb./in.²	Kg./cm.²	Lb./in.²
1:0	40	569	50	711
1:3	15	213	20	284
1:6	6	85	10	142

The Stern cement was altogether outstanding in the 1870s but such was the rate of progress in Germany that by 1885 the *average* Portland cement produced in that country already exceeded in strength the figures given in Table 12. And the same held good in Switzerland (Mangold 1931) where a very rapid advance was made around 1880. By 1875 Portland cement was being manufactured in at least 10 countries, the dates of the first factories outside England, France and Germany being as follows (Quietmeyer 1912):—

Austria 1860, Denmark 1868, Switzerland 1871, U.S.A. 1871, Belgium 1872, Sweden 1873, Holland 1875.

The Early Standard Specifications

The first specification for Portland cement was drawn up by engineers of the *Ponts et Chaussées* about 1850, using the briquette test (Fig. 1) as a basis. I have been unable to find the exact terms of this specification, but White (1866) gives details of the rules in force *c.* 1860. These are set out in Table 13.

TABLE 13

FRENCH SPECIFICATIONS FOR PORTLAND CEMENT *c.* 1860

Minimum density = 1·20

Age of specimen cured in water	Tensile strength* (lb./in.)²	
	1:0	1:2 (by vol.)
2 days	95	—
5 days	185	95
30 days	350	185

* Converted to German Standard.

Grant was aware of the variations in strength resulting from the admixture of different types of sand and he therefore recommended that standard tests should be made on neat cement and that, for convenience in practice, the 7-day test was the most suitable. This proposal was accepted by the Metropolitan Board of Works and the Specifications introduced in 1859 called for a minimum density

[1] According to Reid (1879, p. 308) who visited the Stern factory early in 1879 the 14 kilns (see Pl. XXII) were 80 ft. high and unusually large. Eight hundred men were employed and the steam engines totalled 382 h.p. The residence of the Technical Director, Herr Haslinger, can be seen in the factory.

PORTLAND CEMENTS, 1843–1887

of 1·37 (110 lb. per bushel) and a 7-day strength (1 : 0) of 255 lb./sq. in.; increased in 1863 to 320 lb./sq. in. In 1865 the density requirement was increased to 1·40 with a strength of 360 and, in 1867, there was a further increase to 400 lb./sq. in. These latter strengths were rather excessive, and tended to compel the manufacturers to increase the lime content above the limits which could safely be adopted with the kilns of the period. The comparatively high density also discouraged efforts directed towards fine grinding of the clinker.

An important advance in cement specification was initiated in 1876 when Michaelis issued a manifesto emphasising the following points:

(i) tests should be made on 1 : 3 mortar as well as on neat cement;
(ii) the cement and sand should be gauged by weight;
(iii) the briquette shape and the sand and the testing machine should be standardised;
(iv) cement must not leave a residue of more than 25 per cent. on a sieve with 900 meshes per sq. cm.;
(v) the specimens should be cured for 1 day in air and 6 days in water (as specified by Grant) and the minimum strengths should then be:
 1 : 0 = 25 kg./sq. cm. (355 lb./sq. in.)
 1 : 3 = 5 kg./ sq. cm. (71 lb./sq. in.)

These points were widely debated, a committee was formed to study the various modifications, and suggestions were put forward.[1] These included the proposal by Hugo Delbrück (1825–1900) that the particle size of the standard sand should lie between the 60 and 120 mesh sieves (equivalent to 20 and 30 meshes per linear inch), as used to the present day, whilst there was a strong feeling, led by Rudolph Dyckerhoff, in favour of a 28-day test. The committee presented a report on 24 January 1877, and at this meeting the Association of German Cement Manufacturers was established, under Delbrück, with representatives from 23 firms. Their proposals for a Standard Specification were submitted to the Royal Prussian Minister for Commerce, Industry and Public Works on 28 February 1877; the chief difference from Michaelis being the substitution of a 28-day strength of 8 kg./sq. cm. (114 lb./sq. in.) for the 1 : 3 mix, in place of the 7-day test. The Minister set up a Commission to enquire into the matter and on 10 November 1878, he issued the first officially recognised national Standard for testing cement, containing the following rules:

(i) a residue of not more than 20 per cent. should be left on the 900 sieve;
(ii) a good slow setting Portland cement is one with a final set of more than ½ hour and a tensile strength of not less than 10 kg./sq. cm. (142 lb./sq. in.) when mixed with three parts of sand by weight and tested at 28 days;
(iii) the specimens are to be cured for 1 day in air and 27 days in water; their cross-sectional area is 5 sq. cm. (Fig. 2 (*d*)), the sand will pass the 60 sieve and be retained on the 120 sieve, and the water content is to be 10 per cent. (water/cement ratio = 40 per cent.);
(iv) the 7-day strength of neat cement may be used as a routine control on the site for cement already tested in accordance with the Standard.

This Specification of 1878 was admirable in many ways, but the 28-day strength could hardly be considered a stringent test for the better cements then in production. This defect was remedied, however, in the rules adopted only 1 month later in Austria with a 28-day strength of 12 kg./sq. cm. (170 lb./sq. in.). In addition the Austrian Standard specified a 7-day strength of 8 kg./sq. cm. (114 lb./sq. in.), although in both cases the average of the best six of 10 tests could be used.

[1] For details of the discussions and proposals in Germany 1876–78, see Haegermann (1927) who also gives full transcriptions of the official Standards of 1878 and 1887.

PORTLAND CEMENTS, 1843–1887

In 1880 Grant published his third paper, in which he enthusiastically approved the 1 : 3 mix with standard sand, and for cements setting in more than 2 hours he accepted the 28-day strength of 170 lb./sq. in. without qualification. Realising the continued need for tests on neat cement he specified a 7-day strength of 400 lb./sq. in. (significantly the same as in his rules of 1867, but now fully justified) and added a 28-day strength of not less than 550 lb./sq. in. Much data was presented on the importance of fine grinding and Grant went so far as to suggest that a 10 per cent. residue on the 900 sieve might be adopted. Hence by 1880 good practice was represented by the following minimal requirements:—

Final set	= not less than 2 hours.
Residue on 900 sieve	= not more than 10 per cent.
1 : 0 at 7 days, tensile strength	= 400 lb./sq. in.
1 : 0 at 28 days, tensile strength	= 550 lb./sq. in.
1 : 3 (by weight) at 7 days, tensile strength	= 114 lb./sq. in.
1 : 3 (by weight) at 28 days, tensile strength	= 170 lb./sq. in.

Test Results, 1880–1887

Although the briquette test carried out in accordance with the German Standard of 1878 was reasonably reliable, and although it continued in use with only minor modifications well into the twentieth century, the compression test was more satisfactory. Firstly because it is less prone to error and to the difficulties of precise interpretation, and secondly because concrete is essentially a material used in compression, not tension. The very first scientific investigations on the strength of Portland cement, in 1847, were compression tests and they were made on prisms with a height equal to twice their breadth. This is an excellent shape since it provides scope for the natural development of shear surfaces whilst minimising any dangers due to buckling and eccentricity of loading; and it has been adopted in the present century for research purposes and, in the United States, for Standard Specifications. Nevertheless, after a brief period when prisms were used with a height *less* than their breadth the vast majority of compression tests have been carried out on cubes.

Tests on lime and pozzolana mortars, with 2 in. cubes were made by Jean Rondelet (1734–1829) in October 1787.[1] This type of specimen was virtually forgotten, however, until reintroduced by Grant in 1865. He then adopted 12 in. cubes, but 2 years later he used the 6 in. cube which was to become the standard in England for concrete testing. Bauschinger used similar cubes in 1873 (actually 16 cm. = 6·3 in.) and the more convenient 12 cm. (4·7 in.) cubes for cement mortars in 1876, whilst in the same year Michaelis proposed a cube of 7·07 cm. side (2·8 in.) with an area of 50 sq. cm. for testing cements mixed only with sand: not, of course, for concrete. With these small specimens a neat and comparatively inexpensive machine could be employed (Fig. 10).

Under the guidance of Professor Ludwig von Tetmajer (1850–1905) at Zurich, the Swiss regulations for cement were formalised in 1883 and for the first time the compression test was included, the 28-day strength for 1 : 3 mortar being stipulated as 2,130 lb./sq. in. Meanwhile at the Royal Testing Station in Berlin, Böhme carried out tests on the 50 sq. cm. cubes from 1880 onwards;[2] sparingly at first but in considerable numbers after 1883. Thus, from 1884 to 1886 inclusive, no less than 47 cements were tested with the following average results:

1 : 3 (by weight) at 28 days, tension	= 300 lb./sq. in.
1 : 3 (by weight) at 28 days, compression	= 2,600 lb./sq. in.

[1] These were among the first reliable strength tests on any of the calcareous cementing materials. The specimens were prepared in April 1786 (Rondelet 1802).

[2] The results of these tests are systematically tabulated in a series of papers entitled "Cement Untersuchungen"—see list of references.

PORTLAND CEMENTS, 1843–1887

These figures may therefore be taken as establishing the typical strength of German cements with great certainty in the period *c.* 1885; although the best factories consistently achieved results at least 15 per cent. higher than the average. It will be recalled that in 1875 the 28-day compression strength (1 : 3) of the better cements tested by Bauschinger averaged 1,650 lb./sq. in. The comparable cements of 1885 would be represented by a strength of about 3,000 lb./sq. in. But part of this improvement, apparently amounting to 80 per cent., was due to the fact that the standard sand used

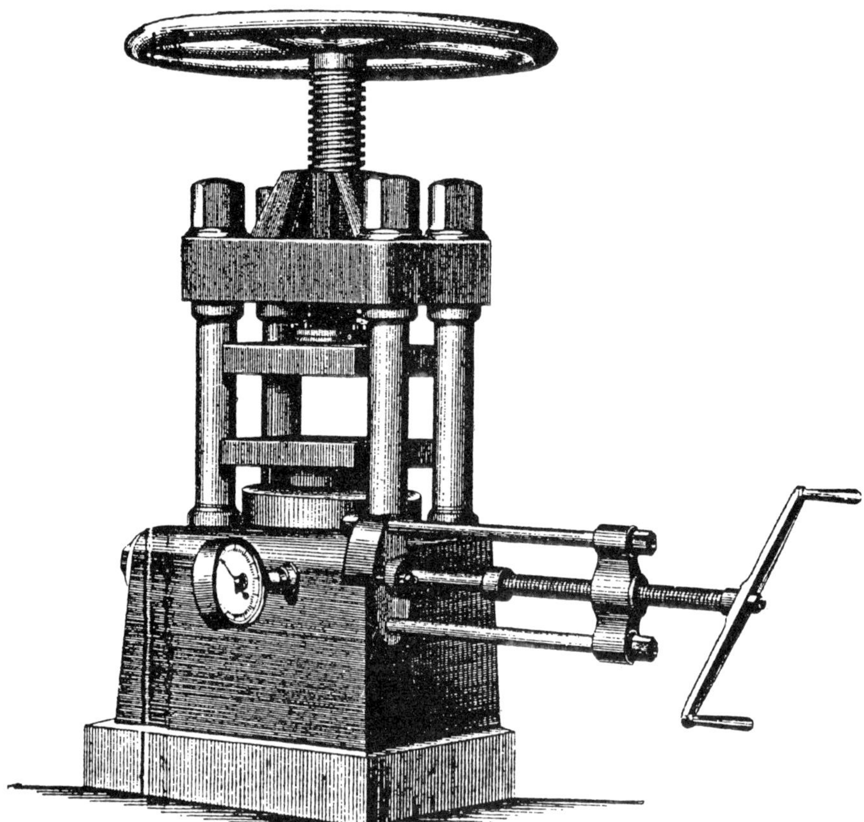

Fig. 10. Michaelis Hydraulic Compression Testing Machine 60-ton capacity, 1876 (Haegermann, 1927, Fig. 137)

in the later tests was superior to the finer grain sand used by Bauschinger. It is therefore more reasonable to compare his strengths with the *average* for 1885, when it will be seen that the increase during this decade amounted to nearly 60 per cent. The earlier tests by Böhme, carried out under conditions strictly similar to those of the 1884–86 tests, give a further indication of the improvements being made. Complete data, including compression strengths, were obtained on eight cements during the years 1880–84. The results are given in Table 10 but here we may note that the 28-day

PORTLAND CEMENTS, 1843–1887

compression strength averaged 2,100 lb./sq. in.[1] The corresponding tension strength was 290 lb./sq. in. and this is practically identical with the average of tests on 11 cements made in 1881 (Böhme 1883). This latter series included tests at 2, 4, 7, 14 and 28 days, which are plotted in Fig. 5.

During this period the English manufacturers were not advancing at the same rate, and a useful comparison between their cements[2] and those made in Germany can be obtained from tests carried out during the years 1878–84 in connection with the Boston Main Drainage Works (Clarke 1885, Grawitz 1886).[3] Average results are given in Table 14, from which it will be seen that the German cements were at least 50 per cent. stronger than the English; and the latter were obtained from highly reputable factories. The foregoing tests have not been entered in Table 10 since there is some doubt as to the correction factor necessary to convert the strengths to the German Standard. Possibly the figures should all be increased by about 15 per cent.; but this would not affect the ratios.

TABLE 14

TENSION TESTS MADE 1878–84 AT BOSTON, MASS. 1 : 3 BY WEIGHT

	7 days	1 month
3 German cements	165	240
4 English cements	100	160
Ratio German/English	1·65	1·50

Standard Specifications, 1885–87

In 1884 the French Minister of Public Works followed the lead of Germany and set up a laboratory in Paris for testing Portland cement, with branches at Boulogne, Marseilles and other ports. After an extensive investigation a Specification was drawn up in August 1885 by Guillain, the chief engineer for the Calais region, and this soon became recognised as a national Standard. For the first time chemical composition was subjected to specification, the stipulations being:

$$\frac{SiO_2 + Al_2O_3}{CaO} > 0{\cdot}44$$

$$Fe_2O_3 < 4\% \text{ and } SO_3 < 1\%$$

The test devised by Vicat (1818) for gauging the time to final set was adopted; i.e. when a needle of 1 sq. mm. area, loaded with 300 gm., just fails to leave an impression on the cement. And the initial set was defined when the needle just fails to penetrate fully through a depth of 4 cm.[4] The initial set must not take place in less than 30 minutes, and the final set must occur within 3 to 12 hours. As in all specifications, Grant's simple test for "soundness," which he introduced in 1859, was again quoted. The strength requirements were expressed in terms of the German briquette and double lever machine. The sand was to be produced by crushing quartz, and defined by the 64 and 144 mesh/sq. cm. sieves. The briquettes were to be prepared with one part cement to three parts sand, by weight, using 12 per cent. water. Tests on neat cement were also recommended, and the minimum strengths for specimens immersed in water after 24 hours are set out in Table 15.

[1] During 1880–84 a total of 23 cements were tested in compression at 28 days. The mean value is 2,080 lb./sq. in. as compared with 2,120 for the eight cements for which complete data are available.

[2] For the difficulties experienced even in 1880 by English manufacturers in attaining the higher lime content, as used by the Germans, see G. F. White's discussion on a Paper by Scott & Redgrave (1880).

[3] Grawitz quotes tests by W. J. Condon of the Boston City Eng. Dept. in 1882 on two German and two English cements. The other cements in Table 13 are reported by Clarke.

[4] These tests are still in use today.

PORTLAND CEMENTS, 1843–1887

The French specification of 1885 marked a further advance in cement testing, but the second German Standard of 28 July 1887 was even more important. Its essential requirements were:

Final set, of slow setting cements, as gauged by Vicat needle[1] >2 hours
Residue on 900 mesh/sq. cm. sieve .. .. <10 per cent.
Strength of 1 : 3 (by weight) specimens at 28 days (water content = 10 per cent.), immersed in water after 24 hours:
tension (5 sq. cm. briquettes) = 16 kg./sq. cm. = 228 lb./sq. in.
compression (50 sq. cm. cubes) = 160 kg./sq. cm. = 2280 lb./sq. in.

Simultaneously the Swiss regulations were modified and the strength requirements were identical with those given above. The residue on the 900 sieve and the final set were also similar (see Table 10).

TABLE 15

FRENCH SPECIFICATIONS FOR THE STRENGTH OF PORTLAND CEMENT 1885

	1 : 0		1 : 3 (by weight)	
Age of Specimen	Kg./sq. cm.	Lb./sq. in.	Kg./sq. cm.	Lb./sq. in.
7 days	20	284	8	114
28 days	35	498	15	213
84 days	45	640	18	256

If we add to the German and Swiss Standards of 1887 the initial set and the 7-day (1 : 3) tensile strength of the French Specifications, and Grant's 7-day tensile strength for neat cement, we have a set of rules which covers all subsequent Standards until 1901. In other words, by 1887 the pattern of modern cement specifications was already established.

Tests, 1887–90

A comprehensive series of tests on French cements was started by R. Feret in 1887 at the Boulogne laboratory. These were carried out in strict accordance with the French Standard of 1885. Of 10 cements six were acceptable on this basis, and the average results for the latter group are given in Table 10 (see also Figs. 5 and 6). Feret also determined the compression strength, using 50 sq. cm. cubes, but none of the cements attained a 28-day strength (1 : 3) equal to that specified by the German Standard. A considerably better position is revealed only a few years later in France by tests reported by Candlot (1891). Of nine cements four passed the German Standard and the average of the best six fell only just short (Table 10). The superiority of German cements at this time is made clear by Böhme's tests during a 24-month period centred on 1888. Compression and tension tests on 36 slow setting Portlands gave the remarkably high average values of 2,800 and 300 lb./sq. in. respectively at 28 days (Table 10) and only five of these cements failed to reach the specified minimum compression strength of 2,280 lb./sq. in.

CONCLUSIONS

By the end of the eighteenth century it was well known that certain limes would set and harden under water, and in 1791 John Smeaton (1724–92) published his classic research proving that this property depends upon the existence of clay in the parent limestone. In examining a considerable

[1] The initial set was also to be measured, exactly as in the French Standard of 1885, but the time for initial set was not specified.

number of these "water limes," as they were then called, he found clay contents in the raw material ranging from 6 to 20 per cent. Limes of this type set quite slowly, taking several days to the final set, but they proved to be useful for making concrete foundations, usually with the addition of pozzolana.

In 1818 Louis Vicat started work on the Souillac bridge, on the Dordogne, which ranks as the first major work with lime-concrete foundations in which no pozzolana was used. In the same year he published his celebrated book, introducing the term "hydraulic" lime for the first time; describing the test for determining final set; giving many strength measurements and announcing the successful manufacture of hydraulic lime from a suitable mixture of chalk and clay. Already in 1813 Collet-Descotils (1773–1815) had shown that the essential component in the clay was silica, which combines

TABLE 16

TYPICAL PROPERTIES OF LIMES AND CEMENTS

Class (kiln temp.)	Approx. proportion of clay in raw materials, per cent.	Description	Date	Specimens cured in water: Compression strength at 28 days 1:3 by weight, lb./sq. in.	Specimens cured in water: Time to final set	Density in loose packing
"Roman" cements (1,100° C.)	30	Ordinary quality	mid-19th century	300	¼ hour	1·0
		First quality		400		
Early Portland cements (1,300° C.)	25	Maude and Aspdin	1843	*c.* 700	½ hour	1·3
		J. B. White & Sons	1847	800		
		Robins and Aspdin	1848	1,000±		
		English cements	1860	900		
		English cements	1865	1,200		
Portland cements (1,400° C.)	23	German cements	1875	1,650	2 hours	1·3
		German cements	1885	2,600		
		Standard Specifications	1887	2,300		
			1909	2,800		
			1918	3,600		
			1932	3,900		
			1942	4,500		
Hydraulic limes (1,000° C.)	18	Metz, Teil, Senonches	Early and mid-19th century	300	3 days	0·8
	15	Lias, Labourgade		200	6 days	
	8	Dorking, Halling		100	12 days	
Limes (900° C.)	1	Chalk	—	0	never	0·5

with lime in the kiln to produce compounds responsible for the hardening of hydraulic limes under water. And in 1822 the distinguished chemist Pierre Berthier (1782–1861) analysed many limes, and also some Roman cements, from which he was able (i) to confirm Smeaton's conclusion that hydraulic limes contain between 6 and 20 per cent. clay, (ii) to show that only those with 15 to 20 per cent. clay are strongly hydraulic, and (iii) that Roman cements contain more than 25 per cent. clay. The strongly hydraulic limes, such as those from Metz, Le Teil and Senoches in France, and the Blue Lias in England, were extensively used for mass concrete from the 1820s well into the second half of the nineteenth century.

Meanwhile James Parker in 1796 discovered the "Roman" cements, which he made from cement stones containing about 30 per cent. clay. These are very quick setting. As such they were valuable

PORTLAND CEMENTS, 1843–1887

for mortar in brickwork and as external stucco, but they could not be used at all conveniently for large masses of concrete. Their strength was typically rather greater than that of the best hydraulic lime (see Table 16).

James Frost, around 1830, succeeded in manufacturing an "artificial" Roman cement from a mixture of chalk and clay, and Sir Charles Pasley in 1836 produced, on a small scale, a cement at least equal to the best natural Roman cement. In all cases, whether "artificial" or "natural", the kiln product at a Roman cement factory had to be ground to a powder, but the kiln temperature was kept below that which would cause clinkering. If, however, clinkers were produced from a mixture rather low in clay content (say about 26 per cent.) the powdered clinker would actually improve the cement. This may have been discovered by Joseph Aspdin, though there is no evidence to show whether or not he had realised the importance of clinkering when he started making his "Portland" cement at Wakefield at some time during the period 1825–28. Owing to construction of the Manchester and Leeds Railway, Aspdin had to move his factory to an adjacent site in 1838. He moved again, in 1848, to Ings Road, Wakefield, and there is an indication that in this third factory his kiln was capable of producing the necessary temperature. Nevertheless, the cement manufactured at the second factory must have been sufficiently good to attract more than local attention, for in 1841 William Aspdin, who had been working with his father, left for London and in 1843 was manufacturing "Portland" cement at Rotherhithe in partnership with J. M. Maude, Sons & Co. This cement was tested in October 1843 by Grissell & Peto, contractors for the new Houses of Parliament, and it was shown to be 80 per cent. stronger than first quality Roman cement of the period. It also weighed 100 lb. to the bushel, in contrast with the 80 to 85 lb. of the Roman cements.

In 1844 J. B. White & Sons, who had taken over Frost's factory at Swanscombe, decided to start making the "new" cement and their works manager I. C. Johnson soon succeeded in this enterprise. He definitely clinkered the materials in the kiln, and we know that the raw mixture contained about 26 per cent. clay; giving a cement with about 58 per cent. CaO. White & Sons went into production in 1845, significantly retaining the name "Portland" cement. Tests in December 1847 indicate that their cement had strengths only 10 per cent. less than those of good Portland cements available in 1860, as tested by John Grant in a notable series of investigations. In 1848 White & Sons started exporting large quantities of Portland cement to France for concrete used in the great breakwater at Cherbourg. Their cement weighed 105 lb. to the bushel. William Aspdin moved to Northfleet in 1847, and his new factory there may have incorporated improvements. He entered a partnership with Robins in January 1848 and tests in September of that year indicate a cement rather stronger than White's although more variable in quality. At this time, it will be recalled, Joseph Aspdin started his Ings Road factory at Wakefield. To these three pioneer firms, a fourth was added in 1849 when Hiltons converted a Roman cement works (previously Sheppard's) at Faversham. In 1850 Johnson left White & Sons to start his own factory near Rochester, followed by another at Cliffe in 1853. William Aspdin left Northfleet in 1851, where Robins continued very successfully, and went to Gateshead. There, from 1852–56, he managed a large factory in partnership with Ord. Johnson took over these works in 1856 when Aspdin went to Germany where he eventually founded important cement works at Lägerdorf. Other firms in England in this period include Thomas Freen's at Wouldham, started in 1853. By 1862 White & Sons were producing 30,000 tons annually and the total output for the whole country amounted to 200,000 tons. The material was now replacing hydraulic lime for mass concrete and Roman cement for mortar and stucco.

The first Portland cement factory in France was started at Boulogne in 1853, and the first in Germany began production at Stettin in 1855. The development in Germany was rapid. Strict chemical control of the raw materials was introduced by Rudolf Dyckerhoff in 1871. Kiln temperatures were raised, and kiln design was improved. The CaO content was increased up to 62 per cent., thereby increasing the amount of tricalcium silicate, and the cement began to be more finely ground.

PORTLAND CEMENTS, 1843–1887

By 1875 a superb series of tests by Johann Bauschinger at Munich University showed that the better German cements were now attaining strengths 80 per cent. higher than those tested by Grant in 1860. By 1885 systematic investigations at the Royal Berlin Testing Station, under Dr. Böhme, showed a further increase in strength of 60 per cent. over the preceding decade. In other words the German cements were now almost exactly three times as strong as the English cements of 1860 (Table 16). The techniques of testing had been improved by Bauschinger, and by Wilhelm Michaelis in Berlin. In Switzerland important work was also carried out by Professor Ludwig von Tetmajer and it was in that country that the compression test was first introduced into official Regulations, in 1883. The German and Swiss Standard Specifications of 1887 are completely modern in form, and required no revision until the first decade of the twentieth century. By 1887 German production had reached one million tons annually.

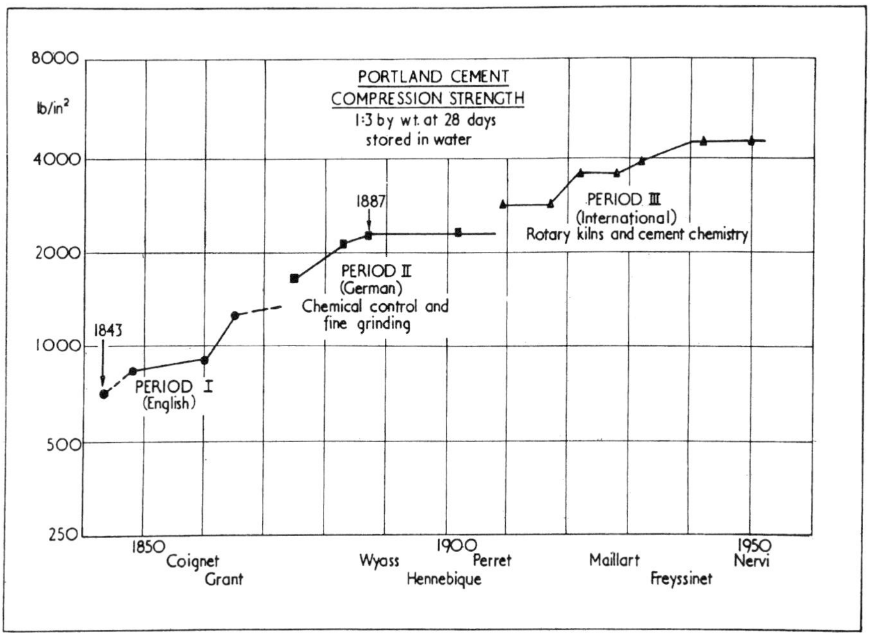

Fig. 11.

With the greatly improved cements of the 1880s, reinforced concrete for the first time became a practical proposition. The idea of combining iron reinforcement with concrete was not new, but it could not be put into effective practice until the strong Portland cements became available from the German factories with their scientific management. By 1900 the English and French, and American, manufacturers had caught up with the developments in Germany. The rotary kilns then initiated a further major improvement and, with important contributions from the science of cement chemistry, the way was open to the present day Portland cements and the associated achievements in reinforced concrete and pre-stressed concrete.

This history of Portland cement may be expressed in the graph shown in Fig. 11. The first break-through was achieved by the Aspdins; certainly by 1843 and probably earlier. I. C. Johnson and

PORTLAND CEMENTS, 1843–1887

George White improved the processes and the latter introduced scientific testing of the material (following the lead of Vicat in relation to hydraulic limes). John Grant was the first systematic investigator of Portland cement and his work led to better control of the manufacture. The Germans then made the second break-through in the 1870s. From their works the first modern Portland cements were produced and from their Specifications of 1887 (associated with the Swiss) we can date the beginning of the modern period, as well as the effective beginning of reinforced concrete.

ACKNOWLEDGEMENTS

I am grateful to Dr. R. W. Nurse of the Building Research Station and to Dr. P. E. Halstead of the Cement and Concrete Research Association for helpful discussions.

REFERENCES

—. Barreau: "Étude sur les qualités et essais des ciments a prise lente, dits Portland," *Anns. Ponts et Chaussées*, 1882 (2eme sem.), pp. 150–225.
J. Bauschinger: "Versuche über die Festigkeit von . . . Portland Cement," *Mitt. Mech. Techn. Lab. München*, vol. 1, 1873, pp. 5–17.
J. Bauschinger: "Untersuchung mehrer Sorten Portland Cemente," *ibid.*, vol. 7, 1877, pp. 1–8.
J. Bauschinger: "Untersuchung mehrer Sorten Portland Cemente," *ibid.*, vol. 8, 1879, pp. 1–28.
P. Berthier: "Analyses de différentes pierres à chaux," *Anns. des Mines*, vol. 7, 1822, pp. 483–510.
—. Böhme: "Beziehungen zwischen der Ergebissen von zwölf deutschen nach der preussischen und russichen Normen untersuchten Cemente," *Mitt. König. techn. Versuch. Berlin*, vol. 1, 1883, pp. 35–44.
—. Böhme: "Cement Untersuchungen," *ibid.*, vol. 2, 1884, pp. 129–43; *ibid.*, vol. 3, 1885, pp. 15–21; *ibid.*, vol. 5, 1887, pp. 120–45; *ibid.*, vol. 8, 1890, pp. 22–49.
J. Bonnin: *Travaux d'Achèvement de la Digue de Cherbourg de* 1830 *à* 1853, Paris, 1857.
R. B. Bradford: discussion on Mather's paper of 1945 (q.v.).
F. J. Bramwell: discussion on Grant's paper of 1866 (q.v.).
E. Candlot: *Étude Pratique sur le Ciment de Portland*, Paris, 1886.
E. Candlot: *Ciments et Chaux Hydrauliques*, Paris, 1891.
E. C. Clarke: "Record of tests of cement made for Boston Main Drainage Works," *Trans. A.S.C.E.*, vol. 14, 1885, pp. 141–70.
A. C. Davis: *A Hundred Years of Portland Cement*, London, 1924.
H. V. Collet-Descotils: "Sur la chaux maigre," *Journ. des Mines*, vol. 34, 1813, pp. 308–309.
Exposition Universelle de 1855 (Report of International Jury on Class 14), Paris, 1856.
Exhibition of 1862 (Report by M. Delesse on Class 10), Paris, 1863.
R. Feret: "Diverse expériences concernant les ciments" *Anns. Ponts et Chaussées*, 1890 (1er sem.), pp. 313–380.
Q. A. Gillmore: *Report on Béton Aggloméré*, Washington, 1871.
Q. A. Gillmore: *Notes on the Compression Resistance of Freestone . . . Mortars and Cements*, New York, 1888.
P. Gooding and P. E. Halstead: "The early history of cement in England," *Proc. 3rd Int. Sym. Chem. Cement*, London, 1952, pp. 1–27.
H. F. Gonnerman: "Effect of size and shape of test specimens on compression strength of concrete," *Proc. A.S.T.M.*, vol. 25, 1925, pp. 237–250.
J. Grant: "Experiments on the strength of cement," *Min. Proc. Inst. C.E.*, vol. 25, 1866, pp. 66–111.
J. Grant: "Further experiments on the strength of Portland cement," *ibid.*, vol. 32, 1871, pp. 266–306.
J. Grant: "Portland cement; its nature, tests and uses," *ibid.*, vol. 62, 1880, pp. 98–179.
G. Gravitz: *The Star Portland Cement Works*, New York, 1886.
F. Guillain: *Devis et Cahier des Charges de Ciment pour les Ports de Boulogne et Calais*, Boulogne, 1885. Reprinted in Candlot (1886) and *Nouv. Anns. Construction* (1886) col. 87–95.
G. Haegermann: "Die Entwicklung des Prüfungswesens" *Die Deutsche Zement Industrie* (ed. Riepert), Charlottenburg, 1927, pp. 265–396.
P. E. Halstead: "The early history of Portland cement," *Trans. Newc. Soc.*, vol. XXXIV, (1961–62) pp. 37–54.
F. M. Lea and C. M. Desch: *The Chemistry of Cement and Concrete*, London, 1935.
E. M. Leblanc: "Note sur les qualités à rechercher dans le ciment de Portland," *Anns. Ponts et Chaussées*, 1865 (2eme sem.), pp. 84–98.
A. Lipowitz: *The Practical Manufacture of Portland Cement*, trans. W. F. Reid and published in H. Reid, *A Practical Treatise on the Manufacture of Portland Cement*, London, 1868.
F. Mangold: *Denkschrift zum funfzigjährigen Bestehen des Vereins Schweizerischer Zement-Fabrikanten*, Basle, 1931.
B. Mather: "Effect of type of test specimen on apparent compressive strength of concrete," *Proc. A.S.T.M.*, vol. 45, 1945, pp. 802–809.
W. Michaelis: *Die hydraulischen Mörtel insbesondere der Portland Cement*, Leipzig, 1869.
W. Michaelis: *Zur Beurtheilung des Cementes*, Berlin, 1876.

PORTLAND CEMENTS, 1843–87

C. W. Pasley: *Observations on Limes, Calcareous Cements, Mortars, Stuccos and Concrete*, London, 1838.
F. Quietmeyer: *Zur Geschichte der Erfindung des Portland Zementes*, Berlin, 1912.
G. R. Redgrave: "The early days of Portland cement," *Building News*, vol. 39, 1880, pp. 725–727.
G. R. Redgrave and C. Spackman: *Calcareous Cements*, London, 1905.
H. Reid: *The Science and Art of the Manufacture of Portland Cement*, London, 1877.
J. Rondelet: *L'Art de Bâtir*, vol. 1, Paris, 1802.
H. Y. D. Scott and G. R. Redgrave: "The manufacture and testing of Portland cement," *Min. Proc. Inst. C.E.*, vol. 62, 1880, pp. 67–86.
J. Smeaton: *Narrative of the Building . . . of the Edystone Lighthouse*, London, 1791.
C. Spackman: *Some Writers on Lime and Cement*, Cambridge, 1929.
A. P. Thurston: "Parker's Roman cement," *Trans. Newc. Soc.*, vol. 19, 1939, pp. 193–206.
L. J. Vicat: *Recherches Expérimentales sur les Chaux de Construction, les Bétons et les Mortiers Ordinaires*, Paris, 1818.
G. F. White: "Observations on artificial hydraulic, or Portland cement," *Min. Proc. Inst. C.E.*, vol. 11, 1852, pp. 478–502.
G. F. White: discussion on Grant's paper of 1866 (q.v.).
G. F. White: discussion on Scott and Redgrave's Paper of 1880 (q.v.).
J. White: *On Cementitious Architecture*, London, 1832.
J. B. White and Sons: *Experiments on the Strength of Portland and Roman Cements*, London, 1852.

PLATE XXII

The "Stern" Portland Cement Works at Stettin, *c.* 1878

(H. Reid, 1879)

5
The emergence of reinforced concrete, 1750–1910

John W. de Courcy

Synopsis
It is suggested that the use of reinforced concrete based on scientific analysis began in the decade 1875-1885. Before that, the medium had been used intuitively, for about 30 years, with a wide range of cements. This use in its turn had developed from the need to provide structural iron with non-structural fireproof encasement. From 1885 to 1910 there was a rapidly increasing perception of the behaviour and potential of reinforced concrete. The paper follows the emergence of this new building medium, from the middle of the 18th century, when the intimately-bonded combination of metal and concrete of any form held no meaning in building, to 1910 when it had become universally known and was being used much in the manner accepted today and in most of the forms now employed.

Preamble
'Reinforced concrete' in the title of this paper implies the deliberate use of long metal bars or similar to add tensile or compressive strength to a structural concrete element in which concrete provides a significant part of the compressive strength and, in certain circumstances, some tensile strength. It requires as essential that the metal be contained in and bonded to the concrete throughout the whole length of the element. Prestressed concrete, a form of reinforced concrete, will not be discussed.

The period before 1830
The use of iron chains to strengthen the dome of St. Peter's Basilica in Rome in 1743 is said by some to mark the start of the present watershed between the structural engineer and the architect. The use of 'chained courses' in masonry was known at that time, as a development of the ancient use of metal ties. Thus, for example, Smeaton used chained courses in the shaft of the Eddystone lighthouse in 1758[1] and Semple used them, run in lead, in the masonry piers of the Essex Bridge in Dublin in 1753[2]. In the present context, however, the most evocative use of iron with masonry is exemplified by Soufflot's design begun in 1757 for the church of Sainte Genevieve in Paris, later to be known as the Panthéon. In this work,

2 EMERGENCE OF REINFORCED CONCRETE

developed by Rondelet and illustrated by him in *Traité de l'art de bâtir*[3], iron rods were threaded through the masonry and grouted in with lime mortar (Fig 1). The principle involved in this work of overcoming the tensile deficiency of the masonry by incorporating iron rods, and the disposition of these rods as illustrated foreshadows reinforced concrete. One should notice here, also, the proposal of Pritchard in 1776[4] to use a permanent soffit ring of iron for a brick river arch at Stourport. In these works, the stone or brick masonry was the primary building material and the iron was added as a structural enhancement. Quite soon, the position would be reversed, with iron becoming, for many buildings, the primary material, with concrete of some form being added as an encasement to protect the iron against fire.

It was in the pursuit of fire protection, rather than of enhanced strength, that our predecessors started the structural odyssey that was to lead, in about a hundred years, to the structural medium of reinforced concrete as we now know it. Fire protection was much in the minds of 18th century builders, with the elimination of timber being a conscious aim. Thus, for example, the Comte d'Espie in 1776[5] advocated the use of brick vaults doubly-curved on the soffits and profiled to a pitch on top, overlain by orthodox tiled roofing as a means of making an incombustible roof. The shape of his vaults avoided the need for iron ties which others had used in stabilising vaults 'armées avec des liens de fer'.

Fairbairn records[6] that 'the first instance on record of the successful application of cast-iron beams to the purposes of building is that of a fireproof cotton mill erected by Messrs Philips and Lee of Manchester in 1801'. The beams and columns were cast iron and the floors were brick jack arches which rested on the exposed bottom flanges of the beams. The filling over the arches was a 'concrete of lime, sand, and ashes'. While Fairbairn advocated wrought iron rather than cast iron, he would always appear to have maintained that such a building was fireproof despite the exposure of the iron flange. Hyatt would later question the distinction between 'fireproof' and 'incombustible'.

Concrete (béton) during the 19th century was a material that might be based on any one of a number of binders, lime, natural cement so-called, or Portland cement. We must wait until perhaps 1900 to be confident that virtually all structural concrete would be based on Portland cement.

1830-1850

During the 1820s, Sir Marc Isambard Brunel, using experience gained on the Thames tunnel questioned[7] whether there might be structural possibilities for the combination of brick in compression and iron or other material strong in tension. As a first step he made a series of pullout tests with various materials to determine their bond in brickwork joints. In several of these, he used iron strip 1$\frac{1}{16}$ in × $\frac{1}{16}$ in (27 mm × 1·5 mm) and found that, when embedded up to 5 ft 0 in (1·52 m) in 2:1 cement : sand mortar, bond stresses of from 1·26 to 1·82 N/mm^2 were obtained. Encouraged by this, Brunel carried out his historic experiment of building a reinforced brick and cement arch by progressive cantilevering without the use of centring (Fig 2)[8]. In the half-arch which constituted the experimental structure, and which attained a cantilever of 60 ft 0 in (18·2 m), the reinforcement was 54 strips of iron arranged in the joints as shown in Fig 3. The structure stood for 2 years before collapsing suddenly in January

1838, as a result, probably, of foundation settlement and a high cross-wind. It had, however, proved an important point, the reality of durable bond between iron and mortar, and at an acceptable level of stress. These studies were not developed, however, as they might have been, towards reinforced concrete. This was still well in the future. When Brunel's half-arch collapsed, François Coignet (1814-1888) was still a young man, Joseph Monier (1823-1906) a boy, and François Hennebique (1842-1921) had not yet been born.

The pigeonholing of Brunel's findings may have been due, in some degree, to C. W. Pasley. During the 1830s, Pasley had made several tests on reinforced beams, using 'longitudinal irons' and cement or lime, which demonstrated the value of reinforcement and the importance of bond[9]. Francis, however, recorded[10] that in about 1840 'Pasley drew the conclusion that concrete was unsuitable for all purposes except foundations and this opinion, coupled with unfavourable reports from other sources, received wide publicity and their overall effect was to delay the more general introduction of structural concrete by as much as 2 or 3 decades'.

Between 1840 and 1850, mass concrete building was being promoted by such as Lebrun, an architect[11] whose béton, in 1843, was based on hydraulic lime. He believed, with many others, in the dependability of mass concrete in lintels, and in suspended slabs and stairways, but he contrived generally, by the use of arches and vaults, to keep all his work in compression. During this time, also, the quest for the fireproof building

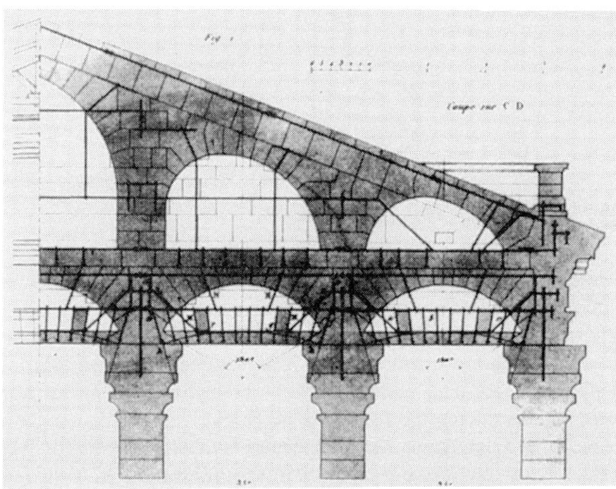

Fig 1. Reinforced masonry in the Panthéon, Paris (1 toise = 1·85m)[3]

ELEVATION.

Fig 2. Brunel's experimental arch at Rotherhithe[8]

4 EMERGENCE OF REINFORCED CONCRETE

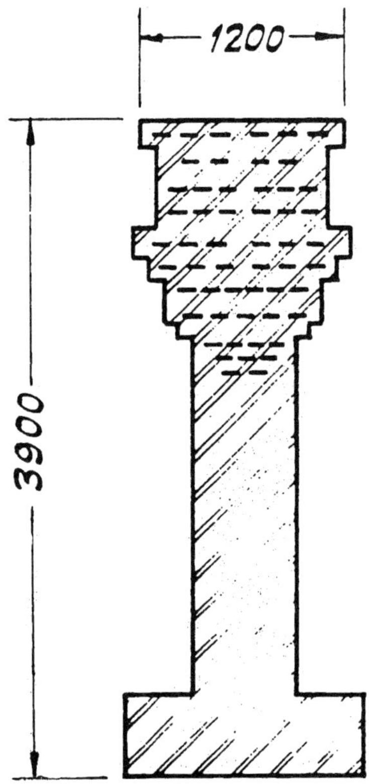

Fig 3. Brunel's arch: arrangement of iron strips in a section near the springing (dimensions approximate)[8]

continued. Thus in 1844, Henry Hawes Fox, a medical doctor seeking safer hospitals, was granted a patent[12] for 'an improved mode of constructing fireproof floors, ceilings and roofs'. As Fig 4 shows, he did make a rudimentary effort to protect the iron loadbearing elements. This combination of iron and 'concrete' was to be replicated, with variations, in many patents during the following 40 years.

In France, two significant ideas, combining iron and concrete for reasons other than fire protection, were realised at about this time. In 1848, Joseph Lambot (1814-1887) used a skeleton of iron rods and mortar to build a boat[13], while in 1849 Joseph Monier also used iron rods as the skeleton in making ornamental flower pots and basins[14]. Both must be seen as in the idiom of reinforced concrete, and possibly the earliest examples of it.

1850-1870

In 1853 François Coignet, who was primarily a chemical engineer, built a house for himself at St. Denis, in which, as Collins records[15] 'the flat roof was strengthened with iron beams'. He had been examining the technology of better concrete for some years and had been building in mass concrete cast *in situ* since 1847[16].

In close succession, William Boutland Wilkinson (2819-1902) in 1854[17], and François Coignet in 1855[18] patented their approaches to this combination of concrete and iron. Both were British patents, and Coignet would take out a further patent in France in 1856. The titles of these patents were revealing. Wilkinson used the description 'construction of fireproof buildings, etc.', while Coignet got his patent for 'artificial stone and cement'. It is of interest to note in passing that Monier, whose name would become very important internationally, did not take out his first patent until 1867[19]. His influence will be discussed later.

Wilkinson claimed his patent on two grounds, the first, which is relevant to this paper, being his innovative slab constructions. He envisaged several types of floor. All were based on a roughly 2:1 rectangular grid of masonry supports (Fig. 5). the 'main beams' spanned the short direction ABCDE. They were a combination of brick arches and concrete, with a wire cable AE placed horizontally near the arch soffits and carefully anchored at both ends. The slabs, which were monolithic with the beams, spanned continuously from HJ to AE to FG, and had horizontal top surfaces.

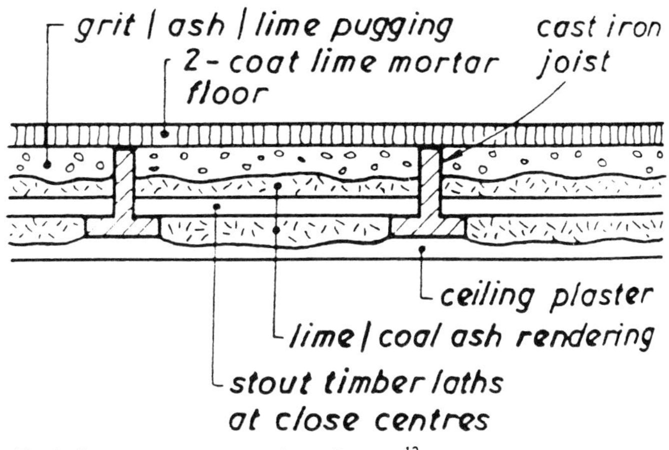

Fig 4. Fox patent: cross-section of system[12]

In some forms the slab soffits were curved to form flat arches. In these forms, the slabs were made with a crude concrete which nevertheless contained 'Portland or other cement of equal quality' and 'strips of hoop iron laid on edge are embedded in the mass across the crown of the arch at about 2 ft (0·61 m) asunder or at greater or less distances according to the desired strength of the floor, and reaching the whole length across the floor or floors, the under edge of the same being at or near the crown of the centring, so that in their low position the strips may act with more power as tension rods to the floor'. From the actual disposition of the hoop iron at several levels as shown on sheet 1 of the patent drawings, however, it would appear that the metal strips as shown would have contributed more to balanced cantilever action than to normal slab action (Fig 6); and the potential strength of the floor slab would appear to lie in the combination of the flat arch and the cantilever action. The small proportion of the 'tension rods' which did in fact reach 'the whole length across the floor' would not be seen today as significant tensile reinforcement for a slab. But the whole arrangement, as with the other inventors, must be seen as foreshadowing modern reinforced concrete.

Another form proposed by Wilkinson had a continuous flat-soffited slab spanning from HJ to AE to FG, or could be used as a single span. This was, in fact, the form (Fig 7) that is generally associated with Wilkinson today. The profiles of the wire ropes and the attention to end anchorage are compelling evidence of his understanding of a problem, but the text of his specification must be read also. 'When ceilings of arch form may be objected to . . . a flat platform of wood is to be erected to the ceiling line and the floor to be composed of plaster, air-slaked lime, cut hay, and ashes and breeze in certain proportions, and wire rope . . . or iron in other forms in a state of tension, hollow bricks, field drain pipes, coke in lumps, etc., being introduced if desired to lessen the weight . . . The wire rope is secured at its extremities by embedding it in the mixture or concrete while in a soft state, and forming the ends into loops or by opening out the strands and hirling them in various directions which renders it so secure as not to be drawn out under any force short of the breaking weight of the rope. In ordinary dwellings I propose placing such wire ropes about 9 in (229 mm) apart and to have a full depth of floor of one-sixteenth the span'. Later,

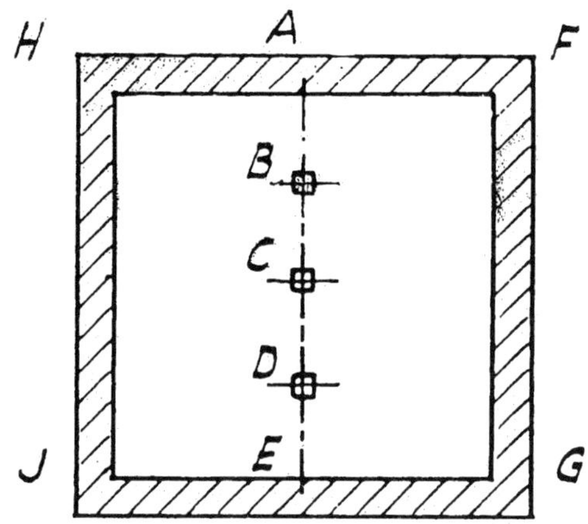

Fig 5. Wilkinson patent: arrangement of floor[17]

in discussing the support of partitions on such a floor he recommended a beam containing wire ropes of similar profile and composed of 'plaster, cut hay, etc.'.

While the profile of the rope, especially in his two-span slab might to some degree suggest reinforcement, it seems more correct to see each rope as a suspension cable (or two such cables conveniently joined in the two-span case) supporting the infill, rather than as the tensile reinforcement in a flexural system. Wilkinson's choice of different materials in compression in the two forms described would seem to strengthen this suggestion.

Fisher Cassie[20], describing a structure of 1865, has proposed that Wilkinson is entitled to 'his proper place in the ranks of the "fathers" of

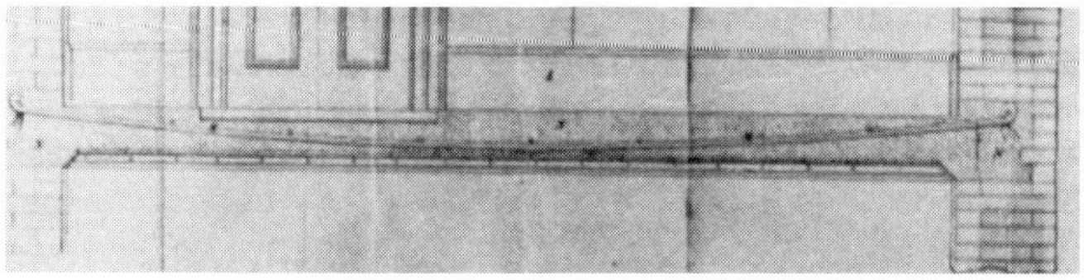

Fig 6. Wilkinson patent specification: section showing straight strips of iron[17]

These descriptions are transposed

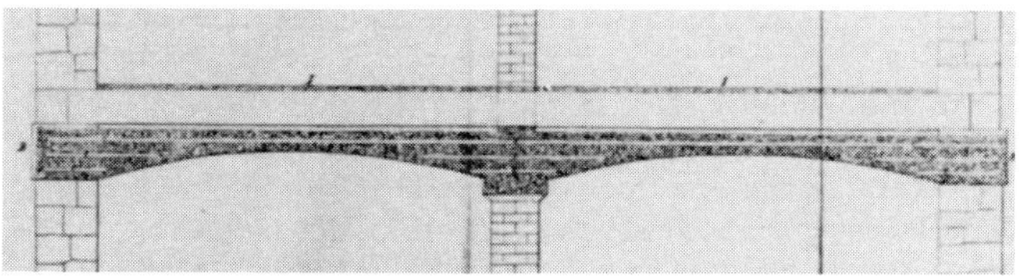

Fig 7. Wilkinson patent specification: section showing curved cable[17]

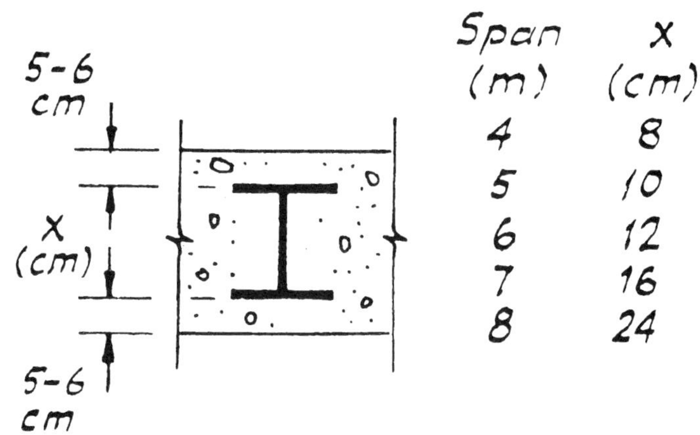

Fig 8. Coignet: span/depth relationship for slabs[21]

reinforced concrete'. This should not be denied him, but possibly more on the grounds of his slabs with arched soffits than for his better known draped wire ropes. It is clear that the house mentioned by Fisher Cassie represents a further development beyond the techniques described by Wilkinson in his patent of 1854.

The Coignet patent of 1855 must be seen as relating primarily to the better making and compacting of a material which he called 'béton aggloméré' (the specification used the term 'concrete'). The four formal claims that he makes for his invention are all based on the nature of this material. His references to the use of iron are incidental and, in the patent specification, rather indefinite. He refers to iron rods placed at convenient distances apart, taking into account their size and 'the area of flooring to establish'. He adds that 'these rods being in the shape of a screw and having a nut at each end will prevent the walls from loosing their perpendicularity'.

8 EMERGENCE OF REINFORCED CONCRETE

There seems little doubt that Coignet envisaged his concrete slabs as adequate for quite large spans, acting as plain concrete unaided in its flexure by tension reinforcement. It is regrettable that this patent was not illustrated with drawings. The text of his book, which will now be discussed, suggests more clearly the 'fail-safe' function that he saw for the iron elements in his slabs.

In 1861, François Coignet published his text *Bétons agglomérés*[21]. The main purpose of his book was to promote 'béton aggloméré' as a material. Any binder could be used at choice in its manufacture, and its merit, as in the patent, lay in its careful mixing and thorough compaction rather than in the precise nature of its constituents. Coignet described it as a 'paste of stone'.

In this book, however, Coignet described two significant combinations of metal and 'béton'. He prefaced the first by noting how costly it was to support the weight of a floor and its loading on iron beams. With a deck of his material, however, the load was carried by the slab to the supporting walls, and iron joists were built in to act as ties between the walls and also to prevent collapse if the floor should fail. The joists should be placed at 1 m centres, and their ends were to be split and opened to provide anchorage. The sizes of the joists, which dictated the thickness of the slab, depended on the span, as in Fig 8. The joists were to be linked by 15-18 mm diameter rods, again at 1 m centres, with nuts at both ends. Coignet's detail generally resembles earlier work in which the 'béton' or concrete (from now on, the term 'concrete' will be used as far as possible) was acting merely as a fireproofing cover, but he is very clear in his mind that it is the concrete primarily that is carrying the load. His recognition, however, of the enhanced resistance of his decks to fire may perhaps be implied in his comment that his system did not incorporate 'one morsel of wood'.

Coignet's second combination of metal and concrete strikes a modern note. If, as he put it, a prodigious resistance were required, either a metal net of large mesh might be added, or 'an infinite number of little fragments of fine iron wire' 4-5 cm in length 'flattened at both ends' mixed and compacted with the concrete would greatly enhance the resistance of the slab to impact and the effect of settlement of the foundations. Interestingly, in the direct context of the present paper, it is this inclusion of discrete wire fragments that he referred to as 'armatures de fer'. It may be mentioned that Coignet in his book used no mathematical analysis and offered no formulae.

In 1862, there occurs an early reference to the combination of concrete and iron in direct compression, with the award to W. H. Wood of a USA patent for filling cast-iron tubes with concrete to form columns[22]. During this time also, from 1856 onwards, steel was being developed by Bessemer, Siemens, and Thomas, and would, during a period of 45-50 years, take over from wrought iron as the metal used in combination with concrete; and in 1863, Bouniceau recorded in the *Annales des ponts et chaussées* the similarity between the coefficients of thermal expansion for concrete and steel rods at, respectively, $1{\cdot}37 \times 10^{-5}$ and $1{\cdot}25 \times 10^{-5}/°C$[23].

As a postscript to this period, it is fitting to note that Thaddeus Hyatt, whose paper of 1877 must be seen as a milestone in reinforced concrete, had already been experimenting for over 20 years. He had, in 1855, designed and made a 9 in × 9 in × (229 mm × 229 mm) beam of flat 9 in × 9in

× 1 in (229 mm × 229 mm × 25·4 mm) bricks 'confined by a tie rod near the bottom of the bricks which passed through holes made for the purpose in them'[24].

1870-1890

In 1870, there was little perception of reinforced concrete. By 1890, it would be clear that a new building medium had been found. In 1870, concrete was substantially in the hands of practical builders. For instance, Potter could say, in criticising even the performance in fire of embedded iron, 'it becomes a question therefore whether iron might not be dispensed with altogether, or at least in many instances where moderate spans from wall to wall occur'[25]; and he could propose 16 ft 0 in (4·88 m) as a reasonable maximum for mass concrete spans, quoting satisfactory load tests on a span of 12 ft 0 in (3·66 m) for a slab 5 in (127 mm) thick at the crown and 7 in (178 mm) thick at the supports, made with 'best Portland cement'. By 1890; reinforced concrete would already have been attracting mathematical analysis for some years.

In 1875, W. E. Ward, using his intuition, supported by tests, had built a house at Port Chester, USA, (Fig 9) in which 'all of the beams, floors and roofs were exclusively made of béton, re-enforced with light iron beams and rods'[26]. (Was this use of the word 're-enforced' in 1883 the first appearance of the English term?.) He recognised that 'the reason for placing the iron beam so near the bottom of the mold was to utilise its tensile quality for resisting the strain below the neutral axis while the béton above this line was relied on for resisting compression from load strain'[27]. Ward recognised the bond between the concrete and the iron; and he recognised that an inverted T-section was more efficient than an I-section, that the composite beam recovered elastically on the removal of load, and that the iron controlled the manner of shrinkage in the concrete. One might note, however, that he had set out in the first place to ascertain 'whether a practically fireproof building could be designed and constructed at a comparatively moderate cost'[28].

In 1877, Thaddeus Hyatt (1816-1901) published his monograph on Portland cement concrete combined with iron as a building material[29]. In 1878 Joseph Monier took out, in France, his first patent for beams containing concrete and iron[30], having taken out several earlier patents for the combination of concrete and iron. In 1879, François Hennebique was already making slabs in concrete and iron, developing the methods he would patent 13 years later[31]. If one were, simplistically, to suggest a starting time for modern reinforced concrete, it would be in these three years.

Hyatt had a broad vision of reinforced concrete. He saw its resistance to fire not as indefinite, but as related to time, from 3 to 48 h, as required. He foresaw T-beam action. He saw the economy of replacing metal by concrete in compression. He had a perception of beam shear—'Were the beams made up of horizontal layers, the effect of the stress would be to cause these to slide on one another, but the beam being solid, the particles are held together by their own cohesion, the shearing strains being thus opposed by cohesive force'[32]. He appreciated the need for bond in this context and tested various mechanical means for securing bars. He recognised the merit of omitting superfluous concrete below the 'neutral line'. He assessed a value for the coefficient of elasticity of concrete. He analysed the strength of combined sections numerically and produced design

10 EMERGENCE OF REINFORCED CONCRETE

Fig 9. Ward's house at Port Chester[26]

tables for solid slabs up to spans of 40 ft 0 in (12·2 m) with *L/H* ratios that would be acceptable today. He designed for ultimate load and applied a safety factor of 4. He was wrong in his analysis, in placing the 'neutral line' always at mid-depth of the slab, and in calculating his resistance moments for the concrete and the metal about that neutral line. He attached a very high capacity for tension to both wrought iron (60 000 lbf/in^2) (420N/mm^2) and concrete (202 lbf/in^2) (1·4N/mm^2) and suggested that the concrete provided part of the tensile resistance.

Hyatt's research did not lead directly to much development, possibly because he himself lost interest in it. He took out a patent in USA in 1878, but there seems little evidence of later researchers acknowledging his study. Ransome & Saurbrey, however, say[33] 'broadly speaking, the Hyatt patent No. 206112 shows and describes everything necessary for the practical use of reinforced concrete'.

From about 1880, Austrian and German interests were negotiating with Monier for his patents[34]. By 1885, Mörsch records[35] that they were largely in the hands of the firm of Freytag & Heidschuch and their engineering consultant, Gustav Wayss (1851-1917). Mörsch observes that 'in his (Monier's) patent drawings are already disclosed all the elements which are today (1907) employed' (Fig 10) and adds 'it is easy to understand how Monier's invention, but little understood and based as it was on an entirely empirical foundation, was destined to develop along entirely different lines in the hands of engineers'[36].

In 1886, Matthias Koenen (1844-1924) published a brief note[37] based on experiments conducted by Wayss. This quite diffident contribution appears to have been the first formal statement of an analysis of the reinforced concrete cross-section. It was reproduced by Wayss in 1887, in *Das System Monier*, the first significant publication in reinforced concrete[38], and would be used by designers for many years. As stated by Wayss, its terms were simple. Let k be permissible pressure on concrete and k_1 be allowable tension in wrought iron. Now, (Fig 11), if Fe is the cross-sectional area of the iron and $M_{max.}$ is the maximum bending moment, $k_1Fe = k\frac{d}{4}$ (b being taken as unity) and $k.\frac{d}{4}.\frac{3d}{4} = M_{max.}$ whence (i) $d = 2\cdot31\sqrt{\frac{M_{max}}{k}}$ and (ii) $Fe = \frac{1}{4}\cdot\frac{k}{k_1}.d$. Again, as with Hyatt, the neutral axis was at mid-depth, the influence of modular ratio not yet

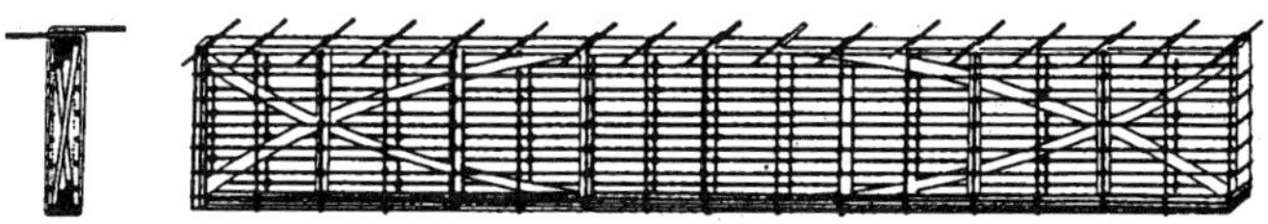

Fig 10. Monier patent specification (1878): elevation and section of beam[30]

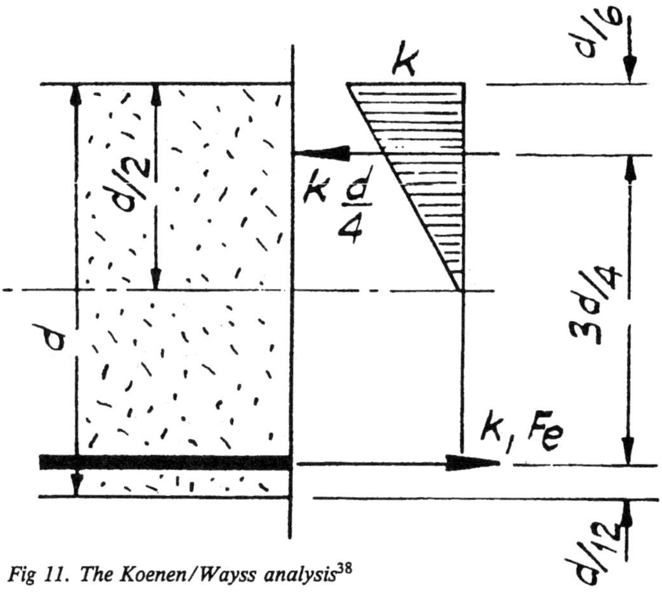

Fig 11. The Koenen/Wayss analysis[38]

being recognised; but the concrete was not used in tension, and the lever arm of the resisting couple was right in principle.

Wayss, in his book, then proceeds to the analysis of arches, which featured largely in the Monier system, and he described in fact most of the applications of reinforced concrete known today. A notable omission was the T-beam. He showed slabs, whether *in situ* continuous, or precast, as supported on iron beams, and in one instance[39] encased timber main beams in reinforced concrete in a composite floor system.

Mörsch records[40] that the first use of reinforced concrete T-beams was in a library building in Amsterdam in 1886. Three years later, in 1889,

12 EMERGENCE OF REINFORCED CONCRETE

Ransome would use this construction in ribbed floors in California[41], using bar reinforcement; and he would observe, about that same project, that the columns used in it 'were probably the first (concrete columns) ever erected'. Whether this claim can be upheld, one is certainly very close to the first use of reinforced concrete columns at that date. The use of columns will be mentioned again later.

1890-1910

Understanding of reinforced concrete developed rapidly during this period. In 1886 Koenen's analysis was published in Berlin. By 1902 Christophe[42] would have assembled a bibliography of over 300 items from 11 countries. This paper can follow only the mainstream of the development. The references suggest further reading.

In 1907 Suenson[43], describing the first course of lectures given on reinforced concrete in Denmark, in 1906, attributed the rapid growth of the medium to two men—'the engineer Wayss in Germany and the contractor Hennebique in France' ('Ingeniør Wayss...og Entreprenør Hennebique . . .').

As Cusack records[44], François Hennebique was both designer and business man. His perception of reinforced concrete was deep and his contribution to its practical development unsurpassed. That the principle of the composite medium continued to be elusive is suggested by the wording in his 1892 British patent[45] 'if a girder be constructed of a bar of iron entirely coated with a layer of cement, to which the iron will thus form a core, a support is obtained of which the resistance to bending will be equal to the resistance to tension of the iron plus the resistance to compression of the layer of cement above the bar of iron. In fact the iron will only feel the effect of the strain by the amount by which the resistance which the cement opposes to compression is exceeded, and only when the cement is crushed'.

Hennebique had the same general understanding that, for instance, Hyatt had earlier shown, for the use of vertical connectors within the concrete section, but he developed the concept further in his patent specification '(these anchors or stirrups) form with the cement the complete and indissoluble union of the lines of tension whilst allowing of there being applied in the construction of the armed cement girders thus formed, the forms and methods of calculation employed for lattice work girders. These stirrups . . . are placed at intervals according to calculation'[46]. In 1909, Mouchel would comment[47] that 'notwithstanding the fact that failure by diagonal tension may be expected to occur in a beam having no web reinforcement if the shearing stress intensity was to attain, say, 100 lbf/in (0·7N/mm²), recent well-authenticated tests have shown that (the use of stirrups and bent-up bars lead to beams) capable of withstanding shearing

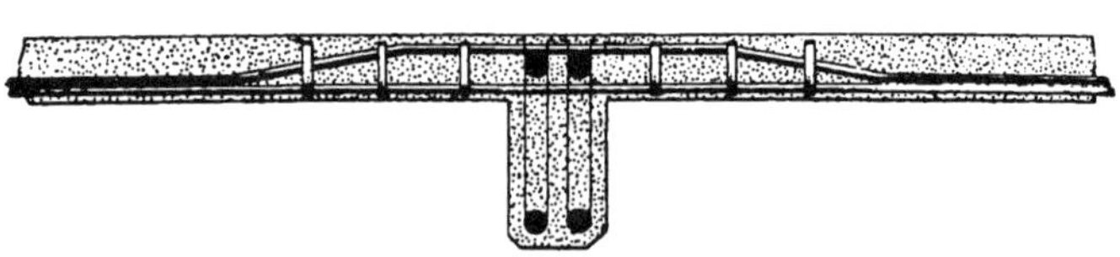

Fig 12. Hennebique detail (c.1900): slab and T-beam[50]

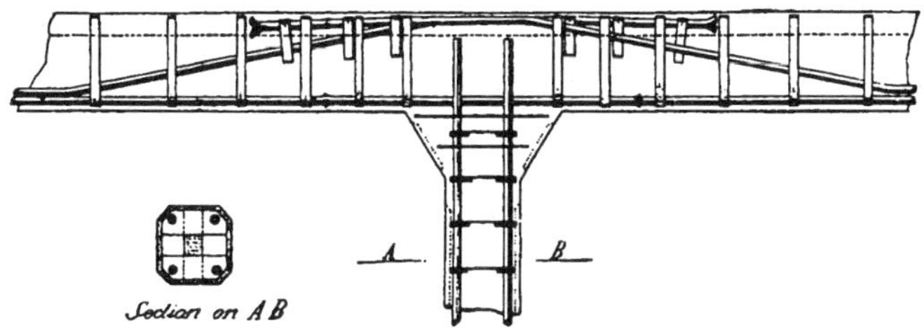

Fig 13. Hennebique detail (c.1900): beam and column[50]

stresses of more than 600 lbf/in^2 (4·2N/mm^2) without giving the slightest indication of failure by diagonal tension'.

The typical Hennebique details of the early 20th century (Figs 12 and 13) were not apparent in his patent drawings. There he saw 'cement of a trapezoidal section as being the most practical arrangement' (Fig 14). The characteristic haunches, chamfers, and bent-up bars, had not yet appeared. One may note in passing that the conscious quest for fireproof construction had not ceased. It was standard practice for Hennebique documents to carry the inscription in one language or another 'Disastrous fires rendered absolutely impossible.'

In 1888, Durand Claye established authoritative values for the modulus of elasticity of various concretes[48]. In 1894, Coignet & de Tedesco[49] formally offered an innovative approach to the design of the reinforced concrete cross-section in bending, Coignet having mooted the idea to the French Society of Civil Engineers some 6 years earlier[50]. They introduced the concept of the modular ratio (N) and exposed the fallacy of locating the neutral axis necessarily at midheight of the element. For $N = 20$, and using acceptable stresses of 40 kg/cm^2 (3·9N/mm^2) for concrete in compression and 1500 kg/cm^2 (147N/mm^2), for metal in tension, and ignoring concrete in tension, they showed that the depth of the neutral axis was at $0{\cdot}4H$ from the compression surface, where H was the effective depth of the section. They then derived, after painstaking analysis, two equations for 1 m widths of slab:

$M = {}^2/_3H$ (M is the cross-sectional area of tension metal in cm^2 and H is in cm); and

$\mu = 6{\cdot}4H^2$ (μ is the moment of resistance of section in kg.m).

It may be observed that this analysis did not consider either under- or over-reinforced sections.

The ability of concrete to resist low tensile stresses continued to be recognised by the theoreticians. Several texts showed the concrete immediately below the neutral axis contributing to the tensile resistance of the cross-section, and based their analysis on the inclusion of this small element of concrete in tension.

The reinforced concrete column came into use during the 1890s. Hyatt, in his US patent of 1878[51], had stated the principle of hooped concrete 'in constructing columns . . . wholly of concrete, I make the structure solid, the concrete then bearing the load, and giving away under compression, would naturally incline to yield in the first place not from absolute crush

14 EMERGENCE OF REINFORCED CONCRETE

of the materials but from want of sufficient tensile resistance at the circumference of the column. But this tendency being resisted by the circular ties, such a concrete could give way only by the crush of its particles'. Hyatt, who proposed longitudinal bars also 'as the case might require', based his hooping on the use of discrete bands. Despite this, Wayss made virtually no reference to columns in 1887 and seemed doubtful of them[52], preferring in this context to confine the Monier system to the fireproofing of cast-iron pillars. Coignet & de Tedesco did not discuss columns in their paper of 1894, and Hennebique did not refer to them in 1892, confining his patent to 'joists, girders, and the like'[53]. It is clear, however, that he did introduce them to his practice shortly afterwards.

In about 1898 Armand Considére (1841-1914) began his significant

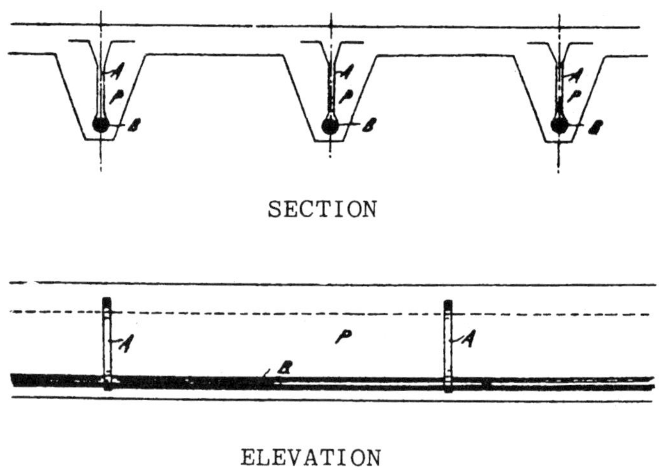

Fig 14. Hennebique patent specification: sections and elevation[45]

researches, one of which was to lead him to accept the concept he called 'béton fretté', 'hooped or specially-armoured concrete'[54]. As he described it, the resistance to crushing of the core of a spirally armoured prism subjected to pressure at the ends exceeds the sum of the three following elements, i.e. (i) the resistance of the concrete itself increased by 50%; (ii) the resistance of the longitudinal armouring taken at its limit of elasticity; (iii) the resistance of an imaginary longitudinal reinforcement taken at its limit of elasticity, the weight of which would be 2.1 times the weight of the special reinforcement[55]. Considére did not claim his concept as original. It had, he said, been examined earlier, certainly by Koenen & Wayss in 1892, and slightly later by Harel de la Noé[56]. In fact, it had also been considered by Hyatt, as we have seen.

The controversy on the merits of spirally—and longitudinally—reinforced columns was spirited. Marsh[57] shows that, for instance, Matrai in Hungary followed Considére. Hennebique, on the other hand, was critical. He saw the merit of 'béton fretté' applying only in the region of ultimate load, and he questioned the ability of a column without longitudinal reinforcement to resist bending. Mouchel would say later[58] 'the conclusions are quite correct so far as ultimate strength is concerned, but ... the added strength cannot be utilised without causing undue shortening, excessive lateral deflection and inordinate strain in the longitudinal reinforcement'. Considére should be allowed the last word in this matter[59] 'it is evident that between this method, supplemented by the use of longitudinal rods, on one side, and the method of reinforcing by longitudinal main rods tied together by belts of lighter material, on the other, there is an intermediate continuous series of methods of reinforcing'. Fig 15 illustrates comparative tests made by Considére 'of longitudinally, and of specially armoured posts'[60].

A new development between 1900 and 1910 was the 'flat slab'. In 1902 O.W. Norcross was granted a US patent for flat slab construction[61]. Its use spread quickly in the United States, and indeed the empirical Chicago rules which governed its design would remain influential worldwide for half a century. 'In 1911 C.A.P. Turner patented a similar girderless system which he called the "Mushroom Slab"'[62] and the American Concrete Institute would later record that it was Turner who, in this matter, had 'initiated one of the most efficient forms of construction in reinforced concrete'[63].

The decade was, however, mainly a time for review and consolidation. While it would, arguably, be dominated by the 'systems' (a witness at the ICE Committee inquiry in 1909, when asked 'how far did you participate in the design of those works?', replied 'they were my design entirely; do you mean, did I ask these reinforced concrete people to do anything?')[64], the subject had become sufficiently known to encourage the publication of formal textbooks. Several appeared in quick succession, as for instance, Christophe in 1900[65], based on three articles published in 1899 in *Annales des travaux publics de Belgique*[66], Mörsch in 1902[67], Marsh in 1904[68], and Taylor & Thompson in 1905[69]. The third edition of Mörsch would be translated by Goodrich and published in the United States of America in 1910[70]. Statutory regulations were issued in several countries and cities, including[71] New York (1903), Germany and Italy (1907), Austria (1908), and France (1906)[72]. In Great Britain, the first formal codification was by the Royal Institute of British Architects in 1907[73], and this would be followed by the London County Council Regulations in 1913.

What then was the understanding of reinforced concrete as the year of 1910 approached? Fashions were numerous, and techniques of design and construction were diverse. Taylor & Thompson listed 37 distinct types of reinforcement being used in the United States[74]. Underlying these, however, existed the fundamentals, all still in debate, modular ratio and the distribution of stress across the section in bending, bond and the resistance to shearing force. The merits or otherwise of compression reinforcement were being discussed, and studies of T-beam behaviour and analysis proliferated.

In 1910, there was broad agreement that the position of the neutral axis depended on the modular ratio, the proportion of reinforcement in the cross-section, and the level and distribution of stress in the section in bending. An earlier view that the concrete did not crack in tension and did contribute materially to the resistance of the bent beam was no longer generally held. While the term 'creep' had not yet appeared in the literature, the experiments of Bach in 1895[75], and of others, had distinguished

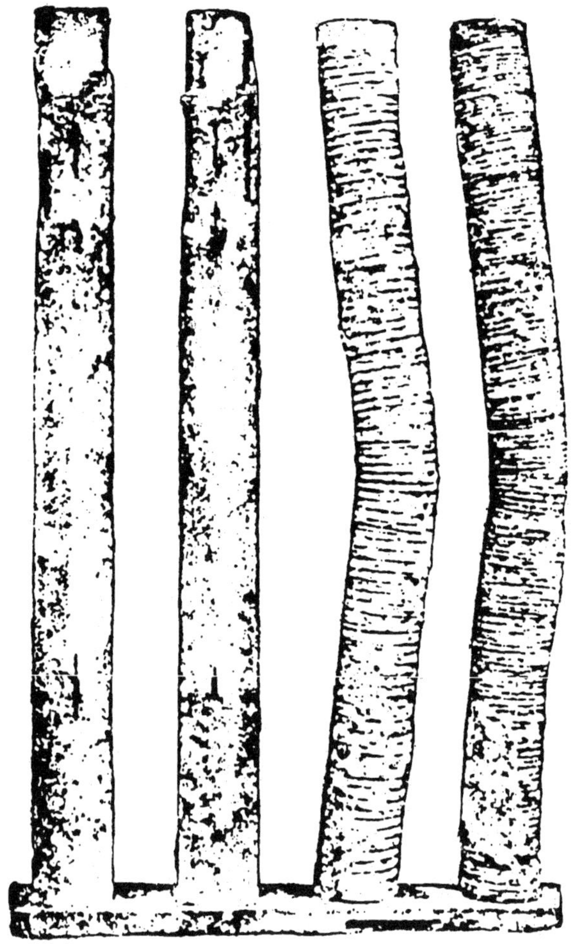

Fig 15. Considére's tests on reinforced concrete columns[54]

between the elasticity of concrete and a phenomenon they called permanent set, and Taylor & Thompson[76] would observe in 1909 that it was the 'total deformation' that 'must be used in reinforced concrete analysis'. As a result, partially of this, the modular ratio, N, for which values as low as 8 had been suggested, was established as 15 in several series of national regulations. This value took into account a 'sustained modulus' for concrete of E_c = 2×10^6 lbf/in^2 (14×10^3N/mm^2). A further reason for choosing N = 15 lay in the general adoption of a straight line triangular stress distribution in the compression zone. This presumed that plane sections remained plane after bending and implied a constant value for E_c. But many studies had shown that E_c in fact varied with the intensity of stress[77], and the continuing planeness of plane sections after bending was challenged[78]. The choice of $E_c = 2 \times 10^6$ lbf/in^2 (14×10^3N/mm^2) offered a safe resolution for these anomalies.

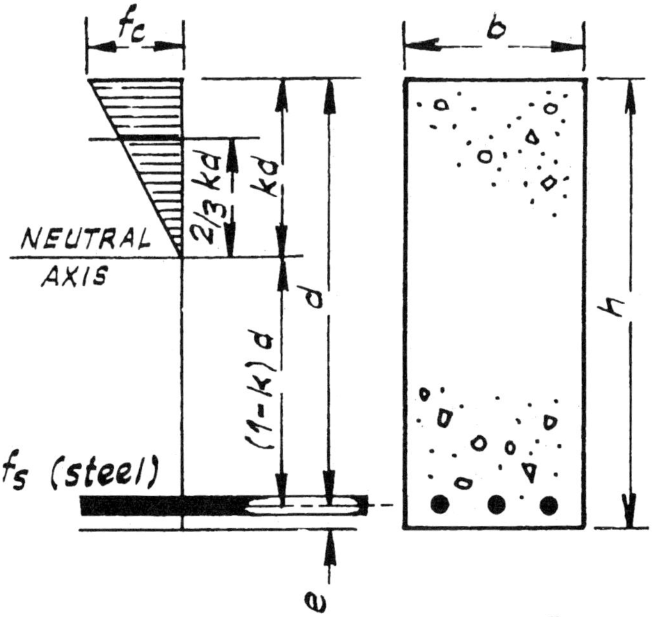

Fig 16. Analysis of the rectangular beam in bending (1909)[69]

The general approach to the design of rectangular beams in bending using triangular distribution[79], as in Fig 16, resulted in an equation which could now link *kd*, the depth of the neutral axis to *p*, the ratio of the tension steel area to the effective cross-sectional area *bd*, and *n*, the modular ratio. This equation,

$$k = \sqrt{2pn + (pn)^2} - pn$$

together with the compatible equation,

$$f_c = \frac{k f_s}{n(1-k)}$$

where f_s and f_c were the stresses in the steel and concrete, respectively, permitted in the design of the basic element, the singly-reinforced rectangular beam in simple bending.

By 1910, the topics of shear and bond had received considerable attention. The bent-up bar was widely used. As Mouchel wrote[80] 'horizontal bars placed near the lower surface of the concrete constitute adequate reinforcement for the middle portion of the beam, but at the ends where tension occurs on diagonal planes, it is necessary to emplov additional reinforcement which in Mouchel-Hennebique practice take the form of inclined bars and vertical stirrups'.

An understanding of diagonal tension had been developing for some time (as indeed it still is, to this day). Many attempts had been made to establish the shearing strength of concrete. Spofford had shown in 1906[81] that the strength of concrete in direct or pure shear was about 64% of the 28-day crushing strength, but it was being accepted that concrete in beams, said to have failed in 'shear', did in fact fail in tension and therefore at a much lower stress, perhaps only 10% of the crushing strength. It was appreciated that this tension developed through the combination of the applied shearing force acting vertically and the horizontal interlaminar resistance to shear of the concrete in the beam; and an earlier theory that the vertical stirrups and bent-up bars themselves acted in shear, in the manner of rivets in steelwork, had been discarded. It was now generally recognised that these elements acted in tension, and the truss analogy, already some 20 years old, continued to be a model in the design process. For convenience, however, the stress in diagonal tension, then as now, was based numerically on the applied vertical shear. Thus $v = \frac{V}{bjd}$,

where *jd* was the lever arm, was an accepted formula, using 'a working stress ... based not upon tests of direct shear but upon tests where the failure was by diagonal tension'[82].

The study of adhesion, which, it has been suggested started with M.I. Brunel in about 1830, had been continuous for orthodox reinforced concrete from 1887. In that year, Bauschinger[83], at the request of Wayss, made tests, frequently mentioned in the literature, that showed the adhesion of the iron and concrete to be 40-47 kg/cm^2 (3·9 – 4·6 N/mm^2). In 1905, Bach reported results of about 38 kg/cm^2 (3·7 N/mm^2)[84]. Marsh[85] describes the work of seven other studies, by Coignet & de Tedesco, Brik, Christophe, Considére, de Joly, Feret, and Hyatt, showing the importance of adhesion and bond in the minds of the researchers of the time. This interest in adhesion would appear not to have lain in the matter of the total amount of embedment required to resist tension at the plane of maximum bending moment. That could readily be resolved by adequate bond length and a variety of terminal details. The problem lay rather in the context of local bond or shear bond. Curtailment of tensile reinforcement was not much used, but the bending up of bars to counteract diagonal tension was widespread, and the ability of the remaining tensile bars to accept the rate of increase of bending moment near the supports was under continual examination. One facet of this study was the danger of lamination of the concrete below the neutral axis under the action of horizontal shear. Lamination was less likely where the full cross-section was available. However, a recommendation existed[86] that there should be a minimum clear distance between bars in a single tensile layer of 1·5 times the bar diameter, since it appeared reasonable to say that the plane at which lamination was most likely to occur was that which contained the horizontal diameters of the tensile bars, the adhesion between concrete and steel being less than the resistance of the concrete to pure horizontal shear.

There has been little reference to concrete technology in this paper. One might conclude, however, by noting in Bach's experimental programme of 1905 his foreshadowing of two practices that would later become basic in the use of concrete. He observed, firstly, that the 'resistance to sliding' of the steel in the concrete was increased, which is to say that the quality of the cast concrete was improved by 'jarring the finished specimens before setting is completed' especially 'when small percentages of water are used', and he observed, secondly, that 'that percentage of water was best with which it was just possible to manufacture the specimens satisfactorily'[87].

Conclusion

The author thanks the very many colleagues and institutions in several countries on whose help he has called, and he would ask them to excuse him for not making individual acknowledgements of their willing and generous cooperation. He would hope to avert the just wrath of any who may disagree with the particular course he has followed, by pleading the limitation of space and, perhaps even more, the subjectivity inevitably attendant on a topic of this nature. He would welcome communications on any point of perceived imbalance or in any matter that will add to this record.

References:

1. Smeaton, J.: *A narrative of the building and a description of the construction of the Eddystone lighthouse with stone*, London, Longman, Hurst, Rees, Orme, and Brown, 2nd ed., 1813, plate 11
2. Semple, G.: *A treatise on building in water*, Dublin, J.A. Husband, 1776
3. Rondelet, J.: *Traité théorique et pratique de l'art de bâtir* (A theoretical and practical treatise on the art of building), 6th ed., Paris, Chez M. A. Rondelet Fils, architecte, 1832, plates 150-151
4. Cossons, N., and Trinder, B.: *The iron bridge*, Bradford-on-Avon, Moonraker Press and Ironbridge Gorge Museum Trust, 1979, p18
5. D'Espie, Le comte: *Maniere de rendre toutes sortes d'edifices incombustibles* (On the manner of making all kinds of buildings incombustible), Paris, Chez la veuve L.Chesne, 1776
6. Fairbairn, W.: *On the application of cast and wrought iron to building purposes*, London, John Weale, 1854, p3
7. Beamish, R.: *Memoir of the life of Sir Mark Isambard Brunel*, London, Longman Green Longman and Roberts, 1862, p285
8. Anon.: 'Mr Brunel's experimental brick and cement arch', *The Civil Engineer and Architect's Journal*, **1**, 1837-8, London, pp119-120
9. Ransome, E.L., and Saurbrey, A.: *Reinforced concrete buildings*, New York, McGraw Hill Book Company, 1912, p20
10. Francis, A. J.: *The cement industry 1796-1914: a history*, Newton Abbot, David & Charles, 1977, pp55-56
11. Lebrun, F.M.: *Traité pratique de l'art de bâtir en béton* (A practical treatise on the art of building in concrete), Paris, Chez Carilian-Goeury, 1843
12. Fox, H. H.: *Fireproof roofs, floors and ceilings*, British Patent No. 10 047, 1844
13. Haegermann, G., Huberti, G., and Möll, H.: *Vom caementum zum spannbeton* (From opus caementitium to prestressed concrete), Wiesbaden-Berlin, Bauverlag GmBH, 1964, **1**, pt. B, pp32-35
14. Ibid., pp64-71
15. Collins, P.: *Concrete, the vision of a new architecture*, London, Faber & Faber, 1959, p28
16. 'Cent ans de béton armé' (100 years of reinforced concrete), *Proc. Centennial Conference*, Paris, November 1949, La Chambre Syndicale des Constructeurs en Ciment Armé de France et de l'Union Française, p34
17. Wilkinson, W. B.: *Construction of fireproof buildings, etc.*, British Patent No. 2293, 1854

18. Coignet, F.: *Artificial stone and cement*, British Patent No. 2659, 1855
19. 'Cent ans de béton armé', op. cit., p35
20. Fisher Cassie, W.: 'Early reinforced concrete in Newcastle-upon-Tyne', *The Structural Engineer,* **33**, 1955, pp134-137
21. Coignet, F.: *Bétons agglomérés appliques a l'art de construire* (The use of bétons agglomérés in construction), Paris, E. Lacroix, 1861, p113
22. Ransome, E. L., and Saurbrey, A.: op. cit., p30
23. Mörsch, E.: *Concrete-steel construction* (first published as *Der Eisenbetonbau* in 1902), 3rd edition (1908) translated by Goodrich, E. P., New York, Engineering News Publishing Company, 1910, p3
24. Hyatt, T.: 'An account of some experiments with Portland-cement-concrete combined with iron as a building material' (1877), *A selection of historic American papers 1876-1926*, Newlon, H. Jr. (ed.), publication SP-52, Detroit, American Concrete Institute, 1976, p73
25. Potter, T.: *Concrete: its use in building*, London, E. & F. N. Spon, 1877, p115
26. Ward, W. E.: 'Béton in combination with iron as a building material' (1883), *A selection of historic American papers 1876-1926*, op.cit., p106
27. ibid., p110
28. ibid., p105
29. Hyatt, T.: op.cit., pp53-99
30. Monier, J.: *Un systeme de traverses et supports en ciment et fer applicables aux voies, chemins ferrés et non ferrés*. Brevet d'invention Français, N.120989: 1878 3 me Addition: construction de poutres, poutrelles (a system for beams and slabs in cement and iron, applicable to floors, railways and other roads. French patent No. 120989: 1878: third addition: construction of beams and joists)
31. Draffin, J. O.: 'A brief history of lime, cement, concrete and reinforced concrete' *A selection of historic American papers 1876-1926*, op.cit., p32
32. Hyatt, T.: op.cit., p75
33. Ransome, E. L., and Saurbrey, A.: op.cit., p35
34. 'Cent ans de béton armé', op.cit., p35
35. Mörsch, E.: op.cit., p204
36. Mörsch, E.: op.cit., p204
37. Koenen, M.: 'Für die berechnung der stärke der Monierschen cementplattes' (Strength calculations for Monier cement slabs), *Centralblatt der Bauverwaltung*, Berlin, Ernst & Korn, 20 November 1886, p462
38. Wayss, G. A.: *Das system Monier: eisengerippe mit cementumhüllung* (The Monier system: an iron skeleton with cement filling), Berlin, publisher not stated, 1887
39. ibid., p78
40. Mörsch, E.: op.cit., p206
41. Ransome, E. L., and Saurbrey, A.: op.cit., p294
42. Christophe, P.: *Le béton armé et ses applications* (Reinforced concrete and its applications), 2nd ed., Paris and Liège, Librairie Polytechnique, 1902

(NB This book contains an extensive bibliography, pp727-739)

43. Suenson, E.: *Jaernbeton* (Reinforced concrete), Copenhagen, P.E. Bluhmes Boghandel, 1907
44. Cusack, P.: 'François Hennebique: the specialist organisation and the success of ferro-concrete: 1892-1909', *Transactions of the Newcomen Society*, **56**, 1984-85, pp71-84
45. Hennebique, F.: *Improvements in the manufacture of joists, girders and the like*, British Patent No. 14 530, 1892
46. Ibid., p4
47. Anon.: *Hennebique ferro-concrete, theory and practice*, London, L. G. Mouchel & Partners Ltd., 1909, p21
48. Durand Claye: 'Procédés d'essai des materiaux' (Procedures for testing materials), *Annales des ponts et chaussées*, Paris, 2e Sem. 1888, p173
49. Coignet, E., and de Tedesco, N.: 'Du calcul des ouvrages en cement avec ossature metallique' (Calculations for works in cement with metallic skeletons), *Memoires de la Société des Ingénieurs Civils de France*, Paris, **1**, 1894, pp282-363
50. Marsh, C. F.: *Reinforced Concrete*, 2nd ed., London, Archibald Constable & Co. Ltd., 1905, p39
51. Ransome, E. L., and Saurbrey, A.: op.cit., p33
52. Wayss, G. A.: op.cit., p102
53. Hennebique, F.: op.cit.
54. Considére, A.: *Notre sommaire sur le béton fretté et ses applications* (Notes on hooped or spirally armoured concrete and its applications), Paris, Publications du Journal Le Génie Civil, 1907 (translated into English: translator unknown)
55. Considére, A.: op.cit. (English translation), pp11-12
56. Considére, A.: *Experimental researches on reinforced concrete*, translated and arranged by Moisseiff, L. S.: authorised version, New York, McGraw Hill Publishing Co., 1903, p123
57. Marsh, C. F.: op.cit., p81
58. Anon.: *Hennebique ferro-concrete, theory and practice* op.cit., pp38-39
59. Considére, A.: *Experimental researches on reinforced concrete*, op.cit., p119
60. Considére, A.: *Note sommaire sur le béton fretté et ses applications*, Op.cit. (English translation), p9
61. Draffin, J. O.: op.cit., p32
62. Turner, C. A. P.: 'Concrete steel construction: part 1: buildings' (1909): *A selection of historic American papers 1876-1926*, op.cit., p274
63. Sozen, M.: *A selection of historic American papers 1876-1926*, op.cit., p243
64. The Institution of Civil Engineers: *Preliminary and interim report of the committee on reinforced concrete*, London, 1910 p164
65. Christophe, P.: op.cit., 1st ed., 1900
66. Christophe, P.: op.cit., 2nd ed., 1902, piii
67. Mörsch, E.: *Der Eisenbetonbau* (Concrete-steel construction), Berlin/Frankfurt, Wayss & Freytag, 1902
68. Marsh, C.: op.cit., 1st ed., 1904
69. Taylor, F. W., and Thompson, S. E.: *A treatise on concrete plain and reinforced*, New York, John Wiley & Sons, 1905
70. Mörsch, E.: op.cit., 3rd ed. (English translation)

22 EMERGENCE OF REINFORCED CONCRETE

71. The Institution of Civil Engineers: *Preliminary and interim report of the committee on reinforced concrete* London, 1910, pp50-63
72. 'Cent ans de béton armé', op.cit., p38
73. The Royal Institute of British Architects: 'Report of the joint committee on reinforced concrete', *Journal of the Royal Institute of British Architects*, Third Series, **XIV**, No. 14, pp515-541
74. Taylor, F. W., and Thompson, S. E.: op.cit., 2nd ed., 1909, pp504-505 (NB This book contains an extensive bibliography, pp725-744)
75. Marsh, C.: op.cit., 2nd ed., pp212-215
76. Taylor, F. W., and Thompson, S. E.: op.cit., 2nd ed., p404
77. Mörsch, E.: op.cit., 3rd ed. (English translation), pp22-23
78. Taylor, F. W., and Thompson, S. E.: op.cit., 2nd ed., p402
79. Ibid., p752
80. Anon.: *Hennebique ferro-concrete, theory and practice,* op.cit., 4th ed., 1921,p11
81. Taylor, F. W., and Thompson, S. E.: op.cit., 2nd ed., p382
82. Ibid., pp446-447
83. Bauschinger, J.: *Bericht über versuche an verschiedenen nach dem system 'Monier' hergestellten objekten* (Report on various tests relating to the Monier system), Berlin-Munich, 1887 (printed by Becker & Hornberg, Berlin)
84. Mörsch, E.: op.cit., 3rd ed. (English translation), p46
85. Marsh, C.: op.cit., 2nd ed., 1905, pp224-225 and pp238-239
86. Taylor, F. W., and Thompson, S. E.: op.cit., 2nd ed., p459
87. Mörsch, E.: op.cit., 3rd ed. (English translation), p45

6
The origins of reinforced concrete

Cyrille Simonnet

It is hard to establish the birthdate of reinforced concrete as a building material: its invention occurred when a builder came up with the idea of sinking metal bars in the mortar in order to obtain a material device that would be rigid and suitable for building. How the invention became official through a legal mechanism (the patent or the licence) that served to validate and sanction it is a different story.

The invention of a building material

Lambot and Monier won their consecration in a period when the material's promotion needed a pedigree myth in order to constitute itself. At the start of the century the material had not yet been baptized as a result of the existence of an entire family of competing systems: concrete with cement and iron, siderocement, iron-cement, ferruled concrete etc. However, a careful investigation reveals that the principle was imagined, described, and even experimented by a number of builders already toward the end of the 18th century. Loriot, who was known in his time for having invented a new kind of mortar (1774), Fleuret, the inventor of the "artificial stone" (1807), Raucourt de Charleville, an engineer of the Ponts et Chaussées and author in 1828 of a long treatise on mortars, and Coignet, the inventor of agglomerated concrete (1852), all had the idea of strenghtening the concrete by means of iron beams, chains or metal nets.[1] Legal registration of the invention in the form of a patent took place about fifty years after its mention in monographies or treatises. Lambot with his boat in 1849 and Monier with his garden boxes in 1865 both earned the title of inventors, while their elder brothers remained forerunners.

Sand, water, cement (or lime) properly dosed and scrap iron placed in a suitable arrangement: these were the ingredients of the new building material. The secret of the mechanical performance of reinforced concrete lied in this harmonious combination, and it was these factors (dosage, type of metal, configuration) that distinguished the various competing systems at this early stage. But is reinforced concrete a building material? This apparently provocative question raises another, more fundamental one: what is a building material? In architecture the material is at the same time the building's physical vehicle and, so to speak, its essential function, which is revealed in the outcome of a union between usefulness and art. As stone or wood, which have, virtually since time immemorial, produced actual building styles based on the techniques of bonding and assembling, also concrete was to be the protagonist of a development of building and architecture that suited its techniques of use, or, more precisely, the methods of handling it. But there is a new element: a great number of professionals, clearly distinguished from one another: suppliers, entrepreneurs, engineers were as many links on an extensive chain, dispersed but in harmony. Reinforced concrete, therefore, is not a raw material that went through a series of transformations in order to reach a final state that corresponded with its being assimilated by the building. On the contrary, it was the product of a process of mixtures, of which "reason" is, unlike wood and stone, not practical but abstract, in that it is governed by a chemical and physical formula, not by the relevance of know-how. In this context the question of the "building material" becomes

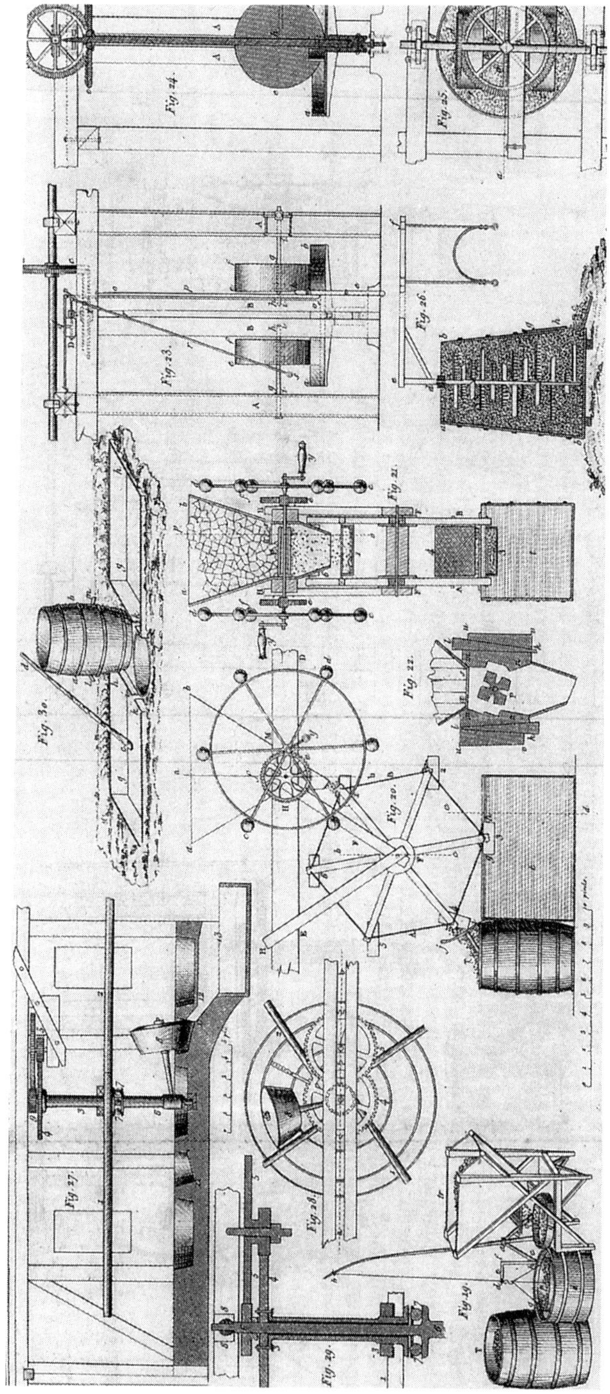

1. Plate illustrating the instruments utilized when preparing lime.
From: Raucourt de Charleville, *Traité sur l'art de faire de bons mortiers* (Paris-St. Petersburg, 1828).

pertinent, and its historical and technical constitution is the subject of our investigation.

Raw material

The development of reinforced concrete is to be interpreted according to the different stages of its history. The first is its production, the second its prescription, the third its consumption. Concrete owes its existence to the elements it is made up of, essentially iron and concrete. Both iron and concrete were protagonists of the absolutely exceptional economic development that took place during the last decades of the 19th century. The role of iron is well known in this context, that of concrete less so. In this sense the dividing line is marked by a date: 1818. Vicat, engineer of the Ponts et Chaussées, defined the formula of artificial hydraulic lime: he "discovered" a process that permitted to eliminate the dependency on the variable and fluctuating properties of the materials extracted from the quarries and at the same time gave the mortar (and later concrete) reliability and quality, in any place and moment.[2] Vicat dedicated his life to visiting all the French quarries, compiling a meticulous table of their geology. This work of patient observation is virtually invaluable: considered an almost anonymous progress, it was to completely revolutionize the economic scenario of building, both in France and in Europe as of 1850. Lime was, up to this time, still extracted from the quarry, often produced by the entrepreneur according to the requirements of the building to be erected. While the stone was extracted, a kiln was built: that of kilns remained a complementary industry, producing mortar in unspecified quantities. In 1850 there were only four cement factories in England (the famous "Portland", a variant of the Vicat process, discovered by Aspdin in 1824), while the only plant engaged in the industrial production of cement in France was Demarle at Boulogne-sur-Mer.[3] The development of the European production did not become significant until after 1880. For example, in 1899 the production of cement products (lime and cement) in France was almost three million tons, 4/5ths of which was manufactured in three manufacturing districts, at Boulogne-sur-Mer, at Theil in Ardèche and in the Grenoble district.[4] This swift expansion, bolstered by a great concentration, coincided with the industrial revolution and was the symptom of an important modification

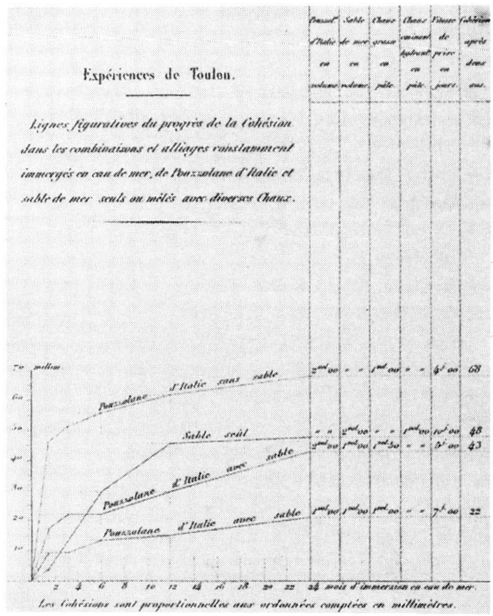

2. Graph illustrating different reaction of pozzolana in mortar.
From: Vicat, ***Résumé des connaissances positives actuelles sur les qualités, les choix et la convenance réciproque des matériaux propres à la fabrication de mortiers et ciments calcaires***
(Paris, 1828).
Bibliothèque Forney, Paris.

of the productive apparatus. In fact, the birth of an industry, in the economic sense of the word, was realized somewhat at the expense of the builders, who made use of equipment and methods practically identical to those of the past centuries.

Cement factories essentially depended on limestone (raw material) and coal (energy). It is not hard to guess their commercial strategy: to produce more, better and at a lower price. The kilns were improved (the rotating kiln, for example, developed in the United States, considerably increased productivity), as well as the treatments (improvement of the sifters, recovery of the roman cement, methods of sack filling etc.). The distribution networks were organized (commercial agencies, advertising), the research was also developed (the cement factories had their own laboratory, to be specific, in Germany). Faced with these developments, the buyers of the products, i.e. the

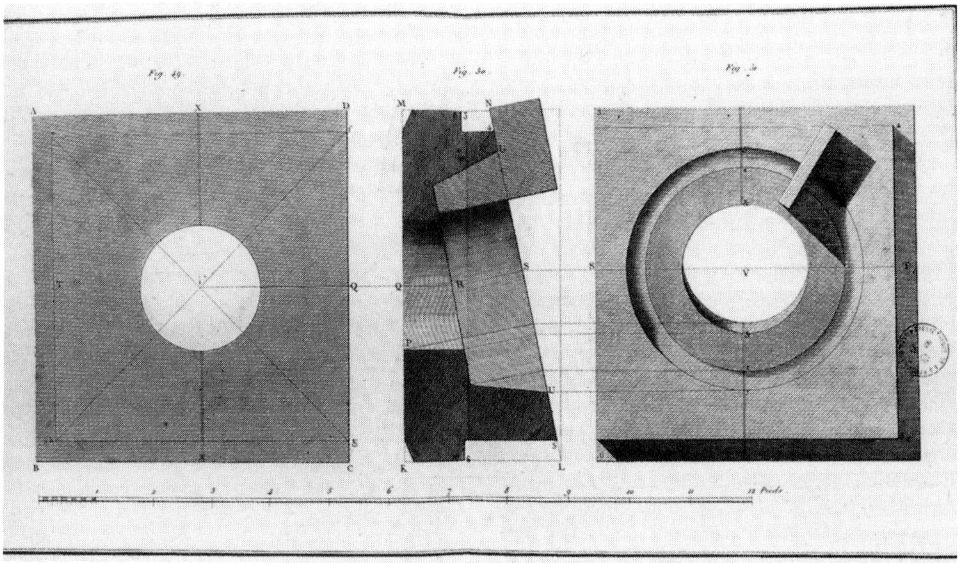

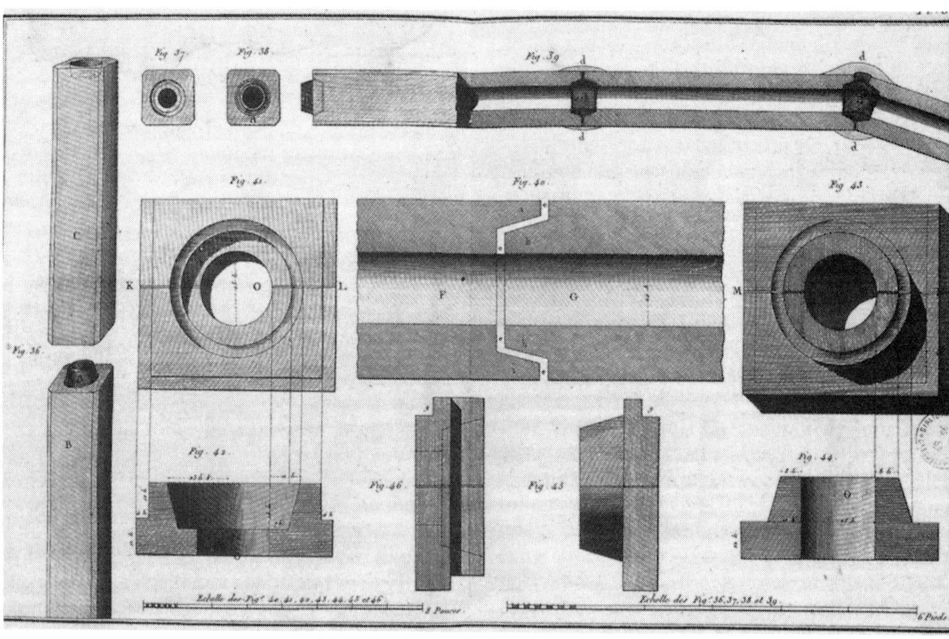

3, 4. The "pierre factice" of Fleuret used for the manufacturing of water pipes. From: Fleuret, *L'art de composer des pierres factices aussi dures que le caillou* (Paris, 1807).

HENNEBIQUE

N'EST PAS ENTREPRENEUR

BÉTONS ARMÉS «HENNEBIQUE», 1, RUE DANTON A PARIS, PREMIER BUREAU D'ÉTUDES DE BÉTON ARMÉ EN DATE COMME EN IMPORTANCE; A ÉTUDIÉ DEPUIS 45 ANS POUR LES ARCHITECTES ET POUR SES 1800 ENTREPRENEURS-CONCESSIONNAIRES PLUS DE 115.000 AFFAIRES, DONT 85.000 EXÉCUTÉES

5. Hennebique trademark from the 1930s.

companies engaged in building and public works, were characterized by a slower and quite inconstant evolution. If we exclude special constructions representing occasions of great consumption of cement products and a strong concentration of labor (such as the excavation of the Suez Canal, with its 13,000 workers), the yards for structures and buildings of average size relied on workers that were extremely scattered, badly equipped and whose approach rather resembled that of craftsmen. As P. Lescure put it, "during the 19th century the building substantially remained the prerogative of the small companies, which considerably privileged the labor factor at the expense of capital investment."[5] As late as 1906, therefore, more than half of the workers in the building profession in France worked in organizations that employed less than ten persons.

From a strictly technical standpoint concrete (and consequently reinforced concrete) derived from two factors: its composition and its treatment, which can be translated into as many industries, cement manufacturers, and building companies, which obviously correspond to extremely different productive and economic logics. In as little as 30 years, during the second half of the 19th century, cement manufacturing made the transition from craft to industry, triggering a commercial dynamism and realizing a geographical concentration without precedents.[6] Only 29 factories were registered in France in 1899, roughly divided into three departments. A similar situation existed abroad.[7] On the other hand, the businesses in the field of masonry or public works remained scattered in tiny units on the entire territory, depending on the fluctuations of the real estate and plants market.

This dual reference (concentration-dissemination) enables us to pinpoint and understand, at least partly, the development of concrete, and more precisely of reinforced concrete. It has been said that iron is the most important building material of the 19th century. True, although this only applies to an extremely specific sector of use, concerning programs (roofing, sheds etc.) rather than the productive infrastructures. In the traditional builder's yard stones were preferred to metal reinforcement, which remained the prerogative of specialized companies. The same companies often acted both as blacksmiths and fitters (for example Eiffel or Moisant).[8] A significant development of reinforced concrete did not take place until after 1890, that is, when the breach between the two sectors on which it relied had become definite. The building material can therefore be considered as the protagonist of a healthy, if not ideal, transition between these two productive sectors, formally speaking antagonistic, but in a close mutual and interdependent relationship. Concrete (and iron), available anytime and anywhere, can sustain the nomadic and scattered work of the builder's yard. Even more so, the use of iron and concrete certainly contributed to transforming the various crafts related to the builder's yard into a simple manufacturing system where the traditional qualification attributed to a "know-how" was replaced by a monofunctionality of the tasks (pouring into forms, grinding, reinforcement forming, crushing etc.).[9] The development of reinforced concrete, from an exclusively socio-economic standpoint, is more easily understood as follows: as a material system which meets, to an equal degree, the requirements of its users and its suppliers, as a device which unites productive capacity and labor availability, while at the same time guaranteeing the total independence of the two spheres.

The bureau d'études and the patent

This image of a mutually compensating economy obviously does not suffice to explain the success of the material. A third element, placed between the cement manufacturer and the worker, regulated the tasks of both with exact-

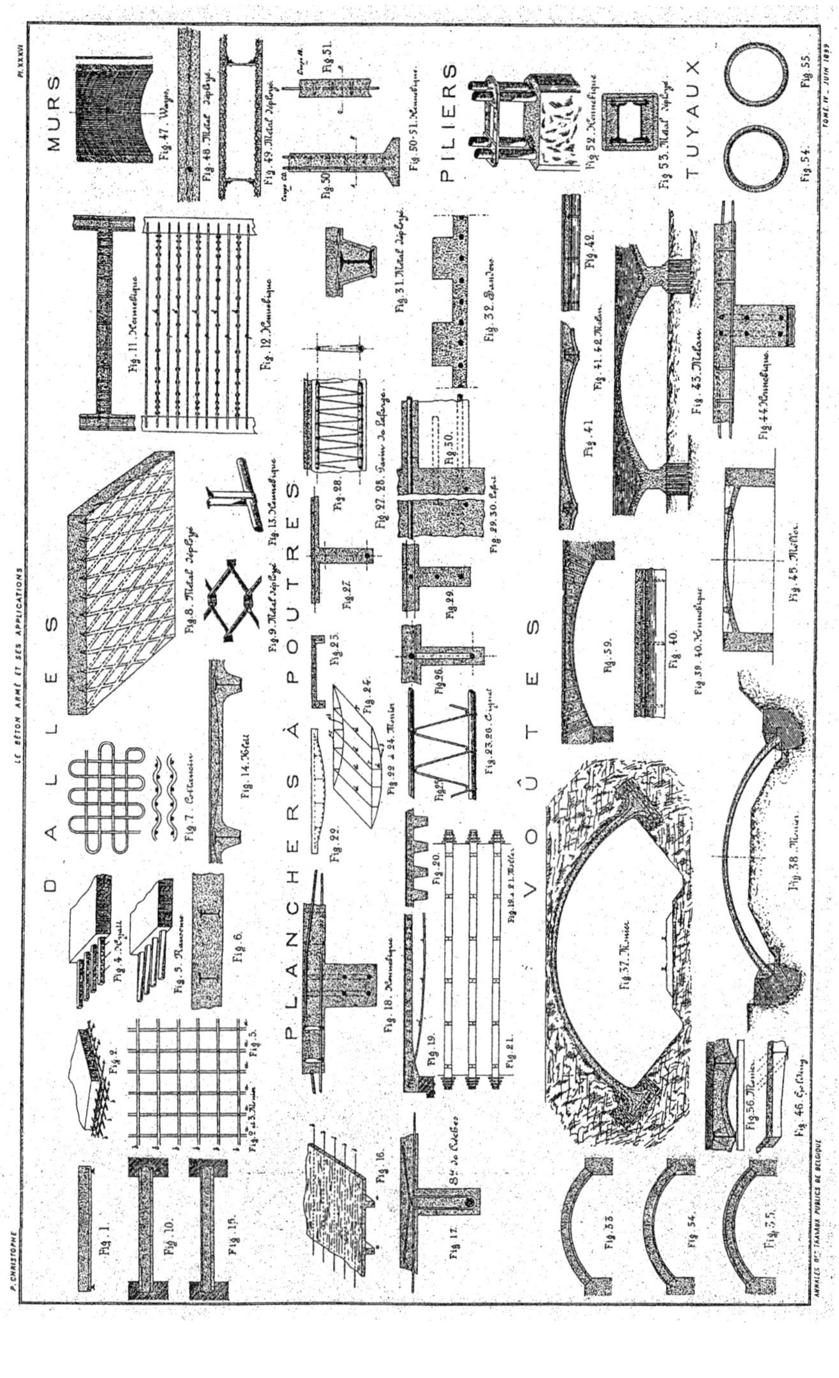

6. Plate published by Paul Christophe in 1899 in which the different systems (Hennebique, Ransome, etc.) are compared.
From: *Le Béton Armé*, 10 July 1899.

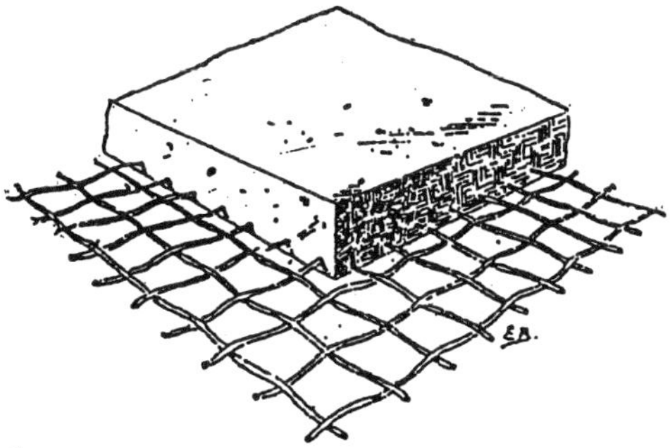

7

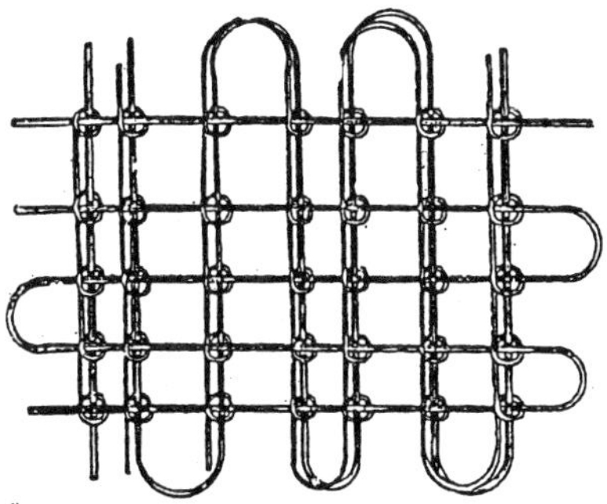

8

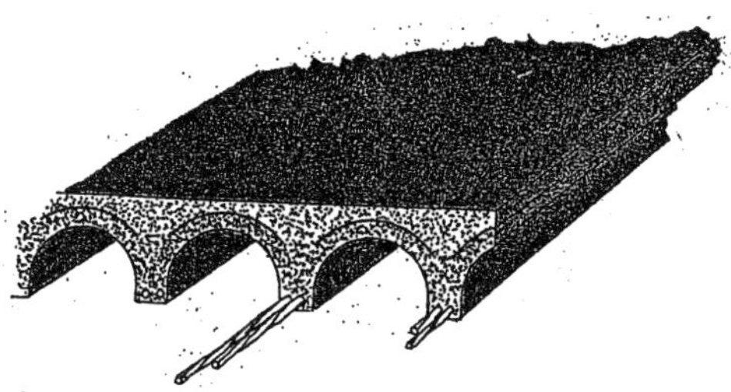

9

7. Beam in reinforced concrete, Cottancin system.
From: *Technique et Architecture*, No. 1-2, 1949.

8. Variations of metal reinforcements of the Cottancin system.
From: *Le Ciment*, No. 1, 25 January 1897.

9. "Twisted" reinforcements by Ransome.
From: *Le Béton Armé*, 10 March 1899.

ness, i.e. the *prescripteur* who was identified by the company under the title *bureau d'études*. The *bureau d'études* acted as an active drive belt between the bulk material and those who used it. Its most defined example, we should say, its very model, was that of the Hennebique company, whose extraordinary success was explained by an unrivalled commercial strategy. The function of the *bureau d'études techniques* (BET) was original and cannot be boiled down to the traditional role of the architect or the engineer, who, even if they were also professionals in the field of *prescription*, performed different functions in the manufacturing process.[10] Reinforced concrete was "theorized"; in other words, before triggering the operative procedures of its use or its installation, simply before existing, a precise description, the creation of which did not depend either on the competence of the workers, or the company, or even on the project for which it was prescribed, but on a conceptual model governed by specific mechanical laws, was required. Although in the decade of 1890 these laws were not yet very clearly perceived, they concerned the strength of the material and the chemical formula of the binders.[11]

The *bureau d'études* knew these laws, these rules: it prescribed the dosage, the configuration of the reinforcement, it calculated the section of the structures, and regulated the manufacturing of the raw materials even before they were transformed. In fact, the basic nature of reinforced concrete is in the end just a principle, almost a "theory", rather than a raw material, physically assimilable by the building industry.

It becomes necessary, at this point, to analyze the patent of invention and its commercial label, the "system". Between 1880 and 1900 a great number of "systems" prospered: Cottancin, Bordenave, Matrai, Bonna, Mélan, Hennebique, Piketty, Hyatt, Möller, Koenen, Coularou, Wünsch, Ransome, etc. These are as many processes, consisting in sinking iron in the concrete, in some cases to form beams, walls, floors, in others to manufacture specific structures (pipes, vaulted ceilings, pillars) in others again to ensure precautionary guarantees (against fires, earthquakes etc.).[12] Their

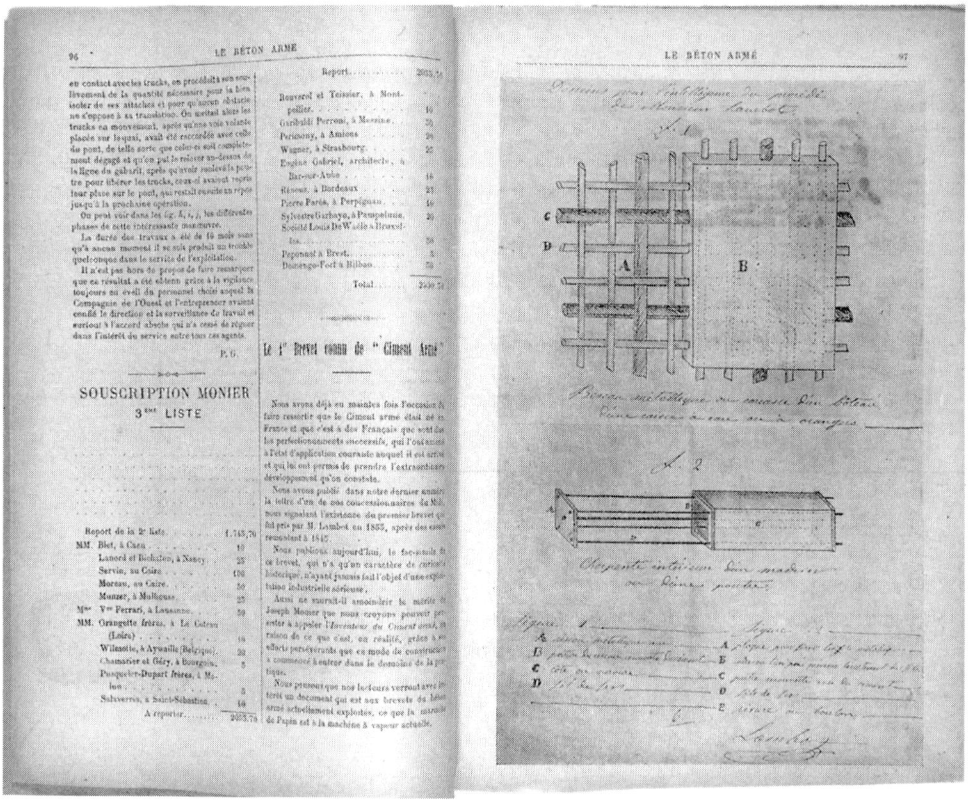

10. Facsimile of the Lambot patent of 1855.
From: *Le Béton Armé*, No. 54, November 1902.

description essentially consists of their configuration in sections (transversal and longitudinal), according to the types and sizes of the work. This is a fundamental point: the concepts of solidity and stability gained new relevance through the only reference to the section gradually eliminating references to proportion. Calculation of size and placing of the reinforcement, methods of building and sometimes of removing formwork, dosage and use of concrete are as many criteria which only those who make use of the system are actually able to describe and prescribe after having adapted it to the project. For example, Cottancin presented, on the occasion of the first "Congrès international des procédés de construction" (Paris, 1889) a system "consisting of a chain woven in a woof" to make watertight pipes; Hennebique invented stirrups to control shear in an economical manner; Ransome coiled an iron spiral to improve adherence with the mortar; Koenen prepared prefabricated non-load-bearing slabs (a forerunner of slab-bracing) in an attempt to eliminate the formwork; Mélan composed vaults based on metal arches sunk in concrete; Considère ferruled his reinforcements by spiralling them in their formwork, thinking of the market of pillars and so on.

Some systems were quite similar to others, and the similarities sometimes gave rise to clamorous disputes (Hennebique-Cottancin). The differences (as well as the similarities) of the processes witnessed a yet disorganized maturation, where the tradesmen sought their

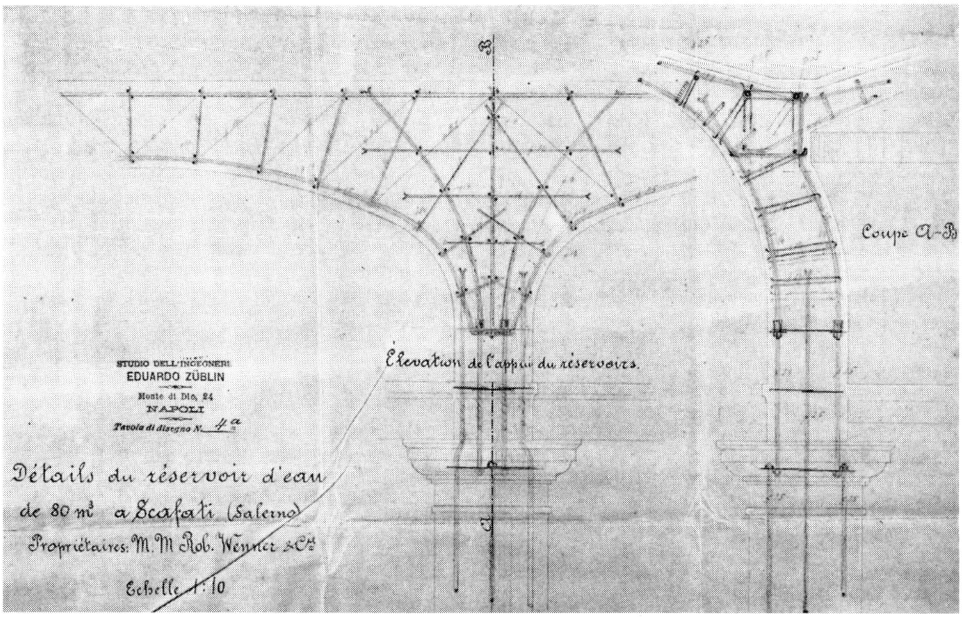

11. Waterworks of Scafati (Salerno).
Constructive detail.
Hennebique archives, IFA, Paris.

trade mark.

Some attempted to specialize in particular types of work: Monier, at the beginning of his activity, chose reservoirs, Matrai water pipes, Wünsch barrel vaults, Ransome silos... The universality of the process (or commercial dynamism) quickly cancelled these functional specifications: after World War I two great families remained: the solid slabs (Monier's heirs) and the truss system (Hennebique), which were, however, soon confused with the same lateral panels: at that point architecture had prevailed over the technical types.[13]

Reinforced concrete therefore revealed a double origin: on the one hand, an economic breach that gave rise to the original function of the engineer-consultant, or the *bureau d'études*, and its possibilities of use; on the other, it was defined by processes of recording controlled (at least before 1906 and in France) almost exclusively by its inventor. The fulcrum of the matter was the *patent*. It is worthwhile to dwell on its particular characteristics. The legal definition of the patent was established in Europe as of 1884, when the "International Convention for the Protection and the Industrial Property" came into effect. In 1881, a jurist wrote

> The patent is never granted for an idea or for a thought, but for the sincere and complete realization of the means or the processes that allow to achieve an industrial progress which society lacks.[14]

The patent, therefore, is an industrial privilege and in fact constitutes an efficient link between the technician and the economist, between invention and use. The building industry, characterized by a craftsman's and manufacturer's approach, scattered in its builder's yard, is a field which is not quite suited to the industrial exploitation of the invention. Apart from the metal construction (controlled by the metallurgic entrepreneurs who rapidly formed important industrial groups) and certain finishing elements (glazed frames, galvanization, heating systems...) we only have a few patents before 1880 that are, strictly speaking, connected to the building industry (i.e. the foundations and the main wall).[15] Coignet is an exception, but he was an industrialist (he produced matches, glue and gelatine), who came from a family of industrial manufacturers, and who in any case exploited his agglomerated concrete in a nearly marginal way.[16] The industrial growth resulted in an increase of patents registered.[17] Reinforced concrete followed this evolution and was promoted by builders of the second generation, of a hybrid identity: engineers who were sons of entrepreneurs, Polytechnic graduates who were real-estate developers, public-works entrepreneurs etc. As products of the companies (such as Boussiron) or of the category of engineers (Piketty), or of the industrial environment (Coignet), they were not as much connected to their origin as to the future of the profession which they contributed to create: the *bureau d'études techniques.* The patent, as a legal instrument, catalyzed the potentiality of the use of the invention and its exchange, in that it ordered and regulated the aspects of commercial exploitation. The invention of reinforced concrete is a result of this factor of balance and particular transformation, just as stereotomy derived from the ancient art of drawing and the kinds of apprenticeship through which it was transmitted. However, concrete is part of another technical and industrial picture. We have already seen how initially it was not the material of any "craft" in the sense of a traditional craftsmanship that still characterized the building industry. Concrete was the result of a coming together of certain ideas, experiences and applications. The vicissitudes of the initial phase of exploitation of reinforced concrete, until the end of the century, should be interpreted in this context.

Expansion, dissemination

In November 1877 Monier deposited a patent in France, which was, one year later, accompanied by an addition, in which the first explanation of the function of iron in the mixture was generally recognized.[18] Monier had already built a number of water-tanks according to the principles of his garden boxes of 1855 (Fontenay, Bougival, Alençon). In these applications, where, above all, the watertightness of the new building material was exploited, the geographical development of concrete was increasingly intense. Although the experts did not heed him, Monier was able to elicit the interest of a number of companies and negotiated his patent in Belgium and in Holland (Picha & Frères), in Austria (Schüster) and in Germany (Martenstein & Josseaux).[19] Due to his promotional efforts the premises for a profitable alliance were quickly created by the Germans: G.-A. Wayss, a German engineer who "discovered" the Monier system on the occasion of the Antwerp industrial exhibition of 1885 and who founded in 1887 a joint stock company for the exploitation of Monier concrete,[20] and C. Freytag, an entrepreneur who gained access to the patent almost by chance when he joined in an association with Martenstein & Josseaux (1884).[21] After some time the two reached an agreement, and the company Wayss & Freytag was founded in 1893, a year after a highly competitive patent had been taken out: that of Hennebique.

An investigation on two important strategies that were to dominate the market of reinforced concrete during the first decade of the 20th century would be appropriate. Hennebique's strength lied in the double practice of centralization (management of the works) and decentralization (market research), with a system of agents and concessionaries which was perfectly organized on a juridical basis.[22] The German power was based on a play of redemption and alliances: in a way it was more "capitalistic" and intercepted the market by means of a stratagem of industrial concentrations (in particular with respect to Dyckerhoff, a big cement manufacturer).[23] Geographically speaking, their areas of influence were different: Hennebique disseminated his products in the entire world, while Wayss & Freytag were active in Germany, in Austria and in central Europe. Technically, also the services rendered were different: Hennebique's stirrups derived from the simple beam, and its efficiency lied in the extraordinary simplicity and variability of the configurations which its handiness allowed. The calculations which he made use of were more economic than theoretic. Hennebique indeed accused science which he suspected as threatening his monopoly:

Gentlemen, do I have to confess it? I am dead shocked by all this accumulation of science... The factors which intervene in [our] formulas are the loads, the capacities which constitute the lever arm of these loads, the strength of the materials used, the height of the couples formed by the solids and the lever arm of the strength of the materials; this constitutes a small and very simple kitchen, in which all the elements are comprehensible and

12. Lighthouse of Nicolaief built by Hennebique.
From: *La Nature*, 1 October 1904.

sufficient to compose, in concrete of cement and iron, combinations of reinforcements and solid and economical floors.[24]

On the contrary, in the case of Wayss & Freytag, calculation and the materials science were mentioned at a higher level: J. Bauschinger, director of the Munich Institute and a great theoretician on elasticity, made his laboratories available for the company. F.-E. von Emperger (theoretician and disseminator in the Usa of the Mélan system), Mathias Koenen (who developed the principle of the *Voutenplatte*, capable of eliminating a great part of the dead load) and Wayss did much for the theory of reinforced concrete.[25] This attitude derived partly from the originality of the Monier system inasmuch as the mechanical behavior of the slab is more complicated to mould than the composite beam.[26] But Bismarck's Germany was characterized by an industrial dynamism without precedents, and was certainly not unfamiliar with these synergies.

The "theories" of reinforced concrete during its initial stages were not yet to prevail over those of the strength of the building materials (elasticity theory), which reached maturity after Navier (1821) and above all Lamé (1852) who simplified the enunciations.[27] Incidentally, it is known that the iron-concrete mixture puzzled the theoretician for a long time, as the components do not appear to be very compatible from a mechanical standpoint.[28] Also W. Fairbairn, an English engineer known for his important experiments on the strength of cast iron and iron, built a spinning mill at Manchester in 1845, where he combined metal and concrete.[29] His intuition ended there, as concrete only served as a firewall.

W. Ward, an American engineer, also sought a method for planning fireproof buildings. Collins[30] reported that he was struck by the sight of the workers busy with the difficult task of cleaning their tools stained with dried concrete when he visited England in 1867. The idea developed, and in 1872 he built his house by sinking metal braces into concrete, recording the experiment with fastidious observations. In that same directions, Thaddeus Hyatt did analogous experiments and published in 1877 what is considered the first real treatise on the subject: *An Account on Some Experiments with Portland-Cement-Concrete Combined with Iron as a Building Material*, in which he justified the mechanical complementarity of the two materials.

In 1887 ten thousand copies of the *Monier Broschüre* were published in Germany by Wayss and Koenen.[31] It would be extremely interesting to know how this booklet was diffused and what effects it had on the milieu. As of this moment the ways of exploiting and disseminating the know-hows were to take on a new direction in Europe. One vehicle acted as a link in this sense, making the immense building potential of reinforced concrete known through its circulation. This is a matter of that principle of amplification represented by advertising and information, synthesized in an exemplary way by the first press organs dedicated to the building material: *Le Béton Armé* in France, organ of the agents and the concessionaries of the Hennebique company, created in 1898; the *Beton und Eisen* magazine in Germany, founded in 1901 by Fritz Edler von Emperger; *Il Cemento Armato* in Italy (1904); *Concrete and Construction Engineering* in Usa (1906); and *Ferro Concrete* in Great Britain (1912). Before a specific knowledge actually developed on the subject of reinforced concrete, relayed by teaching, by the publication of manuals and by institutional research, it was through the columns of these magazines that a true know-how on reinforced concrete was made known to entrepreneurs and *prescripteurs*. Between 6,000 and 10,000 copies of Hennebique's magazine were printed, and in the course of a few years it was disseminated in France, Europe and soon in Buenos Aires, in Indochina and in New York. At once a magazine, a report on experiences and builder's yards, a vehicle for advertising, the journal published in rue Danton provided an efficient promotion that had no precedents and that was to catch many of Hennebique's competitors off guard. The German magazine was more objective, less hurtful (unlike *Le Béton Armé*, no sides were explicitly and openly taken) and more scientific. It appeared in the form of monthly booklets which formed a kind of encyclopedia and proposed a solution to every kind of problem related to concrete, in the attempt to cover all the fields of use of the material.

As for education, Charles Rabut held the first lecture on reinforced concrete at the Ecole des Ponts et Chaussées in Paris in 1897.[32] The education apparatus assimilated the chapter on reinforced concrete, quickly producing an enormous quantity of publications. Three French-speaking publishers promoted the movement and an inspection of their catalogues would be worthwhile: Ch. Béranger (Paris, Liege), Dunod and Eyrolles (Paris). Paul Christophe published *Le béton armé et ses applications* in 1902, immediately followed by Berger and Guillerme with *La construction en ciment armé*: with almost 2,000 pages, the two volumes constitute an impressive bulk. The multiplication of manuals is indisputably a sign of a favorable technological dissemination. In a way, this development followed the period of "patents", which became public domain at the turn of the century and which at this point consisted of improvements rather than of inventions. Besides, the Chambre Syndicale des Constructeurs en Ciment Armé was founded in 1903, a structure of convenient alliances and a sign of professional maturity. A third aspect of the dissemination of the technical culture of reinforced concrete should also be noted: that of its administrative consecration. It took some time before the French administrative authorities admitted the constructional pertinency of the material: the first public structure which officially adopted it was a construction by the Military Corps of Engineers, a shooting gallery in Lyons in 1896. But the admission of reinforced concrete on the part of the administrative authorities was only to take place under the close watch of a laborious technical examination, which was to constitute the famous "circular of 1906", the knowing regulations that governed the conditions of use, calculation and manufacturing of reinforced concrete. The preparation of this document, which required about six years, was also subject to considerable competition, in particular between Hennebique and Considere.[33] The press, the manual and the code: these are three stages of an irreversible evolution. The technical element is here justified, ideologically and scientifically. This is once again a matter of specification: reinforced concrete needed assistance during its "birth". Being a many-sided and plural craft, and as yet not familiar with itself, it was not transmitted through the secret of apprenticeship. It could be imparted from the written page, rather than from experience. Its maturation was governed by the social and economic device that understood it. Such device, as we have seen, was concretized by the patent, whose geared-down efficiency triggered a game that was, at the same time, economic, professional and promotional. The large-scale manufacturing of a raw material, a new way of prescribing, a system of information: these are the active forces that have permitted the use of reinforced concrete on a significant scale, and which have actually created the foundation of a new culture of building.

1. J.-A. Loriot, *Mémoire sur une découverte dans l'art de bâtir...*, Paris 1774; Raucourt de Charleville, *Traité sur l'art de faire de bons mortiers*, Paris-St. Petersburg 1828; Fleuret, *L'art de composer des pierres factices aussi dures que le caillou...*, Paris 1807; Fr. Coignet, *Bétons agglomérés appliqués à l'art de construire...*, Paris 1861.
2. L.-J. Vicat, *Recherches expérimentales sur les chaux de construction, les bétons et les mortiers*

13. Cover of *Le Béton Armé, organe des concessionaires et agents du Système Hennebique*, 1899.

14. Rifle-range of Comboire (Isère), built in 1901 by the Société Anonyme de Travaux et ciments et béton armé (Hennebique system).
Photo A. Michel. Hennnebique archives, IFA, Paris.

ordinaires, Paris 1818. The monography describes the "trick" experimented for the foundations of the bridge of Souillac in Dordogne (1817), which consisted in refiring the lime mixed with tiny clay pellets so as to obtain artificial hydraulic lime. As for Vicat, ref. *Louis Vicat (1786-1861)*, Paris 1962.
3. From "Fabrication du ciment Portland artificiel. Usine de Frangey" in Turgan, *Les grandes usines. Revue périodique des arts industriels*, t. XVII, Paris 1885.
4. Ref. E. Leduc, *Chaux et ciments*, Paris 1902.
5. P. Lescure, *Immobilier et bâtiment en France (1820-1980)*. Paris 1983, p. 15.
6. Ref. for example the history of a company such as Lafarge: L. Dubois, *Lafarge Coppée, 150 ans d'industrie*, Paris 1988.
7. E. Leduc, *Chaux et ciments* cit.
8. Ref. B. Lemoine, *L'architecture du fer. France XIXème siècle*, Paris 1986.
9. Ref. "Entreprises et patronat du bâtiment", *Séminaire histoire des métiers du bâtiment*, Plan Construction et architecture, emploi et valorisation des métiers du bâtiment, MELT, Paris, 28, 29 and 30 November 1989 (typed document), and also P.-W. Kingsford, *Builders and Building Workers*, chapter 6 *Building with Concrete*, London 1973.
10. The drawing of the architect, for example, does not provide indications, in a proper sense, as to the task of the worker (even if it is interposed between the latter and the building material); on the contrary, if it nurtures them, it fully presupposes them. Ref. Ph. Potié, "Un chantier de l'éclectisme: la Préfecture de Grenoble", in *Cahiers de la recherche architecturale*, n. 29, first quarter of 1992.
11. A good picture of these knowledges is provided by E. Candlot, *Ciment et chaux hydrauliques, fabrication-propriété-emploi*, Paris 1898.
12. The taking out of patents concerning reinforced concrete (in Europe before 1900) reaches a hundred. Their use, rather than their technical specifications, would deserve a reasoned classification. A rough outline of this inventory is attempted by M. Dotta Rosso, "Calcestruzzo armato e innovazione tecnologica", *150 anni di costruzione edile in Italia. Il modo di costruire*, international seminar, Rome, 13, 14 and 15 November 1991 (about to be published).
13. Ref. R. Legault, "L'appareil de l'architecture moderne" in *Cahiers de la recherche architecturale* cit.
14. E.O. Lami and A. Tharel, item "Patent", *Dictionnaire encyclopédique et biographique de l'industrie et des arts industriels*, 9 vol., Paris 1881-91.
15. Ref. *Brevets d'invention français 1791-1902*, catalogue of the exhibition, Paris 1958, and also Y. Plassereaud and F. Savignon, *L'Etat et l'invention. Histoire des brevets*, Paris 1986.
16. Ref. C. Simmonet, "Le béton Coignet: stratégie commerciale et déconvenue architecturale" in *Cahiers de la recherche architecturale* cit.
17. Between 1830 and 1840 750 patents were registered in France. Just in the year of 1879, 7821 patents were taken out (12,700 in the Usa, 3500 in Great Britain, 4400 in Germany). From E.O. Lami and A. Tharel, "Patent" cit.
18. J. Fougerolle, "L'évolution du béton armé", *1849-1949, Cent ans de béton armé*, supplement to the *Travaux* magazine, n. 194 bis, 1949.
19. From G. Hüberti, "Die Erneuerte Bauweise", *Von Caementum zum Spannbeton*, Berlin 1964, some sources which contradict those of Collins with regard to this period. Thus the meeting between Wayss and Monier can be interpreted in two ways. Apparently, Collins has used an incorrect information provided by an article of *Beton und Eisen* (IV, January 1905), "Zur Geschichte der Eisenbetons in Belgien", which indicates as 1879 the date of the exhibition in Antwerp, which on the contrary was held in 1885.
20. Literally, *Aktien Geselschaft für Beton und Monierbau*.
21. This same company was to become a concessionary for the Hennebique company in 1898.
22. "We have often said that this organization (Hennebique) which functions so well is almost as beautiful as the invention of the system. I truly agree with this idea!", letter from S. de Mollins to Ritter, on 14th March 1899, *Le Béton Armé*, n. 13, June 1899. On the organization of the Hennebique company, ref. G. Delhumeau, *Hennebique et les architectes*, BRA-MELTM research report, Grenoble-Paris 1989 and, for Italy, R. Nelva, B. Signorelli, *Avvento ed evoluzione del calcestruzzo armato in Italia: il sistema Hennebique*, Turin 1990.
23. Ref. G. Haegermann, "Von Caementum zum Zement", *Von Caementum zum Spannbeton* cit.
24. *Le Béton Armé*, n. 10, 1899, p. 4.
25. Ref. G. Hüberti, "Die Erneuerte Bauweiss" cit.
26. On the original duality of the slab-beam (Monier-Hennebique), ref. R. Maillart, "Le béton armé et son expression", excerpt from *Schweizerische Bauzeitung*, 1 January 1938, reproduced in Max Bill, Robert Maillart, Zurich 1969. On the German developments of the Monier system, ref. A.M. Zorgno, in this issue.
27. Ref. S.-P. Timoschenko, *History of Strength of Materials*, New York 1983, p. 117.
28. The mechanical compatibility of the two materials is a gift of nature: it stems from the consistency of the ratio between the coefficient of elasticity of concrete when compressed and that of metal when in tension.
29. The example has been cited by S. Giedion, *Espace, temps, architecture*, Brussels, edition of 1967.
30. P. Collins, *Concrete, the Vision of a New Architecture*, London 1959, p. 57.
31. G.-A. Wayss, *Das System Monier (Eisengerippe mit Cementumhüllung) in seiner Anwendung auf das Gesammte und Ingenieure herausgegeben*, Berlin 1887.
32. Ex engineer of the Compagnie des Chemins de Fer de l'Ouest, Ch. Rabut was to form the generation of great French builders active until the eve of World War II, as Armand, Freyssinet, Caquot...
33. Ref. G. Delhumeau, "La construction en béton armé autour des années 1900", *Cahiers de la recherche architecturale* cit.

7

Hennebique and building in reinforced concrete around 1900

Gwenaël Delhumeau

For this invention, and this is true for many others, the right moment had to arrive, and not only that, the *right man* also had to arrive! (Paul Gallotti)[1]

When François Hennebique founded his magazine, *Le Béton Armé*, in June 1898, his company had, due to his incessant work, already gained considerable recognition. The increasingly intensive activity of the company, and its efficient network of agents and concessionaries (at that time more than 60 entrepreneurs were using his system all over Europe), seemed to have earned Hennebique a dominant position, at least in France, in the sector of building in reinforced concrete.

The creation of this magazine, a powerful propaganda instrument, constituted an important achievement in the extremely intense expansion policy adopted by the builder in those years. The magazine, which was strategically distributed, starting from the local technical offices, strengthened the impressive organization set up by Hennebique, secured its cohesion, reinforced its efficiency and presented as a model the (new?) functioning of a company dedicated solely to the exploitation of a new building system. This demonstration of strength visibly consolidated the position that the builder had rapidly achieved with respect to a competition that was not too hard in France, but nevertheless solidly established. Perhaps most important, it also tended to modify the relationship of authority with other collaborators, the engineers and architects that Hennebique had depended on until then. *Le Béton Armé* therefore purported to publish their comments on a scientific, technical and artistic nature that they might desire to send, and therefore served as a substitute for the architecture and building magazines which had until then been the privileged media for promoting the Hennebique system. In addition to advertising, the builder also took advantage of the possibility to remove the control of the technical information from the engineers and the architects (who were invited to his conferences), and perhaps, through this act of independence, to assume a dominating attitude toward them. The offensive was also directed toward a territory that is still quite unexplored by historians, i.e. the industrial manufacturers of Portland cement[2] and the prerogatives these sought to retain, especially by controlling *Le Ciment* magazine, at that time the only technical publication in France dedicated to this material and its uses.[3] Discussing, with a very relative objectivity, the value of the different systems of "monolithic construction", *Le Ciment* meant to present itself as a useful instru-

1. Menu of the banquet organized to celebrate the six thousandth project realized by the Hennebique company, May 1899.
Hennebique archives, IFA, Paris.

ment for all those who used reinforced cement, thus united; "the success of the single which is reflected on all the others".[4] *Le Béton Armé* was, in any case, clearly identified with the model which Hennebique, inverting the data, ruthlessly deviated.

It is evident that the company's magazine by itself characterized the Hennebique phenomenon and the concept of monopoly which inseparably accompanied it. Unlike the other producers (and to what an extent!) Hennebique aimed, from the very beginning, at a vast diffusion of reinforced concrete, associated with the typical characteristics of his system, to the point of claiming to have invented the material itself.

From the pages of the magazine, Hennebique and his hagiographers, remembering the origins of reinforced concrete, cited the names of the entrepreneurs presented as "predecessors" (F. Coignet, J. Monier), or some sterile American attempts, while they systematically nullified the competition, as if it did not exist, or nearly. The builders in reinforced concrete were more often than not likened to "travelling salesmen who sell braces and other iron elements", as an inevitable obstacle to any kind of progress, while Hennebique is presented as its privileged representative. Beyond this discourse which is, by virtue of the amount of documents left, still impressing the historian, what ever happened to the "rivalry" which the inventor immediately made synonymous with routine and ignorance? What about the daily struggles for contracts, the strategies developed in order to gain a position on and an introduction to the market? And finally, what are the true implication and the purpose of this kind of message?

The Hennebique strategy

An initial study of the company's archives[5] have revealed some characteristic aspects of the Hennebique strategy, allowing a more thorough understanding of the entrepreneur's career, who in the course of a few years succeeded to command respect and, "thanks to his intelligent activity..., to bring reinforced concrete to most French cities and into some foreign countries".[6] In addition to producing leaflets and brochures, which Hennebique, like the other builders, distributed in order to make his production known (and as many collected references, which he used as a support), his policy has been one of information, based on personal contact and aimed at all those who had decisional powers in relation to building, and in particular at architects. In the offices and on the builder's yards Hennebique engaged in an operation of approach, persuasion and seduction that was far from being a mere visit to the potential customer's home. The exchange of correspondence found not only reveals extremely interesting information on the relationship between architect and entrepreneur; it also contributes to assigning Hennebique a position decidedly beyond the mere role of entrepreneur, to which architects and engineers would prefer to relegate him.

> If you would be so kind as to grant me a two-hour meeting - he writes architect Jules Pray - I would be quite an incompetent if you, after these two hours, should not become convinced that reinforced concrete can almost always be used in a positive and advantageous way from all points of view. In all those areas where I have built, I have set up agents and concessionaries; however, my organization expands very slowly, since my concessionaries must be informed of the work they are to carry out. I leave nothing to chance, I proceed methodically... my works speak for themselves.[7]

Although the Hennebique organization owed its efficiency to its decentralization, it was in Paris that the builder achieved notoriety, during the building of the famous sheds for the Raffinerie Parisienne of Saint-Ouen. On this occasion he personally invited the most important Parisian builders to visit the site and witness the installation (in this case quite complex) of his system and the strength tests that were tried out. It was here, around 1894, on his first important builder's yard in Paris, that Hennebique's most difficult and delicate struggle took place. This was perhaps a matter of converging interests, that tended to coincide, and that in any case gave rise to strategies which, once defined, were to have a decisive effect on the evolution of the Maison Hennebique.

The builder's yard, tactical objectives

The activity of representation related to the builder's yard (while this practice was probably common at the time, Hennebique further developed it) is interesting. Due to his position, the architect had to be perfectly up-to-date as to the progress achieved in the field of building

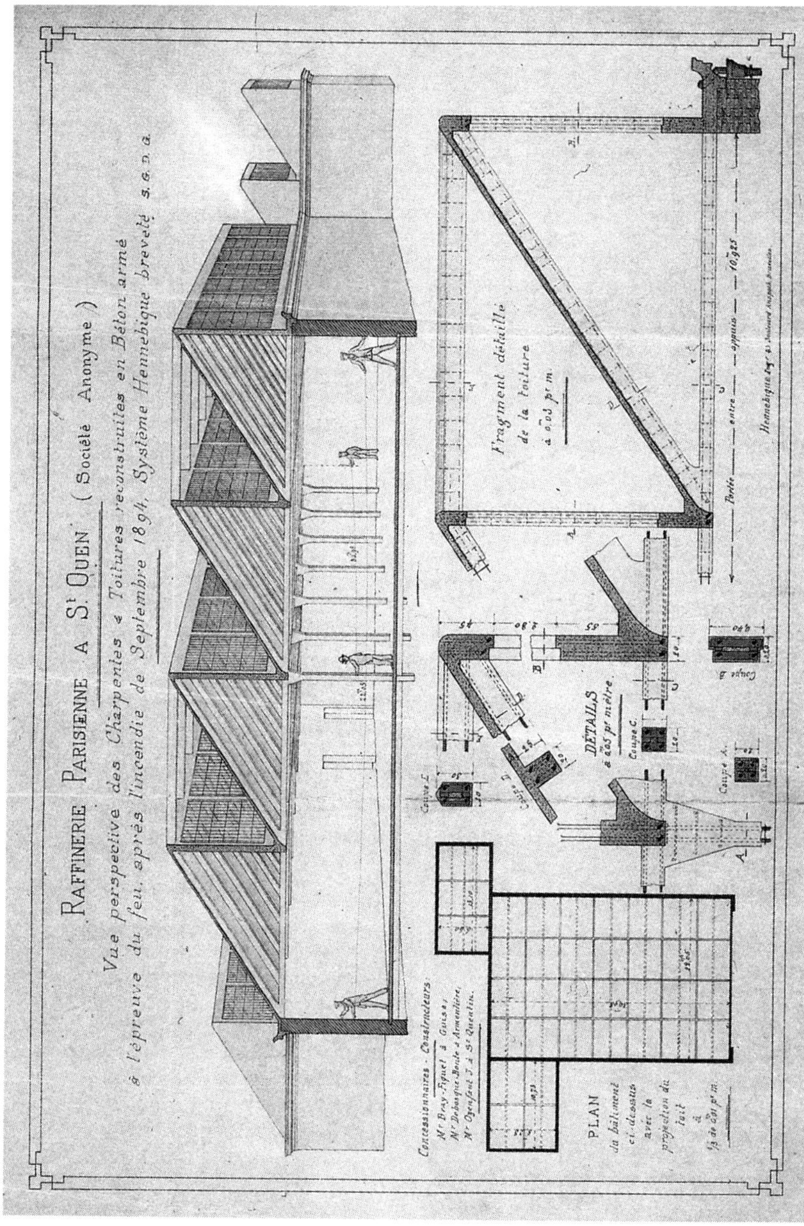

2. The Raffinerie Parisienne of Saint-Ouen, first important Hennebique builder's yard in Paris, 1894. Perspective section, plan and details of the sheds. Hennebique archives, IFA, Paris.

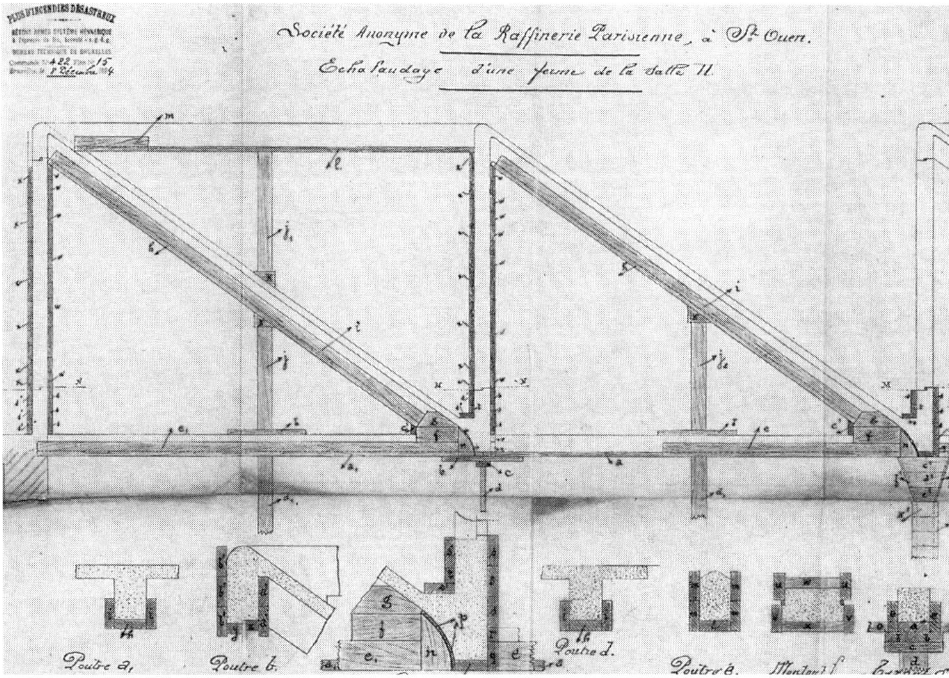

3. The Raffinerie Parisienne of Saint-Ouen. Drawing of the sheds. Hennebique archives, IFA, Paris.

technique, and seek to control its evolution. This is the reason why he should be present on the builder's yard. Hennebique obviously had understood this situation well, and indeed, his existence before 1900 depended precisely on his attitude toward the architects. The builder's yard was a place for demonstration and also a place for comparison, exchange of viewpoints and debates that interested all those involved in construction; an example of this is the series of tests made in February 1896 at Reims. Hennebique wrote Boesch, the regional representative of the Union Syndicale des Architectes, and invited him to attend those tests:

The engineers of the Ponts et Chaussées,[8] of the Eastern railways and of the city will attend it... I also hope to be able to welcome the officers of the Corps of Engineers and some of your colleagues who, like you, are not afraid of studying the new uses of this material, also in face of scientific discussion. I would be particularly delighted if you should accept my invitation to attend this initiative; the theories on reinforced concrete will also be discussed on this occasion. Your application, of a certain importance, of another system (Cottancin?) would allow you to contribute to the discussion with elements of great interest. I take the opportunity to send you, through Mr. Boileau, a brochure describing my system, published by *L'Architecture* magazine.[9]

The architects and the Hennebique bureau d'études

In November 1895 L.C. Boileau, the important secretary of the Société Centrale, dedicated a series of articles to the builder's work in *L'Architecture*. This consisted of a prestigious recognition and a very efficient form of advertisement aimed at the architects. This very detailed study consolidated an affinity that might be compared to the understanding between architect Anatole de Baudot and Paul Cottancin. The stand taken by Boileau around 1900 made him an important figure in the history of reinforced concrete. He is one of those architects who, through his initiatives in favor of the Hennebique system, the applications he had made and the manner in which he had been involved (in all senses), has made a

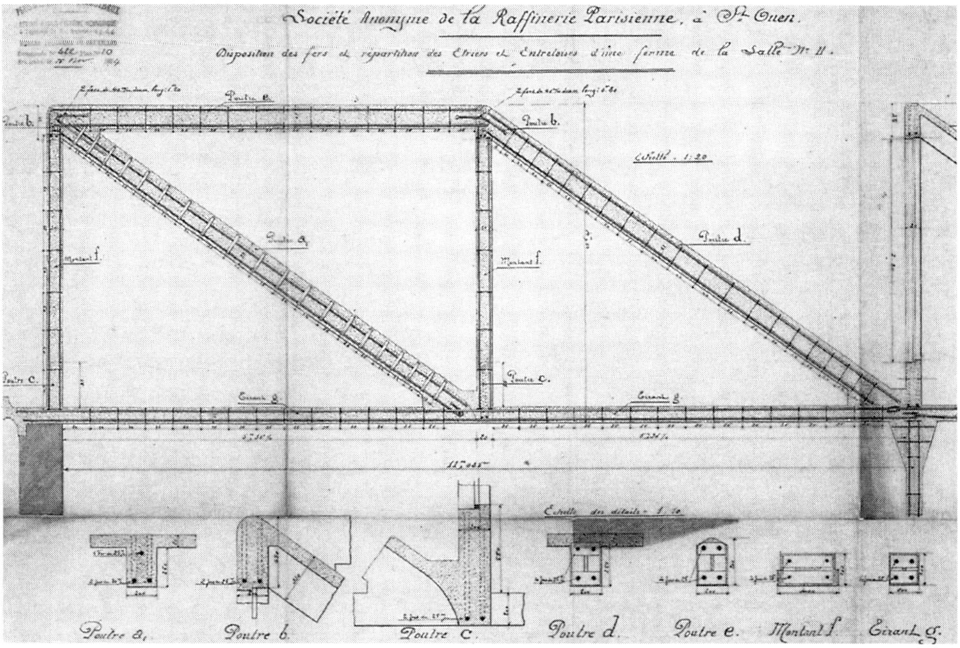

4. The Raffinerie Parisienne of Saint-Ouen. Final drawing for the arrangement of the irons and the stirrups.
Hennebique archives, IFA, Paris.

most efficient contribution to the development of the Hennebique building system in Paris. Until around 1914 he remained the one and only reference for all questions on architecture covered by *Le Béton Armé* magazine. In 1896 Boileau, on behalf of the Société Centrale, invited one of the most important concessionaries of the Maison Hennebique to present, before his colleagues assembled at the Congrès des Architectes Français, the new building methods applied by the entrepreneur. The technical explanations of the Hennebique process were completed by a description of the project of triumphal bridge in reinforced concrete by E. Bérard, which demonstrated the prospects of collaboration between the architect and the Hennebique offices. The latter's system was the only one represented in this meeting dedicated to reinforced cement. Why such elective affinity? Is the choice due to the constructive quality of the system or its producer? It is known that well-known companies prepared equally prestigious projects: suffice it to mention the bridge designed by Edmont Coignet for the Exposition of 1900.[10] The mastering of this new technology by the architect took place by means of the control that he exerted on those who were in a condition to apply this technology. With Hennebique, the architect's relationship and term of measurement were precisely this new entity, the *bureau d'études*.[11] Here he was helped by an interlocutor who was sufficiently elastic and powerful to serve the interests of the customer, who, before 1900, might support him. In the light of the circumstances it seems that it was Hennebique who entrusted Bérard with the task of cladding the reinforced concrete structure that had been studied and calculated by the engineer. Cladding and decorations were designed in copper and lead. The architect, combining new techniques which he coordinated, had at his disposal the collaboration of an entrepreneur specialized in the installation of metal cladding, for the works of plumbing and roofing. An inverted process: Bérard, who during this period participated in the competition for the building of the church of Saint-Jean de Montmartre, entrusted Hennebique with the study of his

5. Advertising for the Cottancin & C. company, diffused around 1895.

6. Advertising for the Hennebique company, diffused around 1895.

project. The architect, who was at first chosen, and at a later stage replaced by A. de Baudot, designed his building entirely in reinforced cement. In his solution he took the structural requirements of the material into consideration; however, in prescribing such material he did not recommend this or that system.[12] Delimiting his field of action, the architect entrusted Hennebique with the technical control of the project, which was thus suited to the specifications of a particular building system. In any case, the as yet obscure history of this famous church can be understood precisely by examining the competition between the Hennebique and the Cottancin systems (and perhaps others). At the conference of 1896, Boileau asserted the position of the architect with respect to the builder; is not this perhaps the sense of the publication, on *L'Architecture*, of an epistolary challenge between Cottancin and Hennebique, the entrepreneurs assigned to their role as "rival tradesmen"?[13]

Systems and strategies

In order to understand what separated and opposed the two builders, a single glance at their advertisements in the press suffices. Hennebique, rather soberly, presented the evidence to the reader: the reality of a network of engineers and entrepreneurs, and its capacity of intervention. Cottancin, on the other hand, explained, with an excessively complicated message, the existence of his system, "the oldest", opposing it to all the other competing processes, to a point where such contrast became the essential character of his principle.

"The only systems featuring flexibility in the architectural forms", the processes used by Paul Cottancin (and by his concessionaries?) - metal framework "without joints and with continuous nets", spike-buttresses, bricks and reinforced ceramics - have been studied, applied and supported at a very early stage by A. de Baudot, and through him, by the Union Syndicale des Architectes Français. As a result of a research carried out by the Master, the Cottancin system became the subject of debate among the members of the Union. In 1893, beyond the "tangible and real" experiences to which Baudot refers, some seem to expect, on the other hand, "detailed and particularly scientific" explanations from the inventor of the process, so that they can express an opinion.[14] During a meeting concerning reinforced cement in December 1895 Cottancin was subjected to cross-examination and provoked, by the uncertainty of his explanations, "a general and very confused discussion, where it seems", as it is put in the minutes, "that one part of the assembly requested a precise formula in order to be able to calculate Mr. Cottancin's slabs". The engineer's stand became even more uncomfortable as, during the same meeting, a debate with a representative of the Maison Hennebique, called to explain the principle of the competing process, had been planned. Since they were based on different criteria the systems were actually not comparable; a difference which was clearly in favor of the Union. Hennebique's representative focused on the rigor and absolute precision of the Hennebique's calculations (who at that time concentrated on transparency and dedicated himself to publishing formulas and coefficients), stressing the evident weak points of the competition.

L.C. Boileau referred to this argument on *L'Architecture* in order to make a distinction and justify his preference of the Hennebique system which, as opposed to that of Cottancin, in his opinion seemed to feature "the so-to-say mathematical certainties which modern architects request" with respect to the metal structure.[15] In 1896-97 Boileau built "the stables and the warehouses" of the Bon Marché. The magnitude of this project (12,000 square meters of floor space distributed on an area of 5,000 square meters), were an opportunity for the architect to experiment the Hennebique system with precision. This building, today no longer existing, was one of the "key features" of the second annual conference of the Maison Hennebique in 1898. It was the most important structure ever realized in reinforced concrete in the capital. The entire structure, beams-pillars-floors, of this modular building was calculated so as to allow for the addition of further storeys. By basing the entire architectural solution chosen on the constructive logic of the material, Boileau attempted to expose the independent reinforced-concrete skeleton, through an efficient play of horizontal and vertical lines that formed the main decoration of this functional, massive and sober building. The architect fully exploited the elasticity of a building material which could be adapted, in an economical way, to the most complex programs. All his concepts related to the subject of reinforced cement, which had here been extensively used, were represented in this work: "Cement everywhere, always cement... not one piece of stone has entered this construction. And yet", as Boileau put it, "this is a reasoned, and not exaggerated, use of reinforced cement".[16] The Bon Marché's builder's yard provided Hennebique with the opportunity to organize, with the collaboration of the Ponts et Chaussées, an impressive program of tests on constructions in reinforced concrete. The inventor proposed to the authorized scientific representatives, behind whom he ably took shelter, his formulas "imbued with empiricism". The inventor of "this scientific feast" (the Congrès du Béton de Ciment Armé of 1898), Hennebique became the manager of an ideal yard for the construction of sheer theory. The experiences realized could refer to another, considerably more important, builder's yard: the Exposition Universelle.

Boileau-Hennebique: a symbolic project

In 1899, on the occasion of the Third Conference of the company's agents and concessionaries, Boileau, "the godfather of the Hennebique system", was "acclaimed" by these and entrusted with the task of realizing a pavilion

7. E. Bérard, project for triumphal bridge to be built above the Seine in Paris, 1896.
From: *L'Architecture*, 26 October 1896.
Photo Bibliothèque Forney, Paris.

8 HENNEBIQUE AND BUILDING

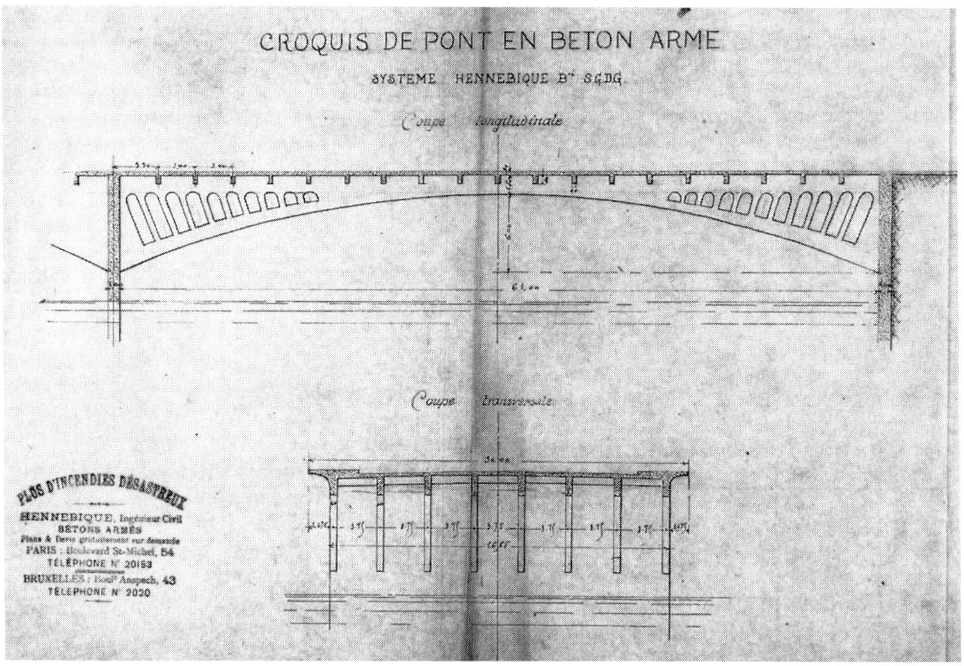

8. Project for triumphal bridge above the Seine, Paris. The structure in reinforced concrete has been studied and calculated by Hennebique. Hennebique archives, IFA, Paris.

destined to represent Hennebique Reinforced Concrete at the Exposition of 1900. The "terrace-gallery in reinforced cement" had to be an occasion to demonstrate interesting applications of the system: foundations on pillars, projections, floors with high loading capacity, hammering of piles, etc.

This was a delicate task for Boileau, who dedicated himself to it completely: not only was it necessary to invent characteristic forms for the material, but, even more so, "to represent in the most exact way possible the inventor's system", by which one had to design, and then to conceive the most appropriate decoration for this singular construction. The "Boileau-Hennebique" project was a clear demonstration of the "mutual support relation between experts, workers and artists"[17] recommended by Boileau. It also symbolized the necessary unity between architects and "expert builders such as our engineers of the Ponts et Chaussées".[18] By this experiment Boileau had to demonstrate the full technical and decorative potential of reinforced concrete to a public that was not expert on architectural matters. The style adopted (simple forms "that are expressed by rational construction") featured all the refinedness of the elements forming this structure, which could be placed between the density of masonry and the slenderness of metal buildings. Every detail expressed a reflection on the system's constructional possibilities and on the solutions that coud be derived from it; the architect thus sought to make the arrangement of the reinforcement typical to the Hennebique system formally understandable, and underscored on some beams the disposition of the characteristic elements: bent bars, stirrups, etc.

Boileau published this project on *L'Architecture* in 1909, in a particular moment of the debate on reinforced concrete: "While the questions verging on the 'new material' are still subject of discussion, the architects might find it rather interesting to discover a study concerning it, conceived from a quite artistic point of view".[19]

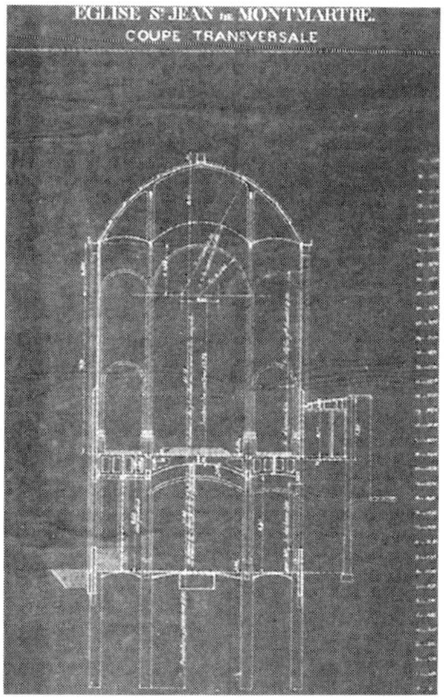

9. E. Bérard, project for the Saint-Jean de Montmartre church, to be built entirely in reinforced concrete, 1896. Transversal section.
Hennebique archives, IFA, Paris.

10. E. Bérard, project for the Saint-Jean de Montmartre church. Longitudinal section.
Hennebique archives, IFA, Paris.

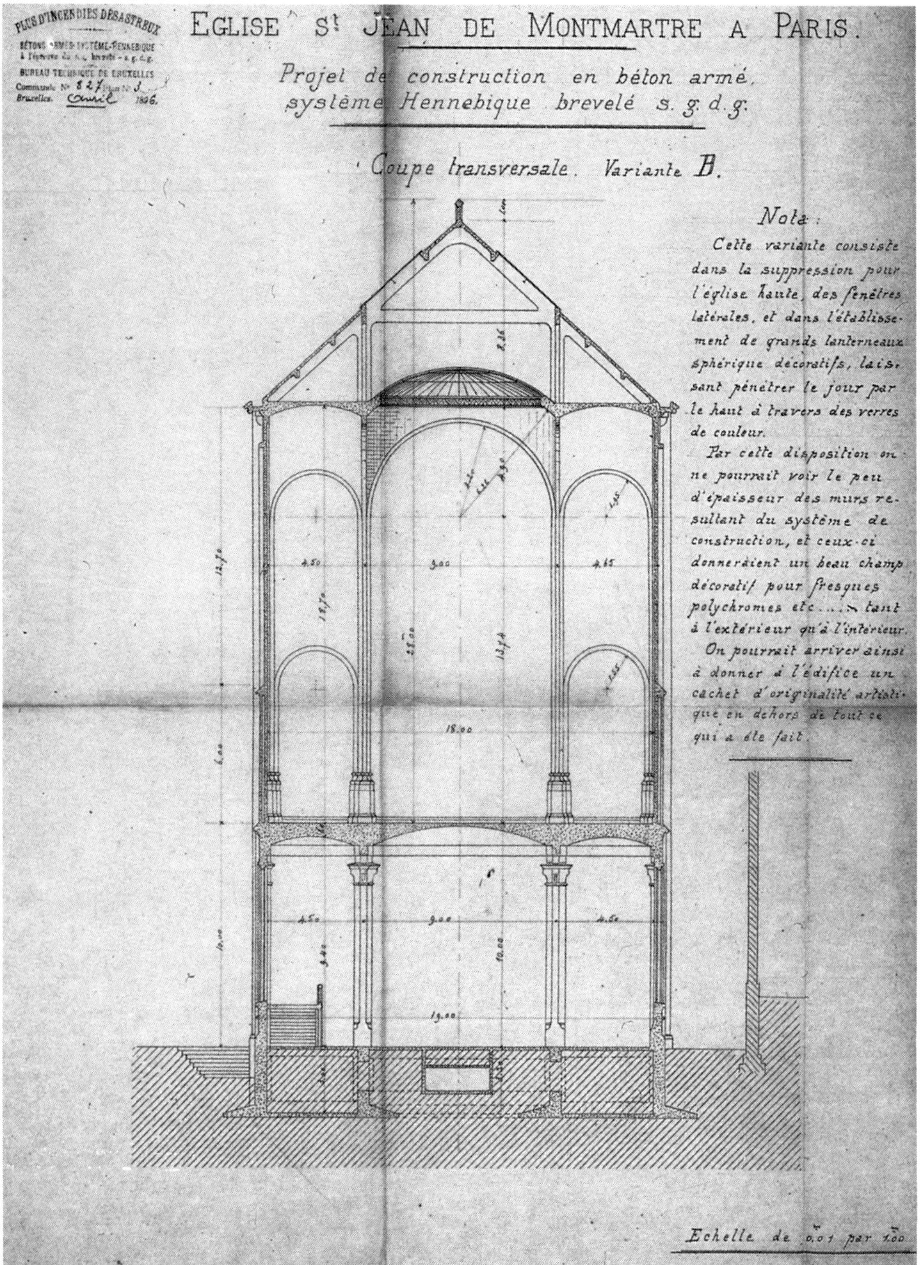

11. Study by Hennebique for E. Bérard's project for the Saint-Jean de Montmartre church. Transversal section.
Hennebique archives, IFA, Paris

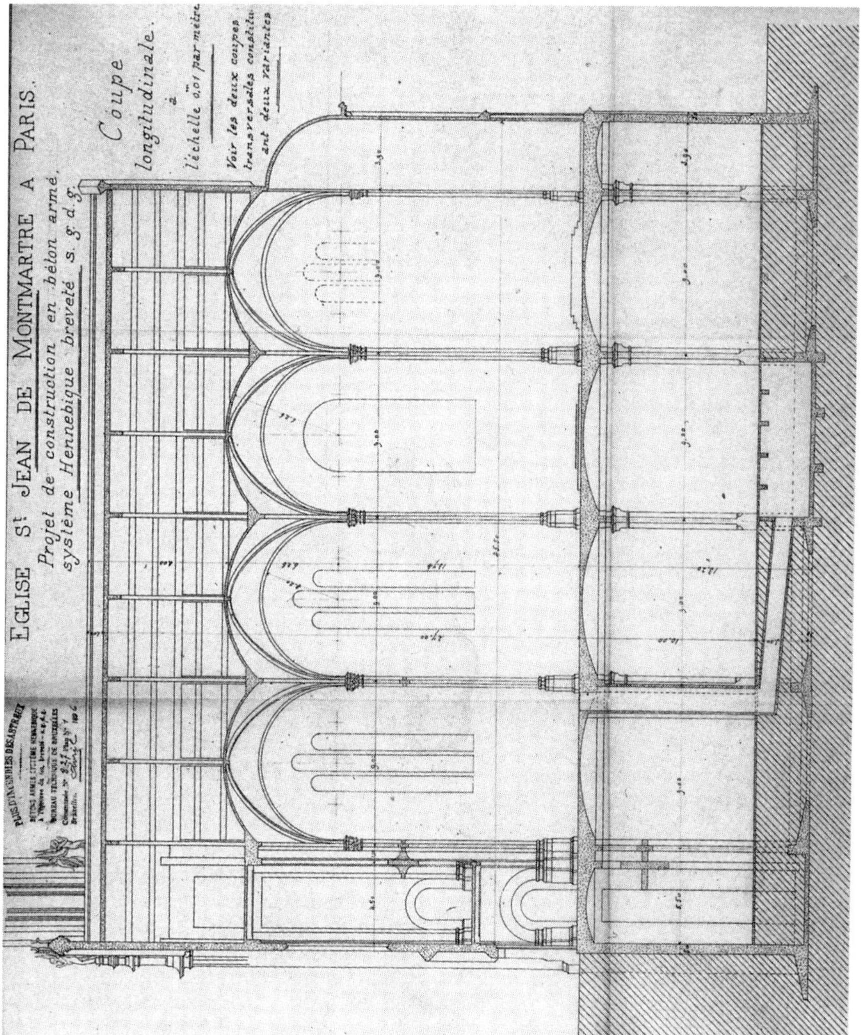

12. Hennebique study for the Saint-Jean de Montmartre church. Longitudinal section.
Hennebique archives, IFA, Paris.

"Sincéristes" and eclectics: speeches and constructional realities

The interest that Boileau showed - regarding the Hennebique reinforced concrete, being in his opinion a prototype of the different systems in use - in reinforced concrete, was indicative of the architects' attempt to master, as early as before 1900, this building novelty and its possibilities, or rather the revolutions which it introduced in the art of building. The challenge consisted for them in knowing all its aspects, being able to define the range of its applications, mastering economic and aesthetic questions: this was the meaning of Boileau's message. A speech in which all the arguments inherent to a debate on the subject of reinforced concrete were, through the experience acquired, under control; this enabled Boileau to face, in particular, the dogmatic stand of the "sincériste" A. de Baudot. The struggle of the latter, as Jean-Pierre Epron pointed out,

> is aimed at the promotion of a new way of considering the structure of a building, which he justifies in relation to the work by Viollet-Le-Duc. He proves the validity of his struggle and completes it by advocating the use of reinforced cement for the building of the church of Saint-Jean de Montmartre... where he undergoes the difficult experience of the architect who claims to formulate the technique rather than compose its elements.[20]

In fact, the hardships encountered during the construction of the church (certainly the building that took the longest time in the entire history of the new building material) becomes utterly evident if compared to the "industrial" function of the Hennebique system. Indeed, instead of limiting its activities to the architectural sector, as the P. Cottancin & Cie seemed to do, this company sought a universal application of the material, from the simple floor of a residential house to works of pure civil engineering. For that matter, the rather bitter remarks by P. Cottancin, published on *L'Union des Architectes*, concerning the speeches held by the students of A. de Baudot and concerning his church, serve to enlighten the divergences of opinion that accompanied the collaboration between the Master and the private "builder of cisterns" of his yard, and therefore of his system.[21] A. de Baudot thus recovered the constructive combinations of the entrepreneur in favor of a discussion on the material.

The completion of the church of Saint-Jean de Montmartre, its publication in the magazines, the publication, finally, of the booklet *L'Architecture et le Ciment Armé*, in August 1905, by A. de Baudot,[22] rekindled the discussion on the new building method and its use in architecture. L.C. Boileau, in parallel with the "conversations", to which he dedicated himself on *L'Architecture*, with the purpose of demolishing, point by point, the arguments supported by de Baudot, presented for the criticism of those at whom his message was aimed (and to others) a project for a church in reinforced concrete which was simply the counter-project of that by A. de Baudot. In this manifesto of eclecticism (on a formal level the architect refers to "modern master" Guadet), Boileau again puts forth his ideas on the use of reinforced cement, a simple and useful building method which, when combined with other materials, allows to perfect interior layouts that the builders of the past would not have been able to realize.

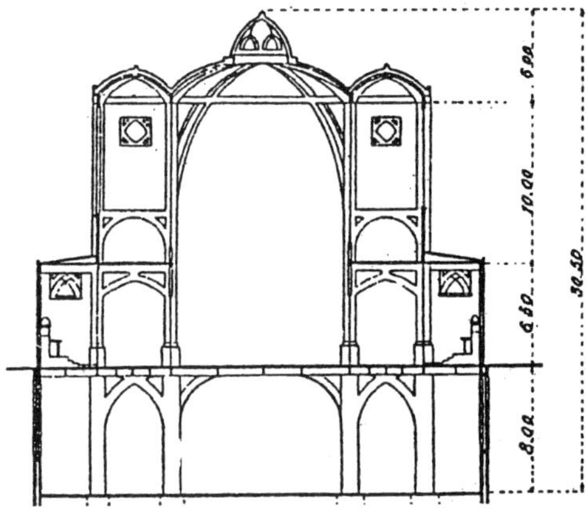

13. Anatole de Baudot, Saint-Jean de Montmartre church. Transversal section.
From: *La Construction Moderne*, 22 April 1905.

These two contrasting conceptions somehow represent the extremes of a debate, centered on precisely the mastering of the material by the architects, and which perhaps in 1906 it had no longer any reason to exist. The use (the correct use) of reinforced concrete in architecture is, in fact, determined by the presence of different systems by means of which

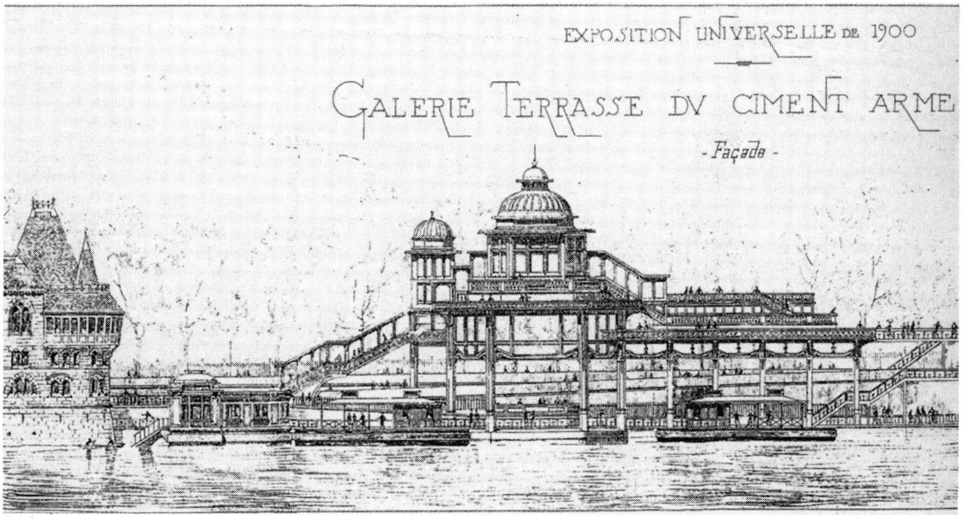

14

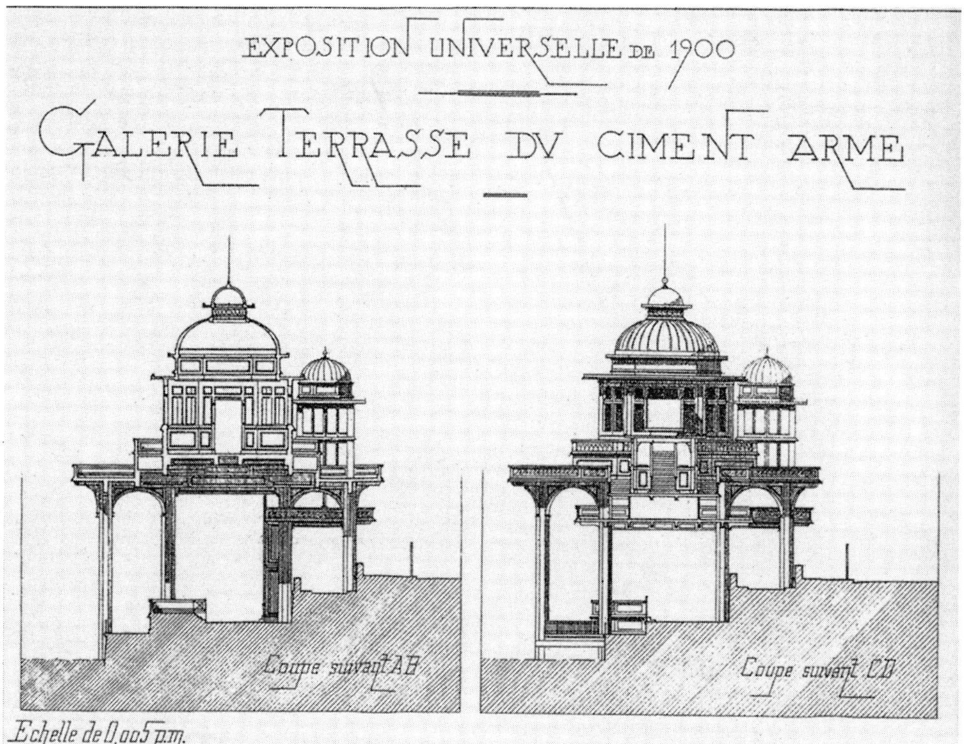

15

14, 15. Boileau and Hennebique, project for "Galerie terrasse du ciment armé" for the Exposition Universelle of Paris in 1900. Perspective, sections. Hennebique archives, IFA, Paris.

the architect conceives the material. Boileau even considered the possibility of combining these different systems (once they had become a prerogative of the public sector) and therefore of inventing some special ones.[23] It was, then, the very concept of systems which was questioned as early as in 1900.

From "system" to material

Evidently, the Exposition Universelle, an officially organized occasion for comparison between the main systems used, was a meeting point for the builders, in which all their efforts, and all the strategies they adopted in order to survive, were presented. However, even if this was the most gratifying certificate for the builders and for their systems, the Exposition of 1900 heralded the end of their dominion over a material that was recovered from on high by the official authorities. On this immense yard (of building as well as of demolition), an opportunity, as Paul Planat defined it, "to make experiences on a large scale that go beyond the resources of an individual initiative", the works, which before the Exposition were subjected to extremely severe tests, acquired their status as buildings in reinforced concrete only after they had been demolished, when they became the occasion for the experiences required by the work of the Reinforced Concrete Commission, which, as we know, served as a base for the elaboration of the Ministerial Circular of October 20, 1906. The ministerial instructions, described in all the specifications of the administration's duties, threatened the very nature of the work of specialists such as Hennebique. Again questioning the existence of their *bureau d'études*, they finally made it possible for anyone to "produce reinforced concrete". This obviously explains the violent criticism that the French circular was subjected to. Its interpretations appear to have radicalized a debate between engineers and specialists, between science and empiricism, focused on the reinforced concrete material. As of 1900 Hennebique had to face an intensified and more aggressive competition, resulting from the several improvements that reinforced concrete had undergone. Thus, the struggle moved to a context far from that of architecture, to the large foundations, maritime constructions and civil engineering yards in general; in any case quite far from the discussions between architects, which *Le Béton Armé* at that time obligingly reported.

This uncertain moment of the history of reinforced concrete (to be precise, the year 1905) was to be defined by Paul Jamot, about twenty years later, "a decisive date for the

16. Girault (architect), Grouselle & C. (builder), the Petit Palais des Beaux-Arts at the Exposition Universelle of Paris in 1900. Photo of the builder's yard during the building of the floor slabs in reinforced concrete. Hennebique archives, IFA, Paris.

17. Thomas (architect), Dumesnil (builder), the stairs in reinforced concrete in the Grand Palais des Beaux-Arts at the Exposition Universelle of Paris in 1900. Photo taken after falsework has been removed. Hennebique archives, IFA, Paris.

architecture of reinforced concrete";[24] a capital date for another history.

1. Paul Gallotti, "1892-1902, dix ans de Béton Armé", in *Le Béton Armé*, No. 55, December 1902, p. 111.
2. With respect to this, see the work of Cyrille Simonnet: *De la pierre factice au béton armé*, report of research BRA-MELT, Grenoble-Paris 1989.
3. *Le Ciment, son emploi et ses applications nouvelles, en France et à l'étranger*, official organ of the Chambre Syndicale des Fabricants de Ciment Portland de France. "*Le Ciment* magazine is sent to the Engineers of the Ponts et Chaussées, the railways, the Plumbing Services, those responsibles of the municipal road networks, the Architects of the State and of the City, the Plumbing Engineers, etc."
4. "It is true, on the other hand, that the differences between the building systems will tend to disappear, and that, subsequently, the different entrepreneurs will have greater difficulty in imposing the superiority of their system. This is not negative for anyone; ...neither for the builders nor for the manufacturers of concrete, as a result of the considerable development of the number of uses", *Le Ciment*, No. 11, November 25th, 1897, pp. 354-355.
5. The Hennebique archives were donated to the Centre d'Archives d'Architecture du XXème Siècle of IFA by CNAM in July 1989. The fund consists of about hundred thousand files, about 7000 photos and the complete collection of *Le Béton Armé* magazine. The pluriennal improvement program of the Hennebique archives has been awarded the financial support of the Getty Foundation.
6. *Le Ciment*, March 1898, No. 3, p. 36.
7. Letter from Hennebique to architect Jules Prey, March 29th, 1895. The document has not yet been included in the inventory.
8. Among which, as Estemble, agent of the Reims technical office, points out to Hennebique: Mr. Maillet Jr. is "... very interested in learning about your system. He is a friend of Cottancin but he hardly trusts the processes of the latter". Letter from Estemble to Hennebique, February 21st, 1896.
9. Letter from Hennebique to architect Boesch, February 25th, 1896. The document has not yet been included in the inventory.
10. "... at the Exposition of 1900, the project for a bridge in reinforced concrete by Mr. E. Coignet had been the subject of much attention, not only because of its lightness, but also for the originality of the processes suggested for its building. Mr. E. Coignet should certainly be flattered since the project which he had drawn has been followed point by point as for rise, capacity, width and system of construction; this is nevertheless hardly a consolation." *Le Ciment*, No. 3, March 1897, p. 80, "Informations" column, comment on the adjudication of the foundations for the building of the Alexandre III bridge.
11. As specified by L.C. Boileau: "We say 'Mr. Hennebique's calculations', since, while it is true that the entrepreneur responsible for our colleagues... is Mr. Dumesnil, the concessionary executor which we mentioned, it should be noted that, according to the treaty ratified between him and the promoter of the new method, it is always the latter who studies the iron and concrete combinations and who prepares the calculations". L.C. Boileau, "Le ciment armé, nouvelle méthode d'application", in *L'Architecture*, No. 45, November 1895, p. 381.
12. Some details as the grooved profiles which constitute the upper floor of the crypt remind, however, of the spike-buttresses used by Cottancin.
13. *L'Architecture*, No. 40, October 3rd, 1896, pp. 299-300 and n. 42, October 17th, 1896, p. 313. "The architects cannot but benefit from these exchanges of explanations", the editors specify, "provided that, to be precise, the latter are presented as little as possible as mere advertisments".
14. The daringness of the intervention is equal only to the irritation of the Master: "What is Mr. Vaillant asking, then? Elements or formulas so that he can himself calculate these floors? Well, in this sense Mr. Cottancin, who is the inventor of the

18. Experimental builder's yard of Bon Marché, 1897-98. Hennebique archives, IFA, Paris.

19. Experimental builder's yard of Bon Marché. Period photo. Hennebique archives, IFA, Paris.

20. Experimental span for the quai de Billy at the experimental builder's yard of Bon Marché. Test with static load. Hennebique archives, IFA, Paris.

21. Experimental builder's yard of Bon Marché. Period photo. Hennebique archives, IFA, Paris.

22. Experimental span for the quai de Billy at the experimental builder's yard of Bon Marché. Test with mobile load. Hennebique archives, IFA, Paris.

process for which he has obtained a patent, confesses that he has also found a method of calculation, but that he desires to keep it secret, something which he is entirely entitled to... what other information could Mr. Vaillant need, apart from the experience acquired, the results of which are available to anyone?" Minutes of the association's meeting of February 18th, 1893. The document has not yet been included in the inventory.

15. "Le Ciment armé, nouvelle méthode d'application", in *L'Architecture*, No. 44, November 2nd, 1895, pp. 369-372.

16. Gustave Olive, "Ecuries et manutentions des magasins du 'Bon Marché' à Paris", in *L'Architecture*, No. 16, April 16th, 1898, p. 139.

17. "Troisième congrès du Béton de Ciment Armé", in *Le Béton Armé*, No. 9, February 1899, p. 7; "Toast de M. Boileau, architecte".

18. "Congrès des architectes français de 1896", in *L'Architecture*, No. 43, October 24th, 1896, p. 321.

19. L.C. Boileau, "Un projet de terrasses en ciment armé", in *L'Architecture*, No. 2, January 13th, 1906, p. 17.

20. Jean-Pierre Epron: *Comprendre l'éclectisme*, research report, BRA-MELATT, Paris-Nancy, January 1991, pp. 64-65.

21. "Essentially, Mr. A. de Baudot has in 1893 acknowledged that Mr. Cottancin has finally had the idea, in order to increase at the same time the size and the strength of his slabs while keeping the same thickness, to reinforce them with projecting ribbing, while Mr. Chaine in 1905 notes that Mr. A. de Baudot is the exceptionally discerning inventor of spikes, that is, of the P. Cottancin ribbings." Letter from P. Cottancin in relation to the publication of H. Chaine's lectures on the church of Saint-Jean de Montmartre. *La Construction Pratique-L'Union des Architects*, No. 25, August 1905, p. 53. This lecture, held at the Union Syndicale on December 17th, 1904, was published in *La Construction Pratique*, as of January 1905. The monthly magazine edited by P. Gallotti, extremely favorable to Hennebique, joins on this occasion the organ of the Union Syndicale, *L'Union des Architectes*. This strategic combination (before Gallotti left the job to critic P. Forthuny) constituted an interesting journalistic bridge between the Hennebique and de Baudot poles.

22. "*L'Architecture et le Ciment Armé* would be interesting also for laymen" it is pointed out in *L'Union des Architectes*. "This gives us all the more reason to believe that the professionals will want to know a detailed report on Mr. A. de Baudot's considerations and will place, among their best bookshelf documents, this book which constitutes, to tell the truth, an historical moment of architecture." *L'Union des Architectes*, No. 8, September 1st, 1905, p. 64.

23. "There are ten or twelve systems of pure building of floors featuring good or less good characteristics and a more or less perfect logic. I am aware that these are patented and that it is not easy to invite builders using different processes to collaborate in the same builder's yard... But the day will come in which the systems will become public. That day, the architects shall not only be able to apply this or that method, but even to invent special ones." L.C. Boileau, "Le ciment armé et l'art de l'architecture", in *L'Architecture*, No. 51, December 23rd, 1905, p. 473.

24. Paul Jamot, "1905, date décisive pour l'architecture du béton armé", in *L'Art Vivant*, No. 41, September 1st, 1926, pp. 641-644.

8

Agents of change: Hennebique, Mouchel and ferro-concrete in Britain, 1897–1908

Patricia Cusack

This paper discusses some of the circumstances of the introduction and establishment of reinforced-concrete framing in Britain [1]. Reinforced concrete existed by the early 1900s as a collection of patented, and a few unpatented, 'systems', with varying dispositions of reinforcement; some systems were designed only for specific structural elements, such as floors or pipes, while others were adapted, not always appropriately, to entire building frames. These systems were commercially exploited by their patentees, but technical details were guarded from public knowledge and protected by vigorous litigation. In 1904, there were over 50 such systems [2]. Among the most flexible and the most widely used was François Hennebique's system, developed in Belgium and France and extended world-wide through a specialist commercial and technical organisation, which was in turn imitated by other major system specialists such as Coignet [3].

The introduction of reinforced-concrete framıng in Britain at the turn of the century was a result of Hennebique's business policy of international expansion. Existing commercial exchanges between Nantes, where Hennebique had an agent, and Swansea, may have facilitated the contacts which led to the commissioning and erection in 1897 of the first fully framed and entirely reinforced-concrete building in Britain, Weaver & Co.'s provender mill in Swansea [4]. During the construction of the mill Hennebique selected a General Agent for his system in Britain, L. G. Mouchel, whose work, until his death in 1908, effectively established reinforced concrete in Britain and especially its use for framed buildings, albeit as a private, commercial product.

This commercial aspect, also the existence of competing methods and self-interest, prompted architects and others to start investigations into reinforced concrete, to broaden professional interests in it and to end the specialists' monopoly [5].

Reinforced Concrete at the Turn of the Century

By the mid-1890s, a number of companies and individuals in Britain were employing reinforced concrete (some since at least the 1860s), but primarily for floors, roofs and beams and probably not for framed construction. Some of these companies, including W. B. Wilkinson & Co., the pioneer patentees of reinforced concrete (1854–55),

* Based on a paper delivered to the Fifth Annual Seminar of the Construction History Society in Slough, England, in September 1986, 'Innovation in Construction'.

62 *Hennebique, Mouchel and ferro-concrete*

continued to design and build in reinforced concrete into the 1900s, but, again, not using complete building frames.

There have been many claims for early designs for reinforced concrete during the nineteenth century, of which only a few are justified; on the other hand, some experiments with reinforcement in Britain have remained little known, such as those by Stuart's Granolithic Co. of Edinburgh in the early 1890s. While W. B. Wilkinson's Newcastle patent of the 1850s remains the earliest ascertained invention of reinforced concrete in the UK, and perhaps anywhere, T. F. Tyerman in Middlesex may have been studying reinforced concrete at the same time, and he filed a patent for bonding metal in concrete before this was included in Wilkinson's own patent. Subsequent inventions in Britain did not improve substantially on Wilkinson's until the early 1890s, with the addition of shear reinforcement in beams, patented in Britain by François Hennebique in 1892.

Throughout the latter nineteenth century in Britain reinforced concrete was not fully distinguished from other combinations of concrete and metal, being advertised under vague descriptions such as F. G. Edwards's 'Fireproof Steel and Concrete' [6]. However, there is some indication that it was becoming fairly widely acknowledged by the time that Hennebique's system was introduced, even if accompanied by considerable scepticism; F. T. Reade for instance in a discussion at the Royal Institute of British Architects in 1907 questioned the sanity of any engineer using reinforced concrete for girders or columns because of the heterogeneity of concrete and steel [7].

The first documented case of the use of Hennebique's reinforced-concrete system in the UK was Weaver's Mill, Swansea, in 1897, the year in which he took out his main British patent. Several other systems applicable to framed-building construction were introduced in Britain during the early 1900s, although the first framed building not in Hennebique's system was probably not until 1905 (using Coignet's system). Systems by Coignet, Considère and Visintini were patented by 1904; by 1907 two further framing systems had been introduced and applied in Britain, one American (Kahn's) and one British (Wells's). Visintini's system acquired a London representative but failed a test by the British Fire Prevention Committee and no framed buildings using it in Britain in this period are known. Documented examples of 'Coignet' framed buildings in Britain prior to 1908 amount to about a dozen (compared to 130 by Mouchel). E. P. Wells, described in 1908 as "one of the few British designers in reinforced concrete" [8], varied his method with circumstances to the extent that he was sometimes said not to have a 'system', but a reinforced-concrete chocolate factory in Portobello, Scotland, was said in 1908 to be "an excellent example of Mr E. P. Wells's system of reinforced concrete construction" [9]. The first major use of Kahn's system was for the headquarters of the Young Men's Christian Association in Manchester (1908), while the first Considère building was not until 1909. Nonetheless, Considère's system had already become a focus for critics of Hennebique's system in Britain, provoking the disparaging comment from Mouchel in 1904 that Considère's method, which employed spiral binding in columns, must "compel us to rob . . . the spring mattress industry of its skilled hands" [10].

Hennebique and Mouchel

Hennebique's contribution to both the development and the propagation of reinforced concrete was considerable. He was the first to make systematic provision for shear reinforcement in beams and to use reinforced concrete in monolithic frameworks. His contribution to the marketing of reinforced concrete was spectacular and in Britain as elsewhere in the early 1900s his system predominated over others, due as much to the organisation of his company as to the intrinsic merits of his system. After superintending his own early works in reinforced concrete, Hennebique established commercial arrangements to ensure the controlled employment of his system, with special training courses for his engineers and selected contractors 'licensed' to use his system in return for royalties.

Between 1892 and 1897, when Hennebique introduced reinforced concrete in Britain, his commercial organisation for reinforced concrete expanded from one technical office, two engineers/draughtsmen (and no licensed contractors yet) to 17 offices, employing 56 engineers and draughtsmen, with 55 licensed contractors. By 1909, there were 62 offices, located in Europe (43), the USA (12), Asia (four) and Africa (three), and Hennebique's system was by far the most popular reinforced-concrete method. By mid-1911 there was a total of nearly 24,000 works executed in Hennebique's system, of which 1073 were in the UK [11].

The uncertainty of knowledge and practice in reinforced-concrete work at this time makes the successful erection of so many large reinforced concrete structures in Hennebique's system remarkable: it seems that Hennebique, and Mouchel too, proceeded empirically, employing a large excess of strength [12] and testing every structure as built.

Among Hennebique's successful early framed structures were a factory in St Michel, Switzerland, and another in Cairo (1895), as well as works in northern France. The commission for Weaver's Mill, Swansea, was obtained for Hennebique in 1897 by his agent in Nantes. Particular care was taken in the design and construction of this mill, which was seen as a preliminary advertisement for the use of Hennebique's system in Britain. The working drawings were executed in Hennebique's office in Nantes; the sand and steel were transported from Nantes (although this was quite an economical matter since coal barges leaving Swansea required ballast on the return journey), and trained French workmen were brought over to help with the construction.

One of the factors in Hennebique's commercial success was his careful choice of agents to head his regional and international offices. As a matter of policy, he accorded the agents a large degree of autonomy. Following initial training and practical support, agents like Mouchel became largely independent of Hennebique, though continuing to pay royalties for the use of the system. Mouchel's organisation, established from 1897, was modelled on Hennebique's in both its structure and its expansionist, entrepreneurial philosophy. Regional engineers were trained, like Mouchel himself, in Hennebique's offices in Brussels or Paris. As Hennebique's agent between 1897–98 and 1908, Mouchel was the central figure, both in the application of reinforced concrete in Britain and in its use for framed buildings.

Mouchel had moved from his home town of Cherbourg to Briton Ferry in South Wales many years previously to work as a mining engineer. At this time, there was a lively coal trade between France and South Wales and Mouchel initiated a number of business enterprises mostly connected with this trade. He became a director of various companies, French Consular Agent for several South Wales ports and an Advisor on Foreign Trade for France. When Mouchel became Hennebique's agent, he was a prospering and well-known businessman in Briton Ferry and district.

Thus there does not appear to be evidence for later claims that Mouchel had been either "Hennebique's partner in France" (Morton Shand, 1932) [13] or one of

Hennebique's "senior engineers" prior to 1895 (Collins, 1959), or that "Hennebique sent Louis-Gustave Mouchel... to Britain in 1895, when he landed at Briton Ferry in South Wales, and immediately proceeded to spread the gospel by erecting Messrs. Weaver's Granary and Flour Mill at Swansea" (Collins, 1959) [14].

Certainly, Mouchel became closely involved in the scheme for Weaver's Mill, since he accompanied one of the directors of Weaver & Co. to France in 1897 to see examples of ferro-concrete construction before the contract for the mill was signed. According to one of Mouchel's former engineers, C. Roch, in 1958, it was while Hennebique was arranging the transport of materials for Weaver's Mill that Hennebique and Mouchel first met, through Mouchel's office as French Vice-Consul [15]. Mouchel's interest in the possibilities of ferro-concrete may have predated Weaver's Mill, however. Hennebique's grandson, Robert Flament-Hennebique, believed that Mouchel himself was Hennebique's first client in Britain [16], and an early short biography (1912) of Mouchel refers to (unspecified) ferro-concrete work, perhaps an extension to his business premises at Briton Ferry [17].

Mouchel's business flair and contacts, as well as his engineering background, made him eligible for Hennebique's agency. He was also a man of personal charm and organisational skill. Mouchel had become Hennebique's 'General Agent' for the UK by 1898. Although Mouchel maintained his commercial offices in Briton Ferry, at least initially, he decided at about this time to make the introduction of Hennebique's system into Britain the "chief object of his professional life" [18].

At the start of his agency, Mouchel coined the term 'ferro-concrete' to describe Hennebique's system in English and set about establishing technical offices for the structural design of reinforced concrete works, although the agreed terms were that Hennebique would provide all working drawings initially. Hennebique helped Mouchel in other ways. Mouchel's engineers either came from Hennebique's offices or attended training courses with Hennebique. Mouchel himself underwent several courses of training at Hennebique's Paris office from about 1899 and subsequently drawings were executed usually by Mouchel, or by other French engineers in Britain trained by Hennebique. Mouchel himself began to patent reinforced-concrete designs which extended Hennebique's system; half of these were for piles, perhaps following a personal interest in underground and underwater construction. Roch in 1958 said that he, Roch, drew up the first projects and working drawings in Britain for ferro-concrete works, which included silos at Birkenhead and a warehouse at Brentford (both of 1899).

Encouraged by Roch, Mouchel moved to London in March 1900, initially to a new but bare apartment lacking either WC or lift (an oversight by the architect), but then to 38 Victoria Street, Westminster, which remained the head office. By 1902, Mouchel had opened offices in Manchester and Southampton and had licensed contractors in a dozen cities in England and Wales. By the end of 1904, further offices had been established in Birmingham, Glasgow and Newcastle-upon-Tyne. British engineers and trainees joined Mouchel's company, including J. S. E. de Vesian who with one of his French engineers, T. J. Gueritte, became Mouchel's partners in 1907.

As well as establishing offices, Mouchel tried to create wider professional and public interest in Hennebique's system by inviting groups of architects, engineers and journalists to observe tests of works, by publishing brochures and by lecturing to professional bodies.

Like Hennebique, Mouchel granted licences to building contractors wishing to employ Hennebique's system. Under the licence, Mouchel received royalties of 12 per cent in return for the provision of working drawings; it was also stipulated that the licensee should not use any system except Hennebique's and should inform the licensor of any infringement of Hennebique's patents. Mouchel licensed building contractors both in the UK and, as a British colony, in Australia.

It was Hennebique's policy to select the most competent contractors as licensees and to ensure they were trained in ferro-concrete construction. However, Mouchel did not limit his licences to established contractors, although he is usually thought to have done so. Nor did he believe that ferro-concrete construction required much training, since workmen of average intelligence could be trained "in a very few days—I was going to say hours... no skill and no mental effort is required" [19]. The brief training given to Mouchel's contractors was sometimes complemented by the use of specially trained French workers, possibly at Hennebique's instigation. Furthermore, Mouchel or Gueritte normally supervised buildings personally, together with the architect or engineer and the contractor's representative.

Mouchel's licensees included a firm of machinery manufacturers, Rose, Downs & Thompson of Hull, who wished to build their own factory extension (1900) and subsequently built a public bridge in Hull, both still in use quite recently (Fig. 1). One of the more prolific contractors was the Yorkshire Hennebique Contracting Co. Ltd., who built a particularly interesting work, the eight-storey, entirely ferro-concrete Lion Chambers in Glasgow, as well as a modern-looking office and pattern shop (1908) in Manchester, both still in use (Fig. 2) [20]. Some of the licensees, such as John Aird & Co. and W. Cubitt & Co., were well-known building contractors.

Cubitt themselves experimented with reinforced concrete. In 1903 they constructed a column on Considère's system for tests by the architect William Dunn and in 1905 they constructed beams reinforced in various ways. In order to 'clarify' the meaning of their Hennebique licence, Cubitt & Co. in 1906 used another system for a small piece of reinforced-concrete work at Whitbread's Brewery, London. Mouchel prosecuted them and won the case, thus successfully preventing licensees from working with systems that competed with Hennebique's. Cubitt's was not an isolated move against the licensing system at this time. Neither was Mouchel's action untypical. In the same year he prosecuted Coignet and his representatives for allegedly infringing Hennebique's patent in constructing ferro-concrete piles in Bristol, thereby initiating a long legal dispute. Although Roch (1958) took a cynical view—that Mouchel anticipated a long lawsuit, while his own piles would enjoy a profitable monopoly—it is more likely that Mouchel's defensiveness arose from his initial struggles in introducing Hennebique's system [21].

The Adoption of Ferro-Concrete

According to L. G. Mouchel & Partners' company records (which may have been begun systematically only after Mouchel's move to London (March 1900), or even later, when L. G. Mouchel & Partners was formed), the first executed Hennebique work in Britain was Weaver & Co.'s provender mill in Swansea, listed as Mouchel's fifth UK project [22]. Between 1897 and 1899 there were only about seven Hennebique framed buildings commissioned in Britain, but by 1908 this had risen to nearly 40 new buildings commissioned or under construction in this year alone. Between 1897–1908 over 130 reinforced-concrete framed buildings in Hennebique's system were built in Britain (Fig. 3) [23]. There were roughly the same number again of contracts for parts of buildings, such as floors, with 89 bridges and a similar number

66 *Hennebique, Mouchel and ferro-concrete*

FIG. 1. Rose, Downs & Thompson, Old Foundry, Hull (1900–01). Hennebique system. Contractor: Rose, Downs & Thompson Ltd (photograph reproduced by kind permission of Simon-Rosedowns).

of reservoirs, tanks and marine and river structures, and some colliery works. In 1909, of about 1000 total reinforced-concrete works in the UK, 700 were in Hennebique's system [24]. Despite Mouchel's somewhat liberal selection of licensees, there do not seem to have been any serious failures with reinforced concrete during his agency, apart from a test failure pending a large contract in Manchester, in 1902, which upset Mouchel but did not prevent him securing the contract.

Contradictory opinions have been expressed about the influence of building regulations on the adoption of reinforced concrete in Britain [25]. Insofar as reinforced concrete was adopted less speedily in Britain than in some European countries, factors other than the provisions of building regulations—such as a distrust of reinforced concrete—were probably more significant, except possibly in London. In 1904 Mouchel characterised London as "the only town in the civilised world where ferro-concrete constructions are actually prohibited" [26]; certainly very little visible reinforced concrete was allowed there. Generally, urban building regulations excluded the use of panel walls, whether with reinforced concrete or steel framing. However,

FIG. 2. Unbreakable Pulley & Mill Gearing Co, Office and Pattern Shop. Manchester (1908). Hennebique system. Contractor: the Yorkshire Hennebique Contracting Co. Ltd.

government buildings and those on railway and docklands were usually exempt from such regulations, while restrictions in rural areas were minimal; furthermore some urban councils such as Newcastle Corporation (but not the London County Council) were willing to waive by-laws for reinforced-concrete buildings.

From about 1903, there was increasing criticism from architects and engineers of the lack of legal provision for the use of frames of reinforced concrete and steel, and especially for the use of panel walls. Following the report of the RIBA Joint Committee on Reinforced Concrete in 1907 and the successful construction of some large reinforced-concrete government buildings in London, from about 1908 more local authorities, including the LCC, began to permit reinforced-concrete framed buildings with reduced wall thicknesses and started the slow process of changing their building regulations in favour of such construction.

The main practical advantages claimed for reinforced concrete by Mouchel and other specialists were fire resistance, imperishability, strength and monolithicism. The chief motives for its use (usually in preference to steel framing) were slightly different: fire resistance; economy of cost and space for heavy-duty structures; and in some cases its superior structural adaptability, relative freedom from vibration and salubrity for food factories. Flexible planning and daylighting were scarcely considered, by either specialists or clients (Fig. 4). The economic advantages of reinforced concrete were greatest in large, heavily-loaded plain buildings, such as warehouses and granaries (where fire-resistance was also a consideration) and in Britain these were the main kinds of buildings erected both by Mouchel and by other specialists. The major

68 *Hennebique, Mouchel and ferro-concrete*

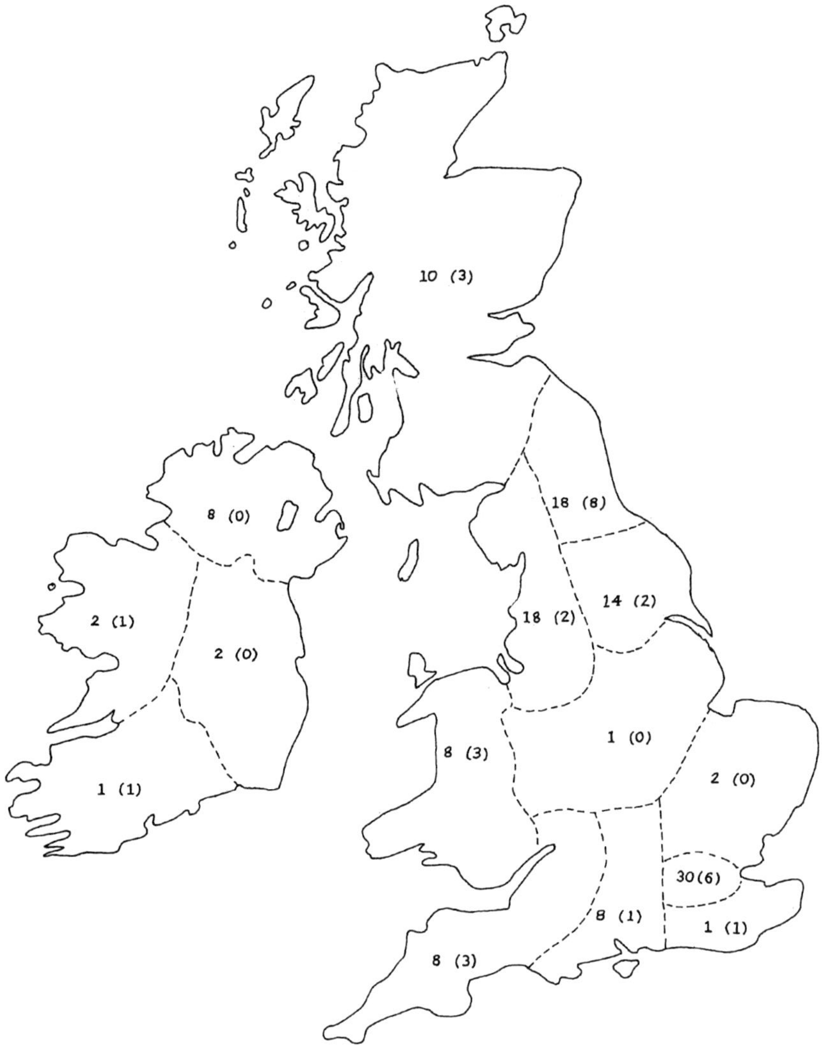

FIG. 3. Geographical distribution of Hennebique framed buildings in Britain 1897–1908. The first figure gives the estimated total number in each region; the bracketed figure gives the number with reinforced-concrete panel walls.

clients for reinforced concrete framed buildings in the UK between 1897 and 1908 can be roughly divided into four groups. In the order of quantity of reinforced-concrete buildings there were:

(1) miscellaneous industrial and commercial companies;
(2) co-operative societies;
(3) railway companies;
(4) grain companies, excluding the Co-operative Wholesale Society.

FIG. 4. Co-operative Wholesale Society Ltd, Grain Cleaning House, Dunston on Tyne (1901–02) (demolished). Hennebique system. Architect: F. E. L. Harris, ARIBA.

However, if clients adopted reinforced concrete for practical reasons, they were also influenced by progressive 'modern' interests. Weaver & Co. were noted for their enterprise and modernity. William Weaver had been chosen as Managing Director because he suited the company's requirement for a 'scientific miller'; he had obtained the "Gold and Silver Medals for Scientific Milling, the highest distinction obtainable in practical modern Milling" [27]. The Co-operative Societies represented a radical tradition of organising commerce and the CWS, who built the Quayside Warehouse, Newcastle, in 1899 and the silos at Dunston in 1901, had a chairman known for his wide interests—"social, scientific and intellectual" [28]. John Buchanan & Bros. Ltd, Confectioners, of Glasgow, who used Hennebique's system for their buildings from 1907, again had a reputation for enterprise and the use of modern facilities in production; their works (1901) displayed a "pantomime of machinery" [29]. The works of Alley & MacLellan Ltd, builders of "portable ships", who in 1905 erected a four-storeyed pattern shop in ferro-concrete in Glasgow, were said to exhibit "originality and energy" (for instance, making use of light railways to convey products between departments) [30]. Other clients too, whether large concerns such as the North Eastern Railway Co. or smaller firms like the Unbreakable Pulley & Mill Gearing Co. Ltd of Manchester, showed a similar 'innovative' tendency remarked upon by contemporary observers.

Although the majority of these reinforced concrete works were commercial and industrial structures and although in this period architects were engaged for only 10 per cent of all buildings, architects were frequently employed along with reinforced-concrete specialists on the early buildings [31]. Architects designed these buildings in detail, then the specialist adapted reinforced concrete to the architect's design, providing all structural details. The results, although perhaps predictable, evidence the dominance of style over material. Since at this time architects, as well as engineers, were usually uninformed about reinforced concrete, the early buildings were designed as if for traditional materials; however, they also tended to conform to a contemporary—and inappropriate—preference for 'Renaissance' styling. The 'thinness' and 'flatness' of reinforced concrete construction had somehow to be given, as Sir Aston Webb put it in 1911, "that architectural character of solidity and permanence without which no architecture can be satisfactory to the eye" [32]. Architects who for reasons of economy used reinforced concrete for the external walls as well as the frame of their buildings tried to give their elevations an identifiable architectural character and to counteract the problematic 'flatness' by means of classical articulation and details imitating stone construction (Figs. 5 and 6). Some of the early buildings were designed in a more functional tradition; for example, Weaver's Mill or Rose, Downs & Thompson's workshop in Hull, on both of which an architect was employed. One distinctive early building entirely in Hennebique's system, the Lion Chambers, Hope Street, Glasgow (designed by the architects J. G. Gillespie and James Salmon) managed to incorporate the characteristic lightness and flatness of reinforced concrete in a hybrid design drawing on Arts and Crafts ideas and related to contemporary Glasgow *Art nouveau* [33]. Another idiosyncratic example was Lynn House, West Hartlepool, designed and built around a Coignet frame in 1906–07 (Fig. 7).

FIG. 5. Co-operative Wholesale Society Ltd, Soap Factory, Dunston on Tyne (1907–09). Hennebique system. Architect: L. G. Elkins, ARIBA.

FIG. 6. Jarrow & Hebburn Co-operative Society Ltd, Drapery Depot, Jarrow (1904). Hennebique system. Architect: J. Cordiner. (Photographed 1977).

Perhaps because of their initial involvement in the design of reinforced concrete buildings, it was architects rather than engineers who first began to question the conditions of reinforced-concrete practice and to criticise the specialists for their monopoly and 'trade secrecy'. (It must be said that some of the criticism levelled at Mouchel was jingoistic.) There was little interchange of experience among specialists, whilst general knowledge about reinforced concrete was seen to be restricted. Consequently, the RIBA established a committee in 1906 to examine and report on reinforced concrete for the benefit of architects; this was accompanied by the launch of a new journal for further debate and publicity and followed in July 1908 by the founding of the Concrete Institute [34].

The main outcome of this activity, apart from provoking the Institution of Civil Engineers into taking a belated interest in reinforced concrete and setting up its own Committee (December 1908), may have been to encourage universities and other educational bodies to take up the subject of reinforced concrete and so continue the process of making the new material more widely accessible.

Conclusion

Mouchel's establishment of Hennebique's system and its application for framed buildings in Britain led to the erection of many Hennebique works all over the UK in this period. The growing awareness of reinforced concrete was indicated by the criticisms levelled at Mouchel for his monopoly of reinforced concrete and by

72 *Hennebique, Mouchel and ferro-concrete*

FIG. 7. M. Robinson & Sons Ltd, Lynn House, West Hartlepool (1906–07) (demolished). Coignet system. Contractor: Watt Bros. Architects: Harry Barnes, ARIBA and Charles F. Burton.

architects' demands for access to what was increasingly recognised as a viable 'new' building material, with definite practical advantages.

The terms of this access were seen to be the freeing of reinforced concrete from patentees' control of its design and execution, as well as the alteration of building regulations to encourage the use of reinforced concrete framing.

The secrecy surrounding the techniques of reinforced concrete obviously restricted opportunities for criticism and correction, and the RIBA Committee and similar initiatives were attempts to open up the subject for study and debate. Yet it is arguable that the conditions of the early employment of reinforced concrete in Britain, largely by Mouchel's company, together with Mouchel's personal dedication to the work, ensured that in a period of rudimentary knowledge about reinforced concrete, Hennebique design and construction was able to work successfully.

The buildings extant from Mouchel's period are in general sound but neglected. Recently several have been demolished for commercial reasons; the former site of Weaver's Mill is now a car park for Sainsbury's. Yet there still exist examples of the mixed results of architects' first encounters with the material in Britain. Architects and their clients are likely to have been working somewhat in the dark, although the architect and client for Weaver's Mill at least had seen Hennebique's mills in northern France. Weaver's Mill and other "functional" examples may well have represented attempts to appear scientifically up-to-date, like the companies they housed. Elsewhere, although there were exceptions drawing upon Arts and Crafts ideas, the predominant architectural symbolism of substance, permanence and classical reference was now cast in reinforced concrete.

Correspondence: Patricia Cusack, Waun-ddu, Dyffryn Crawnon, Llangynidr, Crickhowell, Powys NP8 1NU, Great Britain.

References

[1] The paper is based on material from: P. Cusack, 'Reinforced Concrete in Britain: 1897–1908' (unpublished Ph.D. thesis, University of Edinburgh, 1981). Published articles by the writer which expand some of the themes here are indicated below where relevant. 'Ferro-concrete' was the name given to Hennebique's reinforced concrete system in Britain by Mouchel.

[2] C. F. Marsh, *Reinforced Concrete* (London, 1904), p. 2.

[3] Hennebique's commercial-specialist organisation is discussed in Cusack, 'François Hennebique: The Specialist Organisation and the Success of Ferro-Concrete: 1892–1909', *Newcomen Society Transactions 1984–85*, 56 (1986), pp. 71–86.

[4] Weaver's Mill is discussed and illustrated in: ibid., pp. 76–77; Cusack, 'Weaver's Provender Mill and Silos' (unpublished project, Open University, 1975); Cusack, 'Weaver's Mill', *Concrete*, 10 (March, 1976), pp. 20–23.

[5] See Cusack, 'Architects and the Reinforced Concrete Specialist in Britain 1905–08', *Architectural History*, 29 (1986), pp. 183–196.

[6] F. G. Edwards's advertisement, in Thomas Potter, *Concrete* (2 vols, n.d., *c.*1893), II, p. 290.

[7] F. T. Reade, 'Discussion on the Report of the Joint Reinforced Concrete Committee', *Journal of the Royal Institute of British Architects*, 14 (1 June, 1907), p. 503.

[8] 'Editorial Notes', *Concrete and Constructional Engineering*, 3 no. 2 (May 1908), p. 90.

74 *Hennebique, Mouchel and ferro-concrete*

[9] 'A Reinforced Concrete Factory in Portobello, Scotland', *Concrete and Constructional Engineering*, 2 no. 6 (January 1908), pp. 459–466.
[10] L. G. Mouchel, 'Discussion on Reinforced Concrete', *Jnl R.I.B.A.*, 12 (1904), p. 84.
[11] Compiled from: F. Hennebique, 'Tableau du Développement des Entreprises du Béton Armé', *Le Béton Armé* (Sept 1900), p. 8; *Ferro-Concrete*, 2 no. 3 (Sept 1910), p. 94; *Ferro-Concrete*, 1 no. 2 (Aug 1909), p. 25; *Ferro-Concrete*, 3 no. 8 (Feb 1912), p. 269.
[12] Weaver's Mill, for instance, was apparently 'over-reinforced'.
[13] P. Morton Shand, 'Steel and Concrete. A Historical Survey', *Architectural Review*, 72 (1932), p. 177.
[14] P. Collins, *Concrete* (1959), p. 76.
[15] Mouchel Papers (West Byfleet), letter of C. Roch to C. R. J. Wood, 5 June 1958 (transl. 15 October 1959), p. 6.
[16] Robert Flament-Hennebique, letter to Patricia Cusack, 2 August 1980.
[17] 'The Life and Work of Louis Gustave Mouchel', *Ferro-Concrete*, 3 no. 7 (1912), p. 212.
[18] Ibid., p. 212.
[19] L. G. Mouchel, 'Monolithic Constructions in Hennebique's Ferro-Concrete', *Jnl R.I.B.A.*, 12 (1904), p. 50; L. G. Mouchel, *The Hennebique Ferro-Concrete System* (2nd edn, 1905), p. 20.
[20] Lion Chambers is discussed and illustrated in P. Cusack, 'Lion Chambers: a Glasgow Experiment', *Architectural History*, 28 (1985), pp. 198–206.
[21] Manchester Ship Canal Co. Papers, letter of L. G. Mouchel to W. Henry Hunter, 5 September 1902.
[22] L. G. Mouchel & Partners Ltd. (38 Victoria Street, Westminster), *Project Record No. 1 (Nos. 1–8750).*
[23] L. G. Mouchel & Partners Ltd., *Mouchel-Hennebique Ferro-Concrete. List of Works Executed in the U.K. 1897–1919*, (n.d., c.1920). Also personal inspection of many individual buildings by the author.
[24] 'British Engineers and Ferro-Concrete', *Ferro-Concrete*, 1 no. 4 (October, 1909), p. 65.
[25] P. Collins, *Concrete*, p. 78; Marian Bowley, *The British Building Industry* (Cambridge, 1966), pp. 12–13, 26.
[26] 'Monolithic Constructions', p. 60.
[27] Cusack, 'Weaver's Provender Mill and Silos', p. 6.
[28] *C.W.S. Annual* (1916), preface.
[29] W. S. Murphy, *Captains of Industry* (Glasgow, 1901), p. 200.
[30] Ibid., pp. 123–124 and 126.
[31] R. Norman Shaw & T. G. Jackson, eds. *Architecture: A Profession or an Art* (1892), p. xv; *Jnl R.I.B.A.*, 12 (25 Feb 1905), p. 249 (T. G. Jackson).
[32] Sir Aston Webb, 'Discussion of Sir Henry Tanner's Paper', *Jnl R.I.B.A.*, 18 (January 1911), p. 171.
[33] See Cusack, 'Lion Chambers'.
[34] See Cusack, 'Architects and the Reinforced Concrete Specialist'.

9
The remarkable structures of Paul Cottancin

G.J. Edgell

Synopsis

The paper reviews the work of Paul Cottancin, a French structural engineer working at the turn of the century. The Cottancin system of reinforced concrete, which is based on the use of woven wire placed in relatively thin concrete sections, is described, as is his use of reinforcement in both brickwork and glass. A list of buildings and structures for which Cottancin acted as engineer is given, and five of these are described in some detail. It is concluded that greater study of the history of structural engineering by engineers would be unlikely to familiarise them with the work of Cottancin; also that, although some of his ideas are highly unusual, the concepts he used regularly (e.g. deep beam action, wall-floor combined action) are less well used today than they could be.

Introduction

The developments in reinforced concrete construction that took place in the latter half of the 19th century were diverse and have been described in detail elsewhere[1,2]. The names associated with this period—Monier, Coignet, Considère, Hennebique—are familiar, and it is the aim of this paper to concentrate on the work of an engineer whose contribution has often been overlooked. Paul Cottancin patented a system for reinforcing concrete in 1889, some 3 years before Hennebique, and from then until 1905 acted as both engineer and contractor for a number of buildings that incorporate unusual structural features. Cottancin appears to have stopped designing in 1905, and this is probably associated with his bankruptcy during the construction of the Sidwell Street Methodist Church in Exeter. It seems that his role as contractor was his downfall. However, there are enough surviving buildings and recorded information to show his innovation in the preceding years.

The Cottancin system

The Cottancin system of reinforced concrete is based on the use of a continuously woven mesh-type reinforcement[3,4]. This was used in floors,

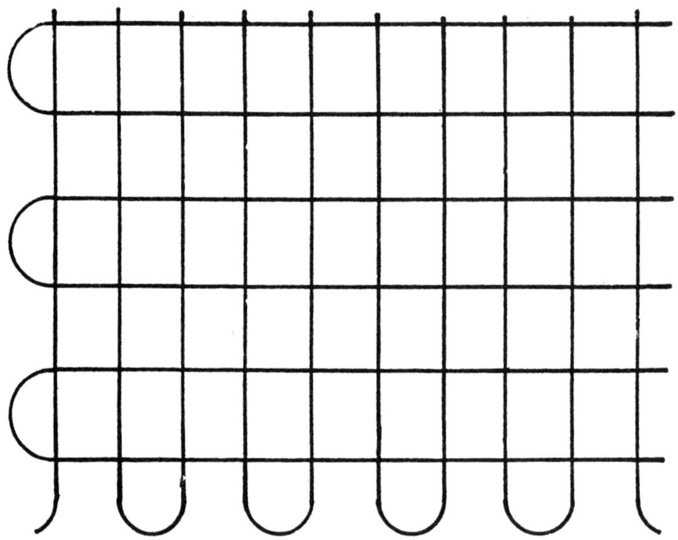

Fig 1. Simple pattern of continuously woven wire

walls or beams. Fig 1 shows the simplest type. In the case of a beam, Cottancin explained the structural performance by analogy with the N-type truss. If we consider the element a b c d, in Fig 2, of an imaginary diamond-shaped woven mesh in a reinforced concrete beam, then, according to Cottancin, the element could be imagined to consist of a number of N-type trusses A B C D as shown. In each case the compression members are formed by the enclosed concrete. Having used the truss analogy to establish that the concrete was in compression and completely restrained by the mesh, Cottancin claimed that the same would be true for a rectangular mesh. As a result of this thinking, Cottancin maintained that, as the wires of the mesh were all in tension,

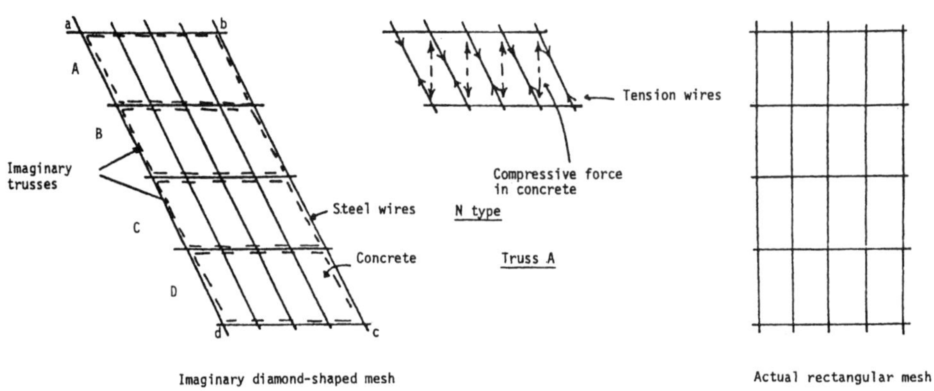

Fig 2. Truss analogy

they could be thin, typically 4 mm, as long as the mesh restrained the concrete in that the rectangles retained their shape. In order not to tie the wires together at the points where they crossed, a single, continuously woven wire was used to form the mesh. Cottancin was convinced that his theory was correct and contrasted his system with the many others available by describing his as one where the concrete was encased by the reinforcement and not vice versa. He reasoned that the absence of cracking in his concrete resulted from the fact that, even at the soffit of a beam, it was working in compression.

A great many patterns of weaving were used for the mesh, although Cottancin felt that they could all be relied on to perform similarly, the choice in any particular situation being based on ease of use.

Similar mesh was used in floor slabs, and Cottancin maintained that the concrete was all retained in compression. So that the concrete was not too far from the mesh, which was at the mid-depth of the slab, the concrete was never more than 60 mm thick and more usually 50 mm. Cottancin believed that it was particularly important for the concrete to be retained in compression, as the bond between the reinforcement and the concrete was not reliable and would be destroyed by elastic or thermal movements or vibration. If this breakdown occurred, the concrete would behave as though unreinforced.

Concrete of this thickness could not span great distances, and a triangulated system of ribs was used, as shown in Fig 3. The ribs, referred to as spinal stiffeners, would typically be 200 mm deep and 50 mm wide and contain at their base a 40 mm × 9 mm steel flat. A flat was also

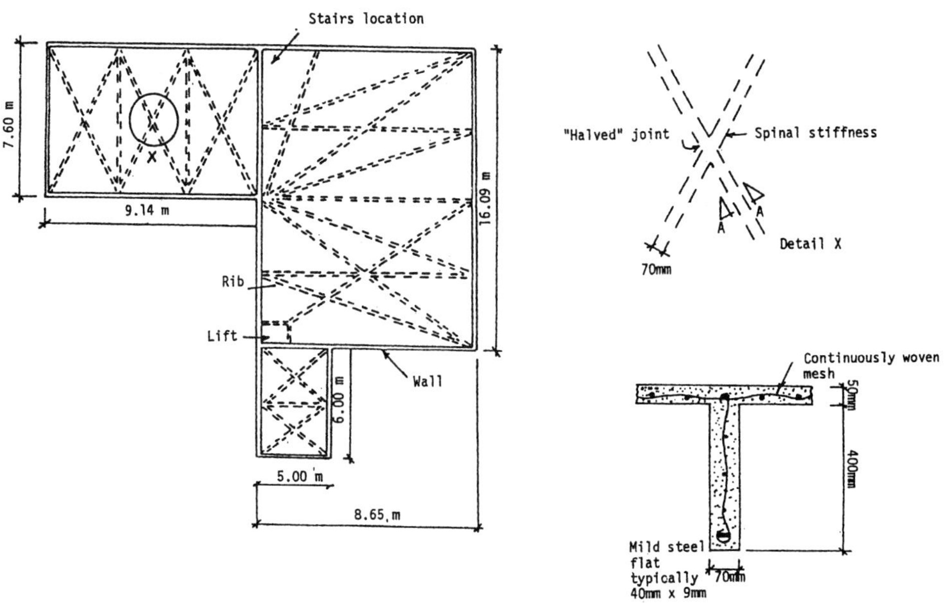

Section AA
(Illustrative - exact reinforcement details not known).

Fig 3. Typical layout of floor ribs (spinal stiffeners) (bus repair and accumulator depot)

generally included at the top, and the reinforcement in the rib was tied to that in the slab.

It is not clear whether any detailed structural calculations were carried out as part of the design. De Tedesco does mention that the system does not lend itself to easy calculation, and Cottancin, in developing his theories for the behaviour of his system, does refer to stresses and forces. However, no firm conclusion can be drawn, and it may be that the designs were supported by the confidence gained from test results. Cottancin did load tests on 30mm-thick, 500mm-square concrete panels in Tunisia[4], and both the French War Department and L'École des Ponts et Chaussées did their own tests[5].

Reinforced brickwork

It is clear from his designs that Cottancin was familiar with using many different materials. However, his use of brickwork was perhaps the most extraordinary. Cottancin treated bricks in the same way as small elements of concrete and retained them in a mesh. A typical design would consist of a wide cavity wall with occasional cross-ribs, both leaves being made of brickwork. At the Sidwell Street church, Exeter, the cavity is 530 mm. The bricks are 215 mm × 75 mm × 73 mm and have four perforations, through each of which passes a 4 mm dia. vertical wire. In each bed joint a horizontal wire threads between the vertical wires. At positions where Cottancin considered there to be particular stress concentrations—e.g. where a rib crosses a cavity or at the sides of openings—he used a larger steel bar, usually a 40 mm × 9 mm steel flat. (See Fig 4.) All the joints and the perforations were solidly filled with 1:4 mortar of an almost liquid consistency[5].

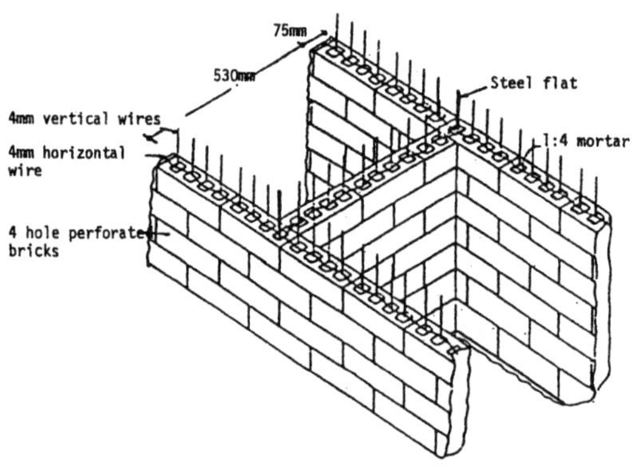

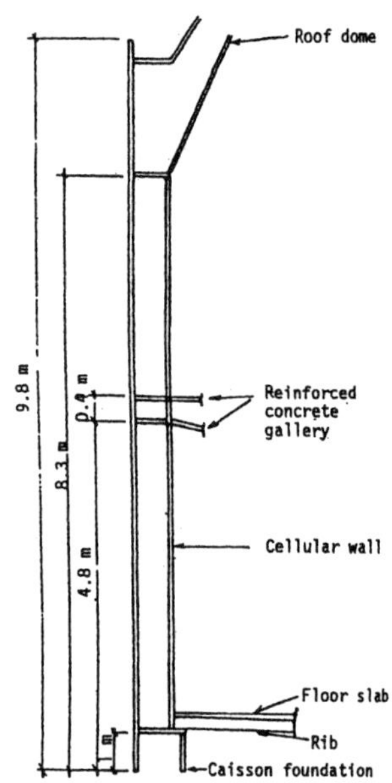

Fig 4. Elevation and isometric of wall at Sidwell St. Methodist Church, Exeter

Glass

In the house[6] in the Avenue Rapp, Paris, it was necessary to light the basement using pavement lights, and although this was common the usual method of supporting the glass panels on the flanges of inverted T-beams led to much of the light being obscured. Cottancin used a design similar to that for his floor slabs. Square glass elements with grooves in their edges were used, and 4 mm steel wires were cemented into the grooves. (See Fig 5.)

In Cottancin's design[7] of the diphtheria pavilion for the Hospital for Sick Children in Paris the structural design and detailing was dominated by the requirements for a hygienic building laid down by the medical superintendent. One small detail that Cottancin devised was for the window units where moulded glass blocks were reinforced with steel wires from the mullions, head and cill. The window panes were set into grooves in the glass blockwork frames and this allowed small pieces of antiseptic felt to be fixed between the frame and the pane, and also between the pane and the fixing bar.

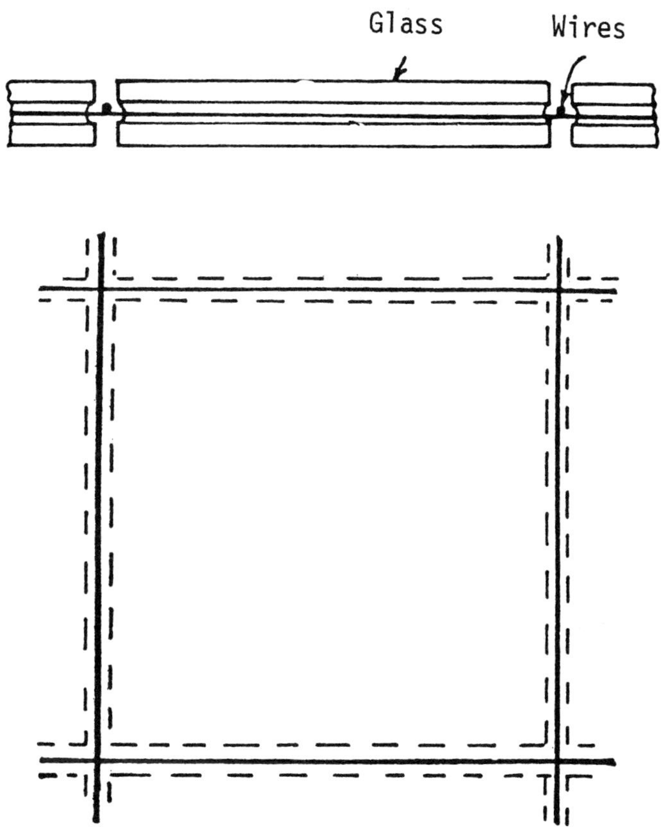

Fig 5. Pavement lights

6 STRUCTURES OF PAUL COTTANCIN

Cottancin's buildings and structures

The buildings and structures for which Cottancin is known to have acted as structural engineer are listed below. The list is almost certainly not exhaustive, but is nonetheless extensive, considering that the activity associated with them was compressed into approximately 15 years.

1899 — 34 Avenue de Wagram, Paris, France (architect M. Lavirotte)
Villa, Cernay, France[5]

1900 — San Marino Pavilion[5,12], Paris Exhibition, France

1900 to 1901 — house in Avenue Rapp[3,5,6], Paris, France (architect M. Lavirotte)

1902 to 1906 — Sidwell St. Methodist Church[3,5,8,9], Exeter, England (architect F. J. Commin)

1903 — Church of St. Jean de Montmartre[3,5,10], Paris, France (architect A. de Baudot)

— Château de Orfraisière[3,5], France

— MM. Ple Frères Shoe Factory[3], Paris, France.

— chimney stack[5], Hirson, France

— highway bridge[5], Montrevil, France

— brickwork water tank[5], Newark-on-Trent, England

— foundations to boilers[5,13,14], pump house, Duck Island, St. James Park, England

— municipal theatre[5], Tulle, Corrèze, France

— foundations to casino and hydro[5,8], Enghien, France

— diphtheria pavilion[7], Hospital for Sick Children, Paris, France (architect M. Belovet)

— foundations to The Government Schools, Tunis, Tunisia[5,8]

— foundations to the palace of the Governor, Tangier, Morocco[8]

— bus repair and accumulator depot, Paris, France[13]

— circular reservoir, Memours, France[3]

Not all of the structures can be described here in detail. It is hoped, however, that enough information is given to interest and stimulate the reader.

Sidwell St. Methodist Church, Exeter, England[3,5,8,9]

Following the completion, in 1897, of a Sunday school in Sidwell Street, the architect Frederick James Commin was commissioned by the trustees to design an urgently needed church.

The plans, which were accepted by the trustees, were for a building not dissimilar in appearance from that finally built (see Fig 6) and which included what appeared to be stone columns supporting a timber framed dome roof, lantern tower, and ventilator turret. The tenders for the building were too high, and subsequently it was redesigned, using the Cottancin system. It seems that Cottancin was introduced to England, if not to this particular scheme, by an English architect Mr Arthur Vye-Parminter who practised in Paris.

The plan view of the walls of the church is shown in Fig 7, and they are as described earlier and shown in Fig 4; they rise to some 8 m above ground level to support the roof dome. The materials were all obtained locally.

Fig 6. External view of Sidwell St. Methodist Church, Exeter

The foundations are typical of Cottancin's work and are cellular, consisting of two parallel leaves of reinforced brickwork connected by a number of intersecting cross-ribs (Figs 8 and 9). The walls were filled with rammed earth, and a 50mm-thick concrete slab was cast onto the top of them, from which the walls were built. This type of foundation seems to have been particularly good in bad ground; at Exeter they were 1 m deep, but for the house in Avenue Rapp, which was some eight storeys high, the foundations extended to a depth of five brick courses only. At the casino in Enghien the foundations were on the mud of a lake bed and the site on Duck Island in St. James Park is clearly far from ideal; although the foundations there extend to a depth of 300 mm only.

8 STRUCTURES OF PAUL COTTANCIN

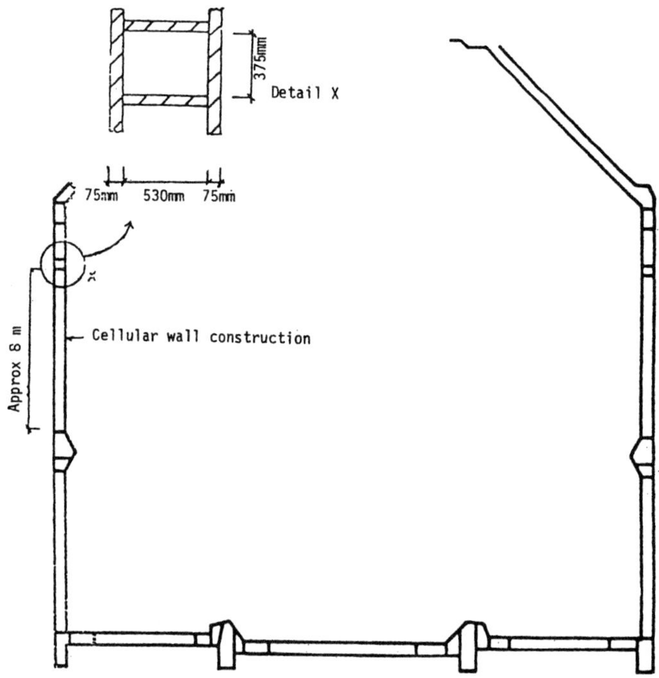

Fig. 7. Plan view of wall layout, Sidwell St. Methodist Church, Exeter

The floor consists of a 50mm-thick concrete slab with a triangulated system of spinal stiffeners.

A cross-section through the church is shown in Fig 10. The main dome consists of two connected domes. The inner is reinforced brickwork and is plastered on the inside. The outer is reinforced concrete, 50 mm thick, covered partly in copper and partly in bitumen, which is visible evidence of the financial problems that arose. The top of the lantern tower leading up to the ventilator turret is reinforced brickwork and the outer surface is curved in both elevation and plan. A fascinating feature is the gallery, which consists of two slabs of reinforced concrete interconnected with ribs, the whole cantilevering some 4·2 m off the walls, the reinforcement of the slabs being connected to that in the two leaves of the walls. The only other supports for the gallery are the staircases at its ends, one each side of the altar. This building did not comply with the local building bye-laws and, as part of the job of persuading the authorities that it was satisfactory, a loading test was carried out on the gallery. It was estimated that the maximum superimposed load in use was 25 ton, and the test consisted of applying a total load of 105 ton, most of which was distributed uniformly over the gallery, although 17 ton was concentrated at the inner edge. The average deflection at this load was recorded as 3/8 in.

It is interesting to observe[9] that all the ornate sculpturing of, for example, window jambs and mullions is, in fact, achieved by using moulded reinforced concrete and not, as is more customary, stonework.

The condition of the building today is encouraging; there has been some corrosion of the reinforcement and subsequent cracking, but the reasons for it are clear.

The building is fairly exposed, Exeter is not far from the sea, and where

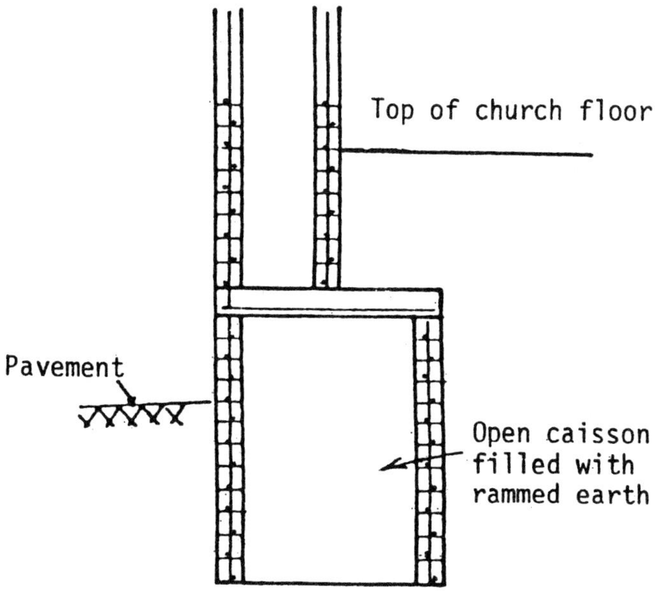

Fig 8. Cross-sectional elevation of foundations at Sidwell St. Methodist Church, Exeter

Fig 9. Top of open caisson foundations at Sidwell St. Methodist Church, Exeter (note the wires protruding from perforations—these to be subsequently connected to steel in floor slab)

10 STRUCTURES OF PAUL COTTANCIN

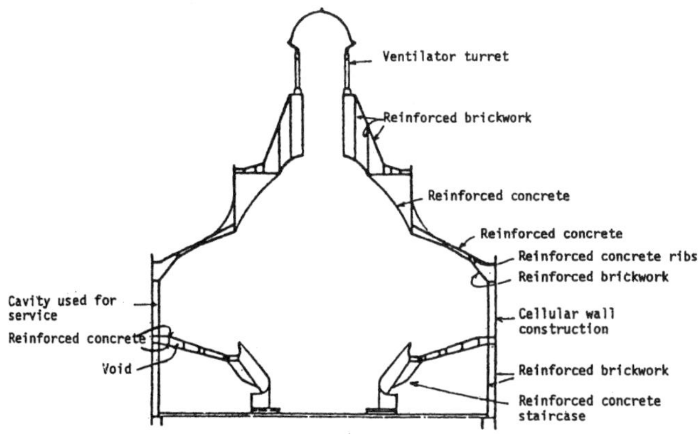

(a) *Cross-sectional elevation*

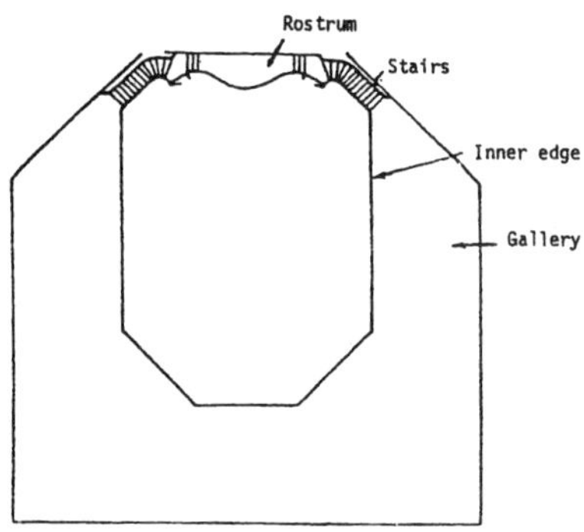

(b) *Simplified plan of gallery*

Fig 10. Sidwell St. Methodist Church, Exeter

the concrete mouldings are damaged the cover to the reinforcement is relatively low (about 15 mm). The brickwork is also damaged, but only at those positions where a large steel flat had been used in a perforation. The steel wires were, where examined, in good condition and phenolphthalein test indicated that, although the mortar filling the perpend joints and perforations was mostly carbonated, that in the bed joints had done so to a depth of 6 mm only.

The Church of St. Jean de Montmartre[3,5,10,11]

St. Jean de Montmartre—probably Cottancin's best-known building—was built in reinforced brickwork as the estimated cost was less than one-third of that for the next cheapest scheme considered, which was steel framed masonry construction. The boldness of the design led to it being called 'the folly of the century'. It has proved to be safe structurally, however, and also durable. The only cracking reported 'is in the belfry which is extremely exposed'. Fig 11 shows an elevation and plan of the building.

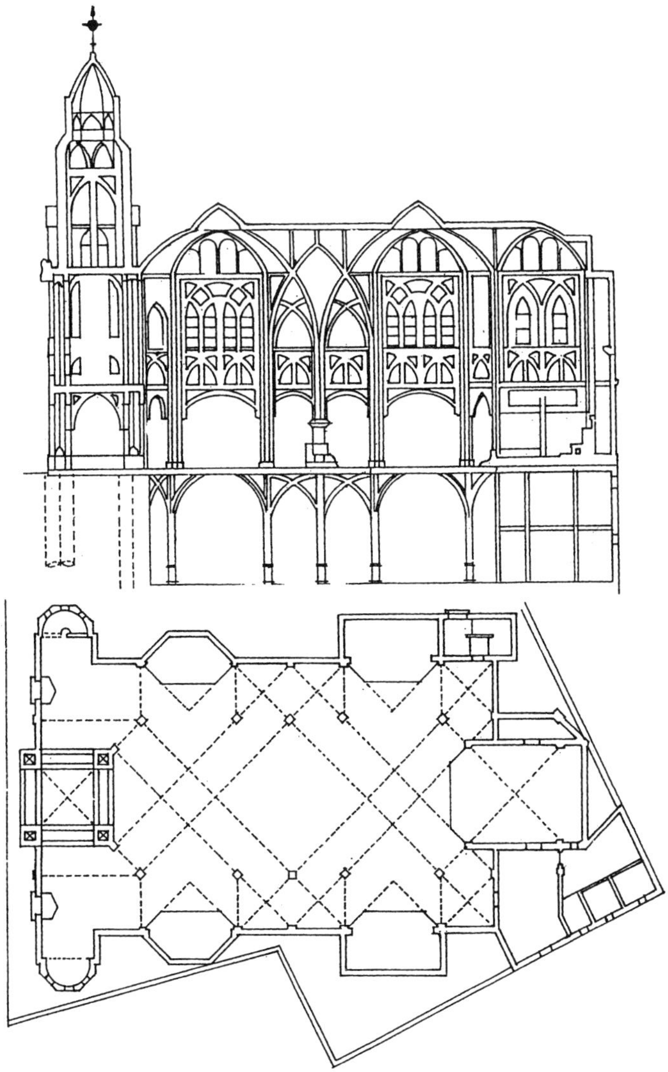

Fig 11. Elevation and plan of St. Jean de Montmartre

The main columns are of interest. These are founded in reinforced concrete cylinders whose overall depth is 12 m. This depth was required because of the poor nature of the ground—fissured gypsum overlain with quarry debris. The columns are made from eight hole perforated bricks, as shown in Fig 12. Steel wires pass up the outer four perforations of each brick with the exception of the corner of the column where a larger steel bar was used. Horizontal wires are again interwoven with the vertical steel. The columns are 440 mm square. At the top of the foundations a

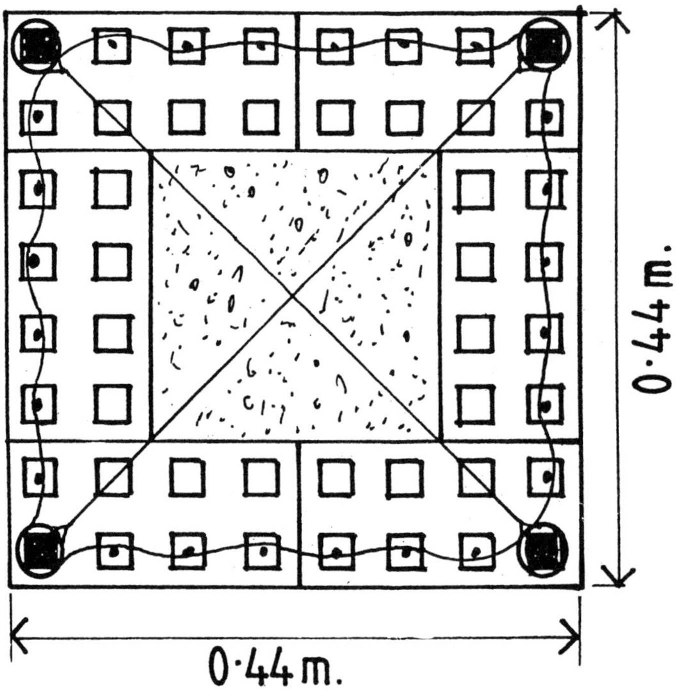

Fig 12. Cross-section of column: St. Jean de Montmartre

number of concrete ribs are connected to the columns to form the crypt floor, which is similar to that at Sidwell Street.

At a height of 10 m a system of arched ribs forms the support for the roof of the crypt, which is the floor of the main part of the church. At the level of this floor the columns then rotate through 45° on plan, all the reinforcement in the perforations following the curves (Fig 13). The columns then rise a further 25 m, and another system of arched ribs—this time at 45° to the aisle of the church, and to those supporting the floor—carries a series of domes forming the roof. The columns also support a number of galleries and a balcony. The external feature that is most often the subject of comment is the slender walls. These are in fact 35 m high and are only 110 mm thick, although they do receive support from the interconnecting reinforcement of the floors and galleries.

As with the Sidwell Street church the building inspectors were sceptical about the strength and stability of the building, and loading tests were requested, this time on the main floor.

One feature that the inspectors found difficult to accept was that, at one corner of the church, there was a passage beneath the crypt and, although there was a column to support the floor and roof, there were no beams to support the brickwork walls. There was at this point a horizontal steel flat 40 mm × 9 mm at the base of the walls which form a sort of giant reinforced corbel 25 m tall, the floor and roof loads going down into the columns and the walls forming part of a very stiff box-like

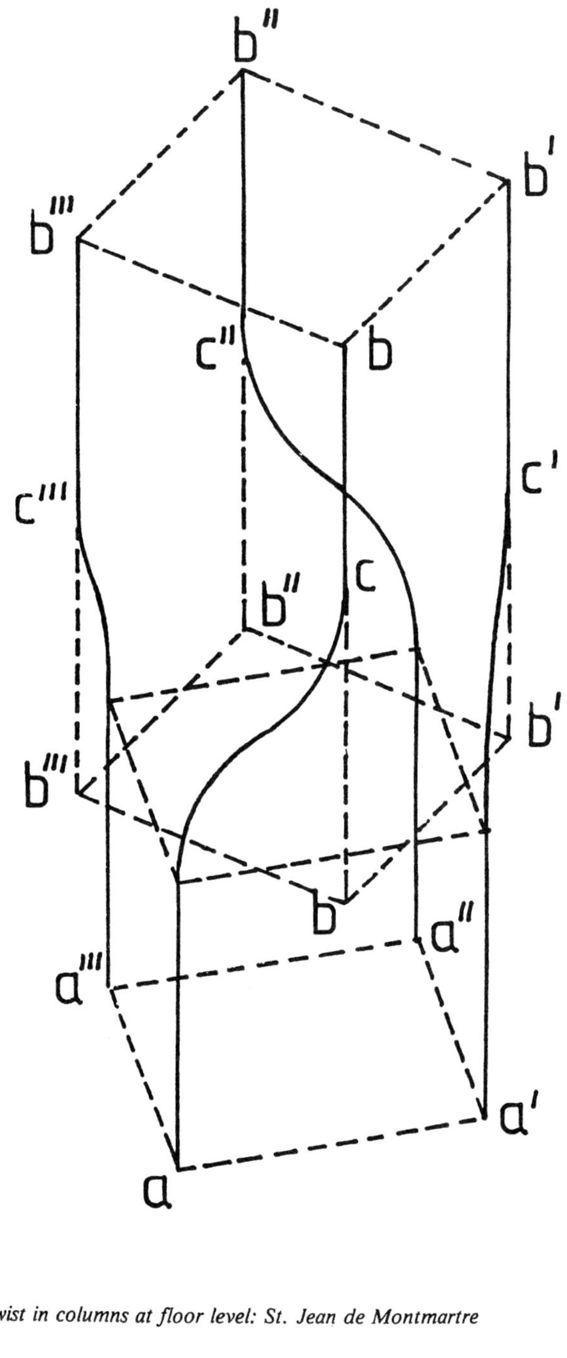

Fig 13. Twist in columns at floor level: St. Jean de Montmartre

structure. In this case the support for one relatively small corner had been removed.

House in the Avenue Rapp[3,5,10]

The house in the Avenue Rapp is shown in Fig 14. It was built in 1902, has eight storeys, including the basement, and on the ground floor are a number of courtyards and coach houses. The building is of interest architecturally and is an example of French Art Nouveau.

The main external walls are of wide cavity construction, all reinforced, and, as was the case whenever this type of wall was used, the opportunity was taken to put services into the cavity. The outer leaf was built with

Fig 14. House in the Avenue Rapp, Paris

bricks having a variety of different finishes and, with the exception of the street facade, where common bricks were used for the inner leaf and the outer, is described as being of moulded stoneware, although reinforced in the same way as elsewhere. The roof is the mansard type and reinforced brickwork of the cavity type was used for the sloping elements. A sort of giant corbel similar to that used in St. Jean de Montmartre was used to, in this case, give a ground floor with unimpeded access to the coach houses. The triangular portion of the building in Fig 15 was designed with no support other than the interconnection with the external walls throughout the remaining six storeys. When completed, however, the visual impression was unnerving and a stone pier was subsequently built; although it does not need to take any load, it inevitably provides some support. The 7 m span arch in the ornate facade is, as with much of Cottancin's work, not what it seems, the reinforced stoneware hanging from the main cavity wall and not needing to transmit any load into the adjacent columns.

It is interesting to note that all the spinal stiffeners for the floors were precast on their sides in the workshop and stored in the basement of the house. Whenever it was necessary for two to intersect to form the skeletal frame to the floor, the precast units were joined by a connection similar to a halved joint in carpentry.

The San Marino Pavilion[5,12]

The San Marino Pavilion was built at the foot of the Eiffel Tower for the 1900 Paris Exhibition. The foundations were of the open caisson type and on top of these was a 60mm-thick floor slab.

The building formed on top of this was 11 m square on plan and 12 m

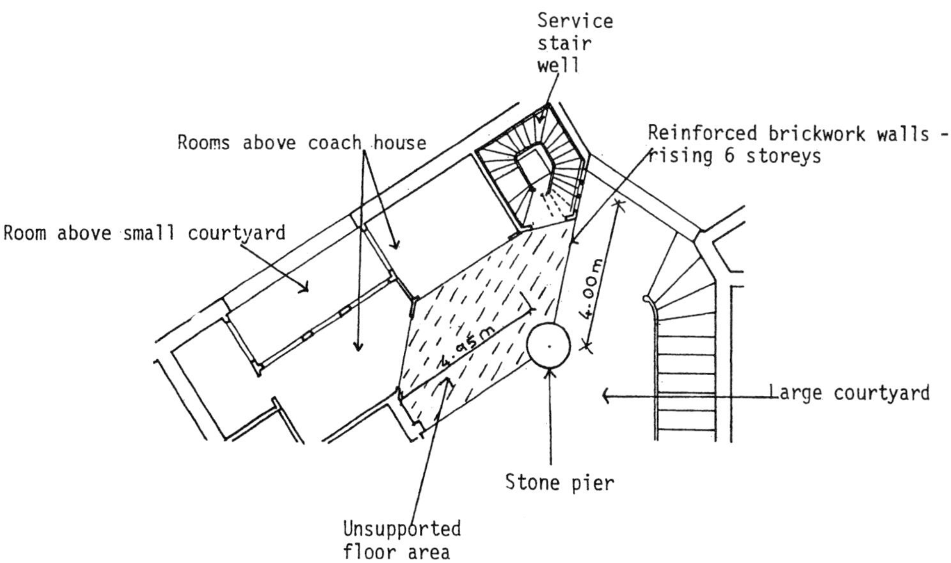

Fig 15. Plan view of 'giant corbel': house in the Avenue Rapp, Paris

high. At two corners was a set of four counterforts to the walls, rising the whole height of the building and forming a sort of hollow tube at these corners. The structure was all reinforced brickwork. From each of these tubes a reinforced brickwork arch sprang and supported a skylight opening at the centre of the building. At the end of the exhibition the engineer from the École Nationale des Ponts et Chaussées who had superintended the construction arranged to have it tested under load. The test consisted of placing an iron band around the circumference of the building about 1 m from the top and using one 50 ton and three 15 ton cranes pulling at the band to simulate a wind load. The horizontal component of the load applied was equivalent to 4 kN/m^2 of wall area and led to no cracking or distress, and no movement could be detected using a theodolite. The building was subsequently fire tested. A wood fire was lighted inside which was sufficient to melt the glass in the skylight and also some of the bricks. However, there was relatively little damage. Expansion of the floors caused some distortion of the walls, but much of the displacement was recovered on cooling. The final test was to, at one corner, remove the counterforts one by one. The removal was by battering the brickwork with a 16 m battering ram swung by eight men. It was calculated at the time that, immediately prior to failure, part of the last remaining counterfort was subjected to a compressive stress of 196 N/mm^2.

Bus repair and accumulator depot[13]

This is a structure which, like the others described, is unusual, although in this case not so many of the facts of the design are clear.

The requirement for the repair bay included a wide door-opening with the accumulator depot above, and this led to the design shown in Fig 16.

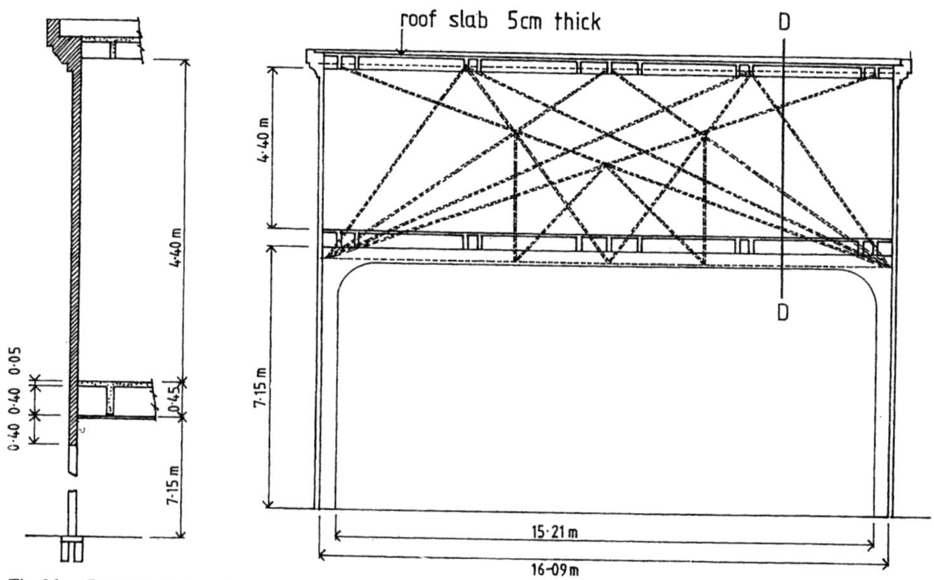

Fig 16. Bus repair depot

The wall above the opening spans the 15 m opening between 440 mm returns at each end and also supports one edge of the upper floor, which is of the slab and spinal stiffener design. The wall is described as being 110mm-thick brickwork and containing the diagonal truss elements shown in the figure within its thickness. The steel elements are 40 mm × 16 mm steel flats which are connected at the edges of the wall to the more usual Cottancin-style reinforcement in the other walls. The wall-truss design was chosen because of the twin requirements of large span and high floor load. When one knows that the floor, which was load tested by the Chief Road and Bridge Engineer, supports a load from the stored accumulators of 35 kN/m^2, the extraordinary nature of the structure may be more fully appreciated.

Other structures

Less is known about the remaining buildings and structures listed in this paper. However, there are enough glimpses in the literature to realise that these, too, are of real interest. For example, the roof structure of the Diphtheria Pavilion[7] consists of two triangular open-section beams; the horizontal and diagonal elements are reinforced concrete with spinal stiffeners; and the vertical elements are reinforced brickwork. These tubes have a clear span of 28 m. Similarly, the dome of the Château Orfraisière[5] is some 17 m in diameter and supported at 42 m above the ground by 450mm-square reinforced brickwork columns.

Conclusion

In order for structural engineering to advance, there must be a continuing development of ideas about structural form and about the use of new and traditional materials. That this can happen means that engineers need to be aware of the history of their subject. Even if the history of structural engineering were to be more closely studied by engineers, it is not likely that they would be familiar with the work of Paul Cottancin. This paper aims to, in some small way, provide a review of his work.

Although Cottancin's ideas would be considered by some as eccentric, there are clearly areas where we can learn from his work. The use of deep beams is perhaps less common than can be justified on economic grounds; the way that wide cavity walls may be used effectively has been demonstrated only relatively recently. The overall advantages of making floors and walls work together are often disregarded. The reasons behind the effective corrosion protection provided in his structures by the use of pourable mortar should be established.

What is to be learnt from Cottancin is encapsulated in a quotation from *The Builder* in 1905, referring to St. Jean de Montmartre—

'The church is certainly one that deserves personal inspection and although our readers may not be anxious to emulate the example presented by St. Jean de Montmartre they may learn from this building how great is the adaptability of reinforced construction and how useful is the aid offered by the system to the architect who is bold enough to break away from stereotyped methods of realising his designs.'

References

1. De Tedesco, N.: 'The history and evolution of reinforced concrete in France', *Concrete and Constructional Engineering,* **1**, nos.1-6, 1906-7, p159
2. Hamilton, S. B.: 'A note on the history of reinforced concrete in buildings', *National Building Studies Special Report no.24,* DSIR, 1956
3. Marsh, Charles F.: *Reinforced concrete,* Archibald Constable & Co. Ltd., London, 1904
4. *Le Ciment,* nos. 1 & 2, 1896; nos. 1 & 2, 1897
5. Galbraith, A. R.: *The Cottancin system of armoured construction,* Cowell, Ipswich, 1904
6. 'Maison de rapport of seven storeys', The students' column, *The Builder,* 25 March and 1 April, 1905
7. 'Diphtheria Pavilion: L'Hôpital des Enfants Malades, Paris', The students' column, *The Builder,* 18 February, 1905
8. 'The Cottancin system of construction', *The Contract Journal,* 31 December, 1902
9. Howard, Margery: 'Cottancin and an Exeter Church', *Concrete,* January, 1977
10. 'L'Église de St. Jean de Montmartre', The students' column, *The Builder,* 11 and 18 March, 1905
11. Bernstien, D.: 'Some early reinforced brickwork masterpieces revisited', *Proc. Vth International Brick Masonry Conference,* Washington, 1979, p231
12. 'Composite structures', *Engineering,* 21 March, 1902
13. 'An omnibus repair and accumulator depot', The students' column, *The Builder,* 25 February, 1905
14. 'Concrete Steel', Twelvetrees, 1905, p165
15. *The Builder,* 21 November, 1903, p526

10

The era of the proprietary reinforcing systems

M.N. Bussell

■ **This paper reviews developments in reinforced concrete construction, mainly in Britain, from the 1890s until the First World War. It focuses on the proprietary systems that offered structural design in this new composite material as an adjunct to supply of the reinforcement or construction of the complete reinforced concrete structure. Of these the Hennebique system, through its British agent, L. G. Mouchel, was the most prolific, with a remarkable record of expansion following its arrival in 1897. A large number of other systems were patented, although only a few found wide use, notably those of Truscon and Considère. Information is provided to assist in identifying proprietary reinforcement, floors and systems of this period in existing construction.**

Introduction: The Early Years

The 1854 patent by William Boutland Wilkinson of Newcastle upon Tyne was the first to propose the use of iron as reinforcement to concrete, recognizing the relative weakness of concrete in tension.[1] Like those of the Frenchmen Lambot and François Coignet in 1855, Wilkinson's patent attracted little interest in the British building industry. It was not until 1892 that François Hennebique, a French contractor, obtained a British patent for his system which, with others later, would find wide use in Britain.[2]

2. The concern was to provide an economical and fire-resistant form of construction—something which was widely sought as a solution to the fatal and expensive fires that frequently consumed mills, warehouses and public buildings in particular. Hurst in Paper 3[26] has reviewed the later nineteenth-century solutions; these were based on beams and columns of iron or steel, and although floors such as brick vaulting and flat and vaulted concrete slabs were themselves 'fireproof', they did not always protect the vulnerable bottom flanges of the beams, nor the columns.

3. Hennebique's patent was based on plain round bars with fish-tailed ends, and stirrups of flat strips, all at that time being of mild steel. The bars provided tensile resistance in beams and slabs, and supplemented the compressive capacity of the concrete in columns and walls. The stirrups provided shear resistance, although they were not mechanically anchored in the compression zone, as is the norm nowadays. Column bars were linked by strips of wire. A later development in 1897 was to provide bent-up bars in beams to provide hogging resistance and supplement shear capacity.

4. At this time, such construction was generally known as 'ferro-concrete' rather than as reinforced concrete.

5. The question arises how such structures were designed. Hennebique carried out tests to establish the strength of beams, and had the technical assistance of a Belgian engineer.[2] His approach was experience-based. Edmond Coignet, son of François, and N. de Tédesco presented a paper to the French Society of Civil Engineers in 1894 which laid down the basis for calculations on the modular ratio method.[3] This remained valid and in use for most of the twentieth century. (This subject is discussed further

by Bussell in Paper 5.)[27]

6. Outside Britain, others were developing their ideas on the reinforcement of concrete. Indeed, it has to be said that, after Wilkinson, the initiative for the development of reinforced concrete came from the Continent (notably France and Germany) and from the USA. Britain lagged behind, and the early development of reinforced concrete construction in Britain was essentially due to the 'import' of Continental and American systems.

7. Joseph Monier obtained a French patent in 1867 for reinforcing concrete plant tubs with wire and rods. His ideas were taken up by the German G. A. Wayss and the 'Monier system'—regarded on the Continent as a synonym for reinforced concrete—was widely used there, but was little used in Britain.[4] In the USA, Thaddeus Hyatt developed reinforcing systems using nuts,[4] or cross-bars placed through holes in square bars,[5] to anchor the steel to the concrete. E. L. Ransome obtained an American patent in 1884 for a square twisted bar which was both stronger (being cold-worked) and had superior bond performance.[4]

8. Numerous other patents were obtained, and systems developed, which made use of various reinforcing section profiles. Expanded metal was introduced as reinforcement to concrete around 1890, and was widely used.[4] The Kahn bar of 1902–3 is of unusual profile, being a square section with two projecting strips on diagonally opposite corners. These are slit to be bent up diagonally in short lengths, forming shear reinforcement. Kahn saw these as forming a 'trussed' beam, and indeed the business which used his system in the USA, and later in Britain, was called the Trussed Concrete Steel Company, affectionately remembered by many as Truscon.

9. A detailed recital of the various patent reinforcement profiles and arrangements, and their related construction systems, would be unduly lengthy. It will, however, be helpful for today's engineer to be able to identify the commonest profiles and systems likely to be found when investigating existing construction of this period, and these are shown respectively in Figs 1–24 and 25–50. Further details of these (and many other) systems can be found in contemporary texts such as Cassell[4] or Marsh.[5] Table 1 lists the common proprietary systems in use in Great Britain in 1907.

10. Hennebique's 1892 patent was followed by the opening of an office in Brussels, and the construction of a refinery in Paris in 1894 and a framed mill in Tourcoing in 1895. In 1897 he obtained a further British patent, and appointed L. G. Mouchel as his agent in Britain, with offices in Victoria Street, Westminster (the traditional home of consulting engineers until recent years). Hennebique's first building in Britain was Weaver's Mill of 1897 in Swansea (Fig. 51). The working drawings for this were done in Nantes in France, from where too were supplied the cement, aggregate and reinforcement; the licensed contractor was also French![2]

11. The expansion of the Hennebique enterprise was remarkable. By 1899, 3061 projects had been undertaken.[2] A major reason for this success was Hennebique's insistence on the use of suitable materials and experienced labour so that the resulting construction was of a good standard. This explains the use of French materials and a French contractor for Weaver's Mill, although it was more common for the company to license contractors whose work would be, in part at least, undertaken and supervised by men who had gained experience with Hennebique. In modern jargon, one would say that the process was one of 'technology transfer', and certainly such was essential, bearing in mind the phenomenal workload in those early years and later. By 1909 the Hennebique system had been used for nearly 20 000 structures, and the company had 62 offices: 43 in Europe (including Britain), 12 in the USA, four in Asia and three in Africa. A list of British Hennebique projects issued in 1911 by what had by then become L. G. Mouchel & Partners totalled 1073.[2] (A second notable factor in Hennebique's success was his company's assiduous attention to promoting its achievements.)

12. A standard specification of 1917 by L. G. Mouchel & Partners[6] is of interest. It covers 'ferro-concrete', and cost 2*s*. 6*d*. (12½p). Steel for reinforcement was to comply with BS 15 of 1912,[7] to be Trisec bars (a patent high tensile steel), or to be shell discard of prescribed quality (the First World War still had a year to run). Concrete proportions by volume were approximately 1·1 : 2 : 4 of Portland cement : sand : coarse aggregate. Provision was

Table 1. The common proprietary systems in use in Great Britain in 1907

No.	Name of system	Form of tension bars	Form of compression bars	Form of shear reinforcement	Method of fixing shear reinforcement	Direction of shear reinforcement
1	Coignet	Round straight bars	Round straight bars	Round rods bent to U-shape	Looped under tension bars and twisted above compression bars	Vertical
1a	Coignet	Round straight bars and round bars bent up near supports	Round straight bars	Bent-up ends of extra tension bars	Continuous with extra tension bars	Diagonal
2	Considère	Round straight bars and round bars bent up near supports	Round straight bars	(a) Bent up ends of extra tension bars (b) Round rods lapped around the main tension and compression bars	(a) Continuous with extra tension bars (b) Bent round tension and compression bars	(a) Diagonal (b) Vertical
3	Hennebique	Round straight bars and round bars bent up near supports	Round straight bars	Steel strip bent to U-shape and made with spring clip	Sprung on to tension bars and bent over for anchorage in concrete	Vertical
4	Improved construction	Round straight bars and round bars bent up near supports	Round straight bars	Round rods wound around the main tension and compression bars	Bent round tension and compression bars	Spiral
5	Indented	Corrugated square bars, bent up near supports	Corrugated square straight bars	Bent-up ends of tension bars	Continuous with tension bars	Diagonal
6	Johnson	Round straight bars woven with wire lattice	—	Trough of wire lattice with rectangular or with diamond mesh	Woven with tension bars	Vertical or diagonal strands, according to mesh used

7	Kahn	Square bars generally straight, sometimes bent up towards supports	Square straight bars	Wings attached to main part of tension and compression bars	Continuous with tension and compression bars	Diagonal
8	Ridley-Cammell	Angle bars straight, and corrugated sheeting	Angle or other bars straight	Trough of corrugated sheeting	Riveted or bolted	Continuous plate
9	Wells	Twin round bars connected by short web	Round straight bars	Steel strip hangers and bonders	Bent round tension bars	Vertical
10	Williams	Rolled steel sections, straight	Rolled steel sections, straight	(a) Round bars (b) Rolled steel sections (c) Spiral coils of steel wire sometimes used in addition	(a) Ends split for anchorage in concrete (b) Riveted or bolted to tension and compression bars	(a) Vertical (b) Diagonal (c) Spiral

Source: *Concrete & Constructional Engineering*, 1907–8, Vol. 2, p. 433

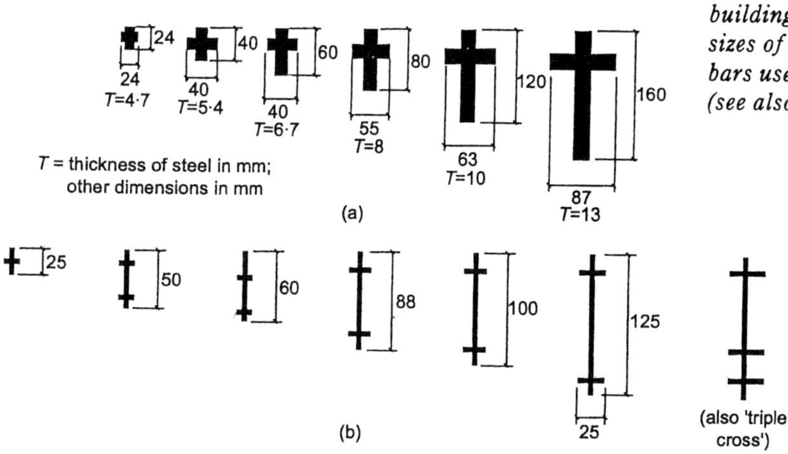

Fig. 1. Bonna (also known as Columbian or Feresco): (a) sizes of Latin Cross sections used in building frames; (b) sizes of Double Cross bars used in floors (see also Fig. 32)

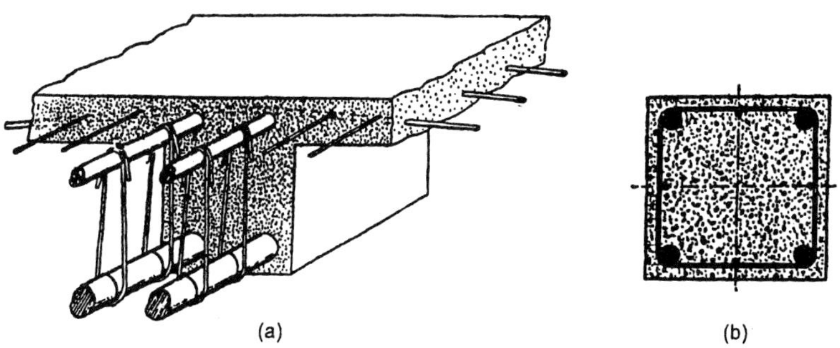

Fig. 2. Coignet system using plain round bars: (a) beam and slab; (b) column (see also Fig. 26)

Fig. 3. Types of Corr Bars

made for load testing if required. (This was common practice at that time.)

13. The deflection criterion for passing the load test was onerous by today's standards: under $1\frac{1}{2}$ times the imposed load, the structure was not to deflect more than 1/600 times the span! It should, however, be borne in mind that such structures were designed using elastic modular ratio theory, with permissible stresses in steel typically of 16 000 lb/in^2 (110 N/mm^2) and in concrete of 600 lb/in^2 (4·1 N/mm^2), so that the section sizes and the reinforcement were more generous than a present-day design would require. A further factor—certainly for the floor slabs—is that the framing plans usually provided beams in two directions (on the precedent of iron and steel frames). Consequently, a floor slab was supported on all four sides and, when loaded, would tend to behave more as a shallow dome in compression than as a slab in flexure.

14. A typical if early Hennebique building structure is the Co-operative Wholesale Society (CWS) warehouse on Quayside, Newcastle upon Tyne[5,8] (Fig. 52). The building dates from 1897–1900 (although the ninth floor was added in 1908). The foundation is a raft in view of poor ground conditions—in reality the raft is a series of haunched slabs spanning between substantial ground beams, these base slabs being 7 in (178 mm) thick at their centres and the building grid being 14 ft 6 in (4·42 m) square. Floors were designed for an imposed load of 6 cwt/ft^2 (32·3 kN/m^2 *sic*), and the 7 in (178 mm) thick slabs were satisfactorily test-loaded to 50% above this.

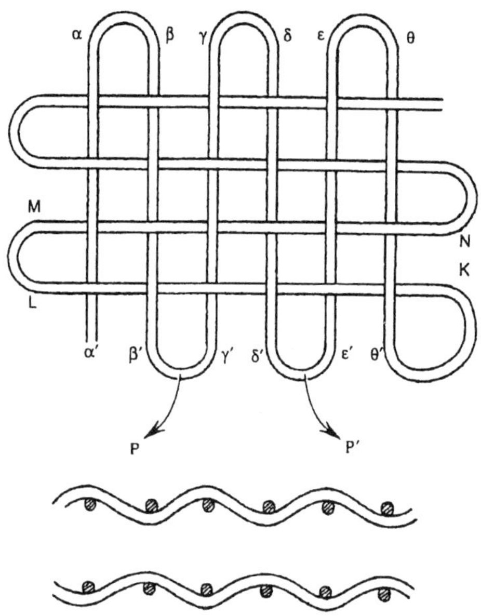

Fig. 4

Fig. 5

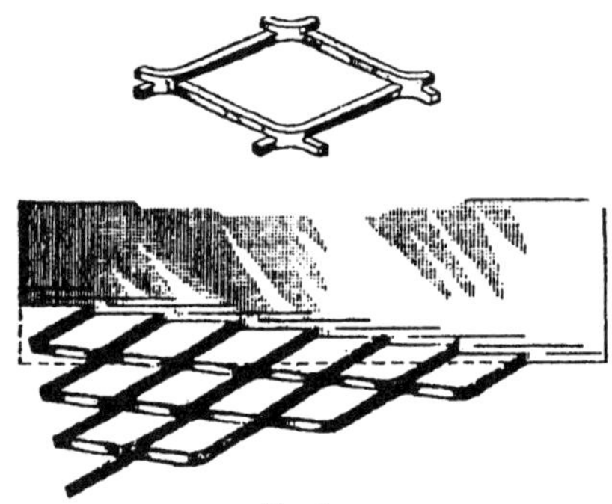

Fig. 6

Fig. 4. Cottancin: woven mat of small diameter rods and flats (see also Fig. 27)

Fig. 5. Diamond: rolled ribbed bar with constant cross-sectional area

Fig. 6. Expanded metal (Expamet): John F. Golding patented 'slashed metal screening' in 1883 in the USA

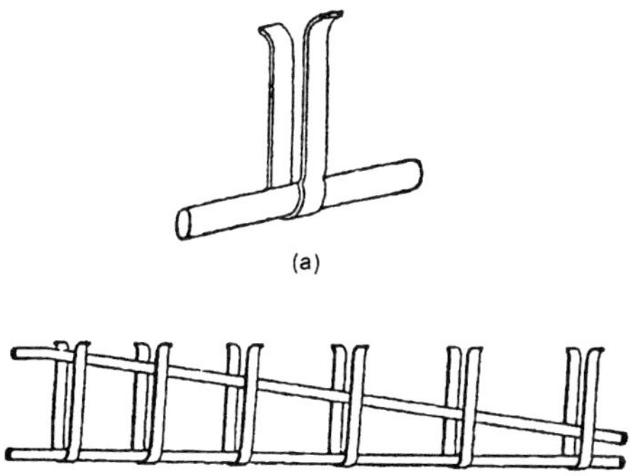

(a)

(b)

Fig 7

(a) (b)

Fig. 10

Fig. 11

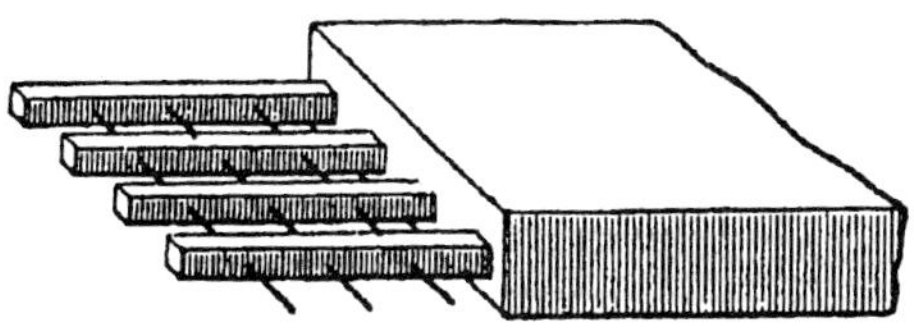

Fig. 8

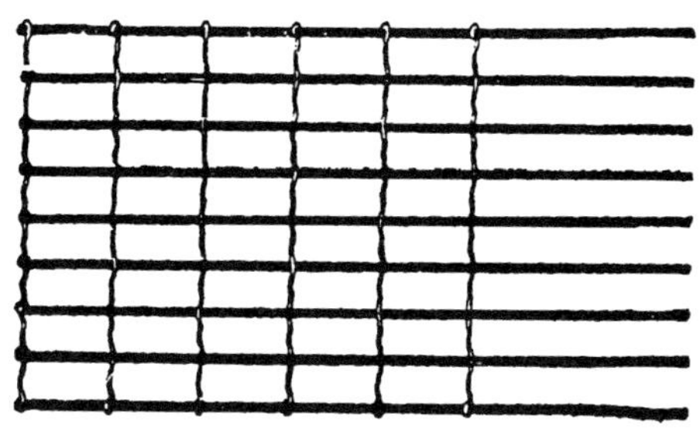

Fig. 12

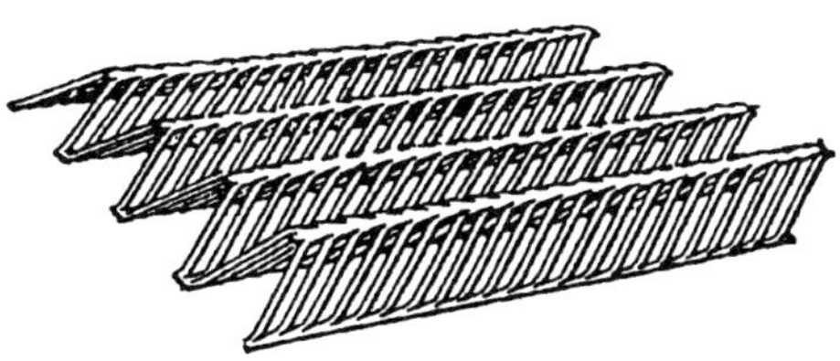

Fig. 9

Fig. 7. Hennebique: (a) stirrup round tension bar; (b) tension bars and stirrup (see also Fig. 34)

Fig. 8. Hyatt: early American reinforcement. Flat bars pierced with round bars, used in slabs and later in beams (also known as Jackson bars)

Fig. 9. Hy-rib (also known as Herring-bone metal lath): folded expanded metal that could act as both permanent formwork and reinforcement

Fig. 10. Indented Bar (see also Corr, Bar, Fig. 3 and Johnson

Fig. 11. Johnson: early deformed bars of high carbon steel with a yield strength of 65 000–70 000 lb/in^2 (448–482 N/mm^2), (also known as Corr, Corrugated or Indented Bars)

Fig. 12. Johnson: triangular or rectangular woven steel lattice

8 PROPRIETARY REINFORCING SYSTEMS

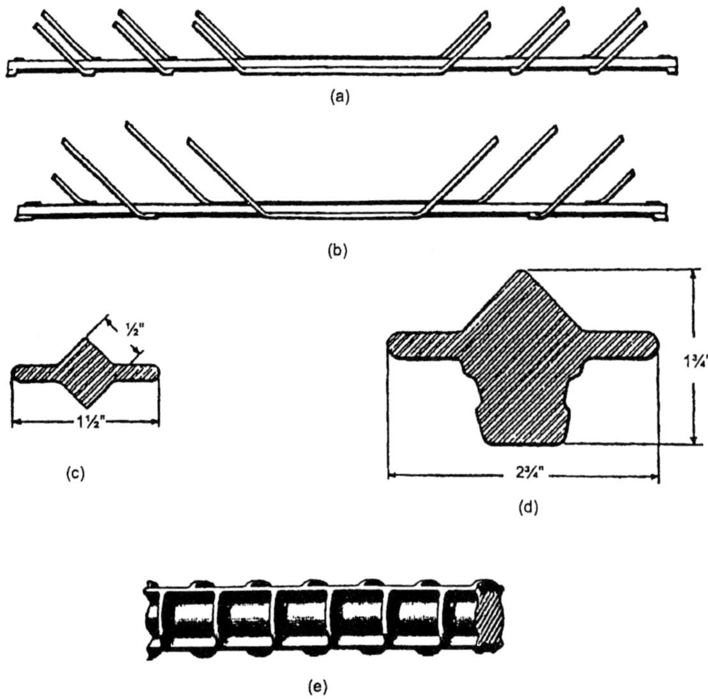

Fig. 13

Fig. 13. Kahn: unique trussed bars with fins rolled opposite corners of plain square bar and bent up to improve anchorage: (a) originally in pairs; (b) later staggered; (c) typical profile; (d) larger section; (e) a later and more conventional profile was the Kahn cup bar. Yield strengths 33 000–37 000 lb/in² (227–255 N/mm²) (see also Fig. 36)

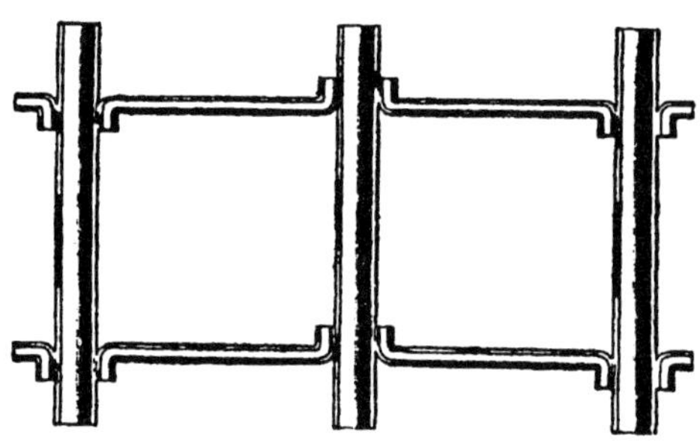

Fig. 14

Fig. 14. Kahn expanded metal: a sheared expanded rolled profile with the main bars normal to the secondary steel (unlike expanded metal)

(a) (b)

Fig. 15

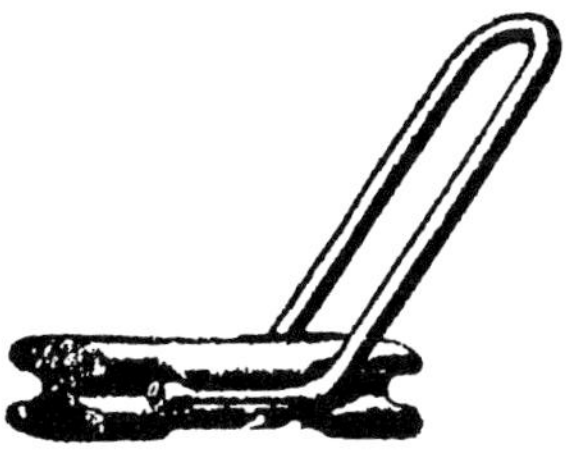

Fig. 16

Fig. 17

Fig. 18

Fig. 19

(a) (b)

Fig. 20

Fig. 15. Lock-woven or lock-tied mesh: (a) wires secured by mechanically-fixed wires or (b) by staples, in lieu of welding to provide an anchored mesh system

Fig. 16. Monolith: a grooved bar into which angled stirrups could be fitted, being held in place by swaging

Fig. 17. Ransome: the archetypical square-twisted cold-worked bar (also hot-rolled), with a yield strength from 53 000 lb/in² (379 N/mm²) (see also Fig. 44)

Fig. 18. Spiral: similar to Ransome, but with corner 'beads', with a yield strength 65 000–80 000 lb/in² (448–551 N/mm²)

Fig. 19. Square-twisted Lug: a further deformed (if not tortured) version of the Ransome square-twisted bar with additional lugs to improve bond

Fig. 20. Thacher: a deformed bar: (a) constant cross-sectional area without sharp re-entrant corners; (b) an early form was of flat strip with indentations

10 PROPRIETARY REINFORCING SYSTEMS

Fig. 21

Fig. 22

Fig. 23

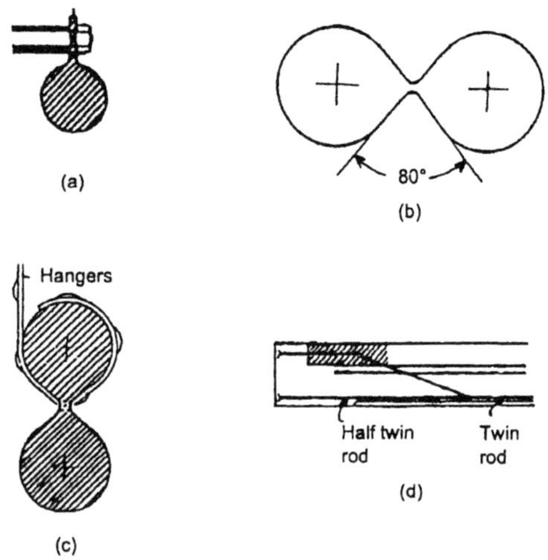

Fig. 24

Fig. 21. Truscon: a deformed bar with notches to accommodate washers to provide mechanical anchorage; also used the Kahn reinforcement system

Fig. 22. Unit: rolled bars (two or four) with linking continuous webs

Fig. 23. Universal: an indented bar not widely used

Fig. 24. Wells: (a) originally a single round bar profile with a flat continuous fin: which could be perforated to accommodate transverse bars (spacers in beams, links in columns, etc.); (b) later developed into twin round bar joined in an hour-glass profile; (c) perforated to receive indented or perforated hanger bars or; (d) slit longitudinally allowing one bar to be cranked up over supports as anti-crack steel

Procurement

15. How were such buildings procured? Clearly Hennebique and his competitors were in effect providing a 'design and build' service, although the basic structural layouts and performance specification would usually be defined by the client and his architects. At the turn of the century there were barely any independent consulting engineers practising, so far as building design was concerned. (Sven Bylander, for example, who had worked on steel framed structures in the USA, was employed by Waring White, its builder, for the structural design of the 1904 steel-framed Ritz Hotel in London—which incidentally had reinforced concrete floor slabs.) Architects would be generally competent to design most elements of the building, including foundations and structural masonry—which were largely sized by empirical rules relating foundation width to wall thickness, and wall thickness to height of building—and timber floors and roofs, for which tables and standard textbook designs were available. When it came to the design of steel frames or concrete, however, it was normal for the design—that is. the calculations and drawing-up of details—to be done by the 'specialists'. For steelwork this would be the steel fabricators, and for reinforced concrete it would be Hennebique, Truscon, Considère and others. Nor was their appointment necessarily a matter of selecting just one such company to implement the architect's scheme.

16. The example of the recently refurbished YMCA building in Manchester is illuminating.[9] This was originally conceived in 1908 by the architects Woodhouse, Corbett & Dean as a steel skeleton with brick walls. Two particular features of the building led to a decision to consider reinforced concrete. One was the provision of a swimming pool on the top floor, the box structure for this logically being best done using reinforced concrete. The other was the need to have a column-free hall at lower level. The original plan was to span this by storey-height steel trusses built into masonry walls on the floor above, but clearly it would be a sensible alternative to design the walls in concrete as deep beams, with considerable savings in steel.

17. The drawings were sent out to five specialists (of which one was the Trussed Concrete Steel Company), who were required to prepare schemes in concrete and to submit these together with tenders prepared by contractors, who were to be proposed by the specialists. So, the 'main' contractors were effectively the choice of the concrete specialists. This apparently back-to-front approach makes sense when the importance of good materials and workmanship, already stressed, is taken into account. Presumably the specialists would propose only contractors licensed by or approved by themselves to use their system.

18. Four of the five specialists submitted schemes with the tenders. It then appears that the architects spot-checked selected typical designed members from each scheme and calculated their strengths using the 1907 RIBA Joint Committee Report.[10] These calculations were compared with the required strengths.

19. Having accepted the Trussed Concrete tender using Kahn bars, the client required the final design by the specialists to follow the RIBA Report methods, with C. F. Marsh, author of the first British book on reinforced concrete,[5] acting as checking engineer. The specialists were required to pay his fee for this service! There were at this time no regulations for reinforced concrete construction in Manchester, and Mr Marsh's 'seal of approval' was noted as being most helpful in obtaining building control consent.

20. The picture one has, therefore, is of the design and construction of reinforced concrete at this time clearly being a specialist activity. Concrete specialists such as Hennebique were anxious, of course, to increase turnover and profits, but were also no doubt concerned that use of unsuitable materials and inexperienced labour, without established standards or codes of practice, could lead to failures and consequent loss of confidence in reinforced concrete as a structural material. They were willing to engage in competitive bidding against each other for work—indeed, they had no choice—and that necessarily meant preparing designs and pricing schemes competitively with no guarantee of payment. Against the previously quoted figure of 3061 projects realized by Hennebique in the period 1892–9 must be set 8078 not realized.[2]

21. The ethical position for a professional

engineer employed by a specialist concrete firm was in those days ambiguous; at least one ICE member was expelled for over-zealous efforts to 'procure' work for his employer. This eagerness may be explained (if not condoned by Edwardian professional men) by the fiercely competitive commercial environment.

The systems develop

22. Hennebique was, with Mouchel, for several years the only major concrete specialist in Britain. The company began working in Britain in 1897. The Trussed Concrete Steel Company took out a British patent in 1903 to cover use of its American Kahn bar.[11] In 1904 Edmond Coignet built a group of tobacco warehouses in Bristol, recently demolished, while the British Reinforced Concrete Co. (BRC) was established in 1905. Armand Considère took out a British patent in 1902 for the use of helical binding in columns, based on tests he had conducted which showed—as is now well appreciated—that adequate restraint to the compression bars is essential, both to restrain them from premature buckling and to enhance the strength of the concrete core. But this was not used until 1907. He later developed the U-hook to anchor main bars. Other forms of reinforcement included expanded metal, as illustrated in Fig. 6.

23. Anchorage and bond were recognized as important design matters, and the disadvantage of the plain round or square bar was one factor in the proliferation of patents for the use of deformed or profiled bars. Thus the Ransome and Kahn bars have already been noted; others include Johnson's indented bar, patented unsurprisingly by the Patent Indented Steel Bar Company, and much used by the British Concrete Steel Company. An advertisement in a contemporary textbook[12] by the former company notes that this bar profile has a continuous mechanical bond with the concrete, without slip. Significantly, it also points out that the bar is offered without fee or royalty, and may be used for any concrete system. In other words, the company was offering a product—the bar—to anyone who would buy it. This is in strong contrast to the 'concrete specialists', mainly from the Continent, who were offering a complete system.

24. The clients for early reinforced concrete structures included, paradoxically, some of the more conservative bodies such as the government, dock and harbour companies, railways, water boards, and agricultural and co-operative societies. Concrete was clearly economically competitive with steel and masonry from the outset, for if it had not been so then it would never have been considered. At the same time it was seen to offer other advantages. The embedded reinforcement appeared to be inherently protected against corrosion by the concrete—only later did it become embarrassingly clear that cover thickness, mix design, compaction and (particularly) water content were vitally important. The cover also fire-protected the reinforcement. Concrete was pourable, so that structures could be built with non-orthogonal forms at only slightly greater cost. The Manchester YMCA swimming pool has already been cited as a case where the mouldable nature of concrete, its structural strength and its potentially water-resistant nature combined to offer the 'ideal' material. Water tanks, silos and bunkers were obviously similarly suited to concrete construction.

25. The government was a quick and eager user of concrete. Crown buildings were exempt from building regulations, and many architects (then as now) were keenly interested in new structural developments. Sir Henry Tanner became Chief Architect in the Office of Works in 1898 and supported the use of reinforced concrete in its projects.[13] The King Edward VII building of the General Post Office in King Edward Street in the City of London[12] (Fig. 53) was built on the Hennebique system in 1907–10. It is no coincidence that Sir Henry took the chair of the RIBA Concrete Committee when it was established in 1905 (see Bussell[27]) and was an active member of the Concrete Institute (see Witten[28]).

26. The ready embrace of reinforced concrete by the public utilities (including railway companies and port authorities) and by commercial bodies was undoubtedly spurred by practical considerations of economy, including what was perceived as minimal maintenance cost. Today we may question this strictly utilitarian approach on aesthetic grounds. Unfortunately, too, durability was inadequately understood (see Bussell[27]). Weaver's Mill in Swansea (Fig. 51) was not a visual delight as it

stood abandoned in the early 1980s, and it was demolished despite being the oldest surviving reinforced concrete structure in Britain. It is not certain which structure now holds that title, although the 1897–1900 CWS Quayside warehouse in Newcastle (Fig. 52) is a contender. (It has recently been repaired externally while a new use is sought.)

27. Many concrete structures of this period were bold in concept. One of the boldest was the 1909 Royal Liver Building at Pier Head, Liverpool, architect W. Aubrey Thomas[4] (Fig. 54). One hundred and sixty-seven feet (51 m) high at roof level, and crowned by the Liver bird with an overall height of 310 ft (94 m), it offered transatlantic voyagers the sight of the nearest thing to a New York 'skyscraper'. The tallest concrete building in the world previously had been the 18-storey Ingalls building in Cincinnati, Ohio, of 1902, at 210 ft (64 m). This was reinforced on the Ransome system of twisted bars, mixed with plain round bars. Column main bars were spliced using early 'couplers'—sleeves of wrought iron tubing filled with grout. Writing of the building, Twelvetrees notes wryly that 'the architectural design is not remarkable for novelty',[12] and indeed it can be said that many early concrete buildings—as opposed to structures—did not advertise their concrete construction. From the outside, and often even from the inside, the structure could be of steel or of concrete—it was hard to tell. The expressionist use of concrete in habitable buildings had to wait for the birth of the modern movement (see Newby[29]). On the other hand, functional structures were usually clear in displaying their concrete form and finish.

28. This review suggests that, until the First World War, the concrete specialists 'had it all their own way' so far as design was concerned. While this is true in terms of the preparation of designs, drawings and schedules, it was often required of the specialists that they also satisfy an independent consultant. The example of Marsh and the Manchester YMCA building has already been described above. As for construction, the specialists either did the work themselves or licensed contractors who were guided by the specialists in sound construction practice, usually in return for paying a royalty for using the system.

Construction practice

29. Actual construction practice was inevitably more basic than today, although the principles of sound practice were little different. Materials were to be clean and properly stored; formwork was to be clean, rigid and secure. It was in the mixing, transporting, placing and treatment of concrete that the main differences were to be found.

30. All concrete was site-batched—there were no ready-mix trucks, no skips manoeuvred by tower cranes, no pumps. Hand-mixing was common, although machines were being used. Concrete was mixed by volume proportions (a practice that still survives today) or in 'recipe' form, as for the 1917 Mouchel specification in which 6 cwt (305 kg) of cement were typically to be batched with $13\frac{1}{2}$ ft^3 (0·38 m^3) of sand and 27 ft^3 (0·76 m^3) of coarse aggregate to give a 1·1 : 2 : 4 mix, with a probable cube strength of around 15–20 N/mm^2.[6]

31. Transporting was likely to be by chute or by wheelbarrow once the fresh concrete had been raised to the required level by a hoist. Placing would be by spade or shovel, working the mix as far as possible into its final location. It would then be compacted, up to a point, by ramming or 'punning' by hand. Punning tools came in a variety of shapes, including a rod with an enlarged box-end and another shaped like a hockey stick.[4] Not until 1917 did Eugène Freyssinet establish the essential importance of mechanically compacting the concrete to expel air and ensure thorough filling of the profiles with complete surrounding of reinforcement by concrete. Even then, it was 1924 before he used vibrating tools in practice. Before then, the importance of filling the shutters and fully surrounding the reinforcement with concrete was understood; but, without mechanical compaction, and with (often) congested reinforcement, the all-too-obvious answer was to make the concrete wetter and more 'self-compacting'. Some specialists, including Ransome, argued in the 1900s that a wet mix would indeed be superior to a dry mix.[11] It may be chance that Duff Abrams completed work in 1918, just a year after Freyssinet's researches on the need for compaction; Abrams's law declares that the water–cement ratio is a fundamental determinant of concrete strength. In plain terms, a

'wetter' concrete will be weaker when hardened than a 'drier' concrete.

Variations of technique and form

32. Various construction techniques and structural forms were introduced during this period which, at first glance, might be thought to be later in origin. Precasting, for example, was the subject of Hennebique patents as early as 1897, both as self-contained units and as permanent shuttering with projecting reinforcement to provide composite action with *in situ* concrete. Hurst[26] has described even earlier examples of precast units.

33. A simple precast floor system has recently been revealed in the 1904–5 Joshua Hoyle building in Manchester, which has a steel frame and, clearly an innovation for this time, an unattended automatic lift.[14] However, it is the floor units that are of most interest. They are precast unreinforced half arches (Fig. 55) that are laid in pairs to sit on the lower flange of the I-beams and to butt at the crown—pure three-pinned mass concrete arches. The coarse aggregate is clinker, and timber shavings as are nowadays found in woodwool slabs. Reuse proposals will have to bear in mind the potential combustibility of such concrete!

34. Hurst[26] has mentioned the filler joist floor slabs in which unreinforced concrete (spanning effectively as flat arches) was supported by I-beams or joists at typically 0·6–1·0 m spacing (Fig. 56). When this construction, also known as joist concrete, was introduced in the second half of the nineteenth century, the joists were of wrought iron, while the concrete would contain either lime or Portland cement as binder, with coarse aggregate of broken brick, slag, clinker or 'breeze' from coal fires, or stone. With the wider availability of steel from the 1890s, this displaced wrought iron for the joists, while the concrete was increasingly made with Portland cement, sand and stone aggregate, typically as a 1 : 2 : 4 mix.

35. This filler joist floor construction was widely used in offices and blocks of flats, notably the 'mansion blocks' of Edwardian times and later, where its high mass gave good resistance to air-borne sound. Timber flooring laid on battens (particularly when carpeted) also gave good structure-borne sound resistance.

36. Tests to assess the enhancement of strength by what is nowadays termed composite action between the filler joists and the concrete were carried out at the National Physical Laboratory as late as 1922–3.[15] The results were incorporated into BS 449[16] in 1932, and indeed remain in the 1969 metric edition still used today, but the wide adoption of orthodox reinforced concrete floors, both *in situ* and precast, resulted in the abandonment of filler joist construction, with its requirement of formwork for the invariably *in situ* infill concrete.

37. A problem often found with clinker concrete in such construction arises when the floor or roof becomes wet, as in neglected bathrooms or kitchens, or on flat roofs with degraded waterproofing. The clinker contains compounds of sulphur, and also of nitrogen and chlorine, which in wet conditions can severely aggravate corrosion of the iron or steel joists. Repair can be expensive and disruptive.

Publications of the time

38. It may be helpful to review the technical information and design guidance available to clients, architects and structural practitioners during this pioneering period of *c.* 1897–1915 (that is, from the first British Hennebique–Mouchel designs to the publication of the London County Council Reinforced Concrete Regulations, which offered the first 'code of practice' for reinforced concrete design[17]). Regulatory practice is covered in Bussell[27]; here attention is drawn to contemporary texts and specialist literature.

39. As already noted, *Reinforced Concrete* by C. F. Marsh appeared in 1904, being the first British textbook on reinforced concrete design and construction.[5] It is a substantial text, well illustrated, and can be fairly called a 'state-of-the-art' review. Later editions followed in collaboration with W. Dunn, which gave increasing space to calculation methods as the use of reinforced concrete widened.[18]

40. A more enduring record of concrete practice began for the general reader with the publication in 1906 of the first issue of *Concrete & Constructional Engineering*, which continued until 1966. This journal is a rich source of information on both individual buildings and practice of the time.

41. Essentially a house journal, but none

the less useful for that, was Hennebique's *Le Béton Armé*, produced from 1898. An English sister journal, *Ferro-Concrete*, began publication in 1909. Today this is a valuable if scarce source of detailed information for British Hennebique buildings such as those mentioned here. The early editor was W. Noble Twelvetrees, also the author of early guides to theory and practice in reinforced concrete.[12,19]

42. The Concrete Institute was set up in 1908 for architects, engineers and contractors, publishing its first *Transactions* in 1909 (see Witten[28]).

43. Other concrete specialists published literature, both technically supportive and openly promotional, during this period, while various 'independent' textbooks started to appear. The libraries of the Institutions of Civil and Structural Engineers (and also of the Royal Institute of British Architects) have good collections of useful material.

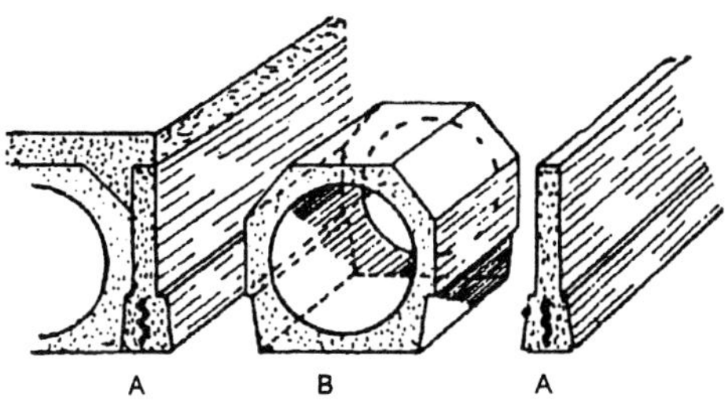

Fig. 25

Fig. 25. Armoured Tubular Floor: (a) precast rib with corrugated strip reinforcement; (b) fired clay, or coke or breeze concrete hollow infill blocks

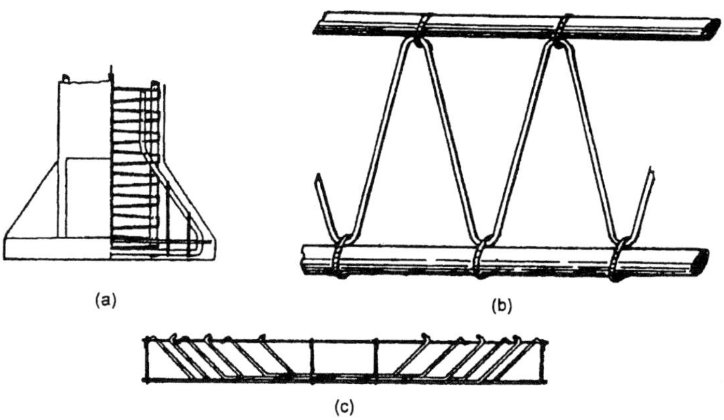

Fig. 26

Fig. 26. Coignet: (a) continuous top bars supporting stirrups with spiral links in columns; (b) stirrups sometimes formed of spiral rod wired to top and bottom bars; (c) bent-up bars used in small diameters as shear reinforcement (see also Fig. 2)

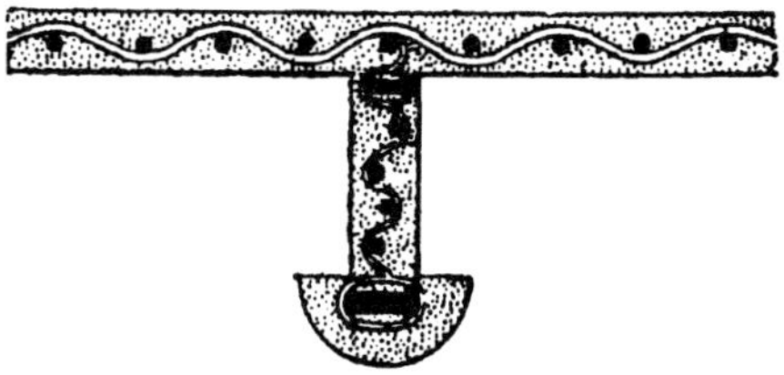

Fig. 27

Fig. 27. Cottancin: interlinked mat of small diameter bars. System closer to ferro-cement than 'true' reinforced concrete, with thin slabs reinforced by ribs often acting as folded plates (see also Fig. 4)

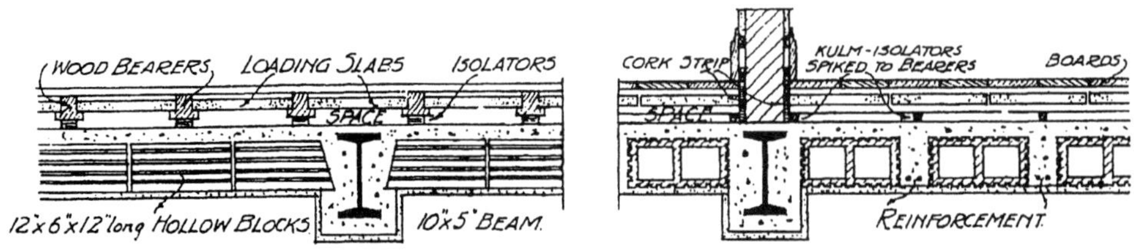

Fig. 28

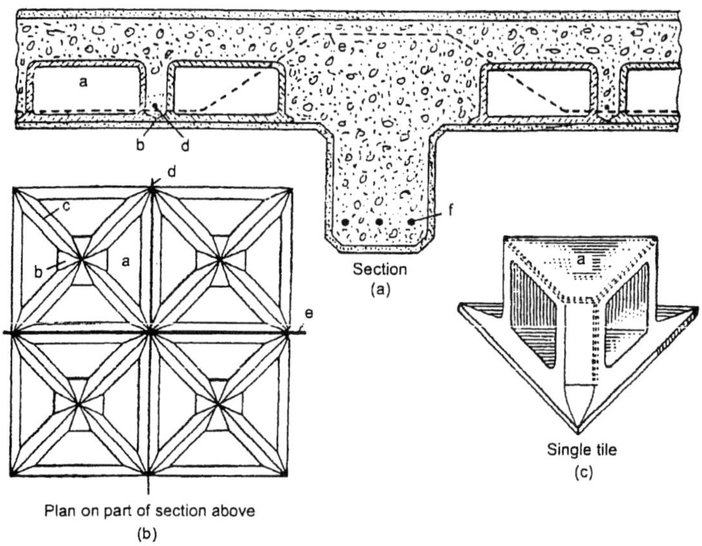

Fig. 29

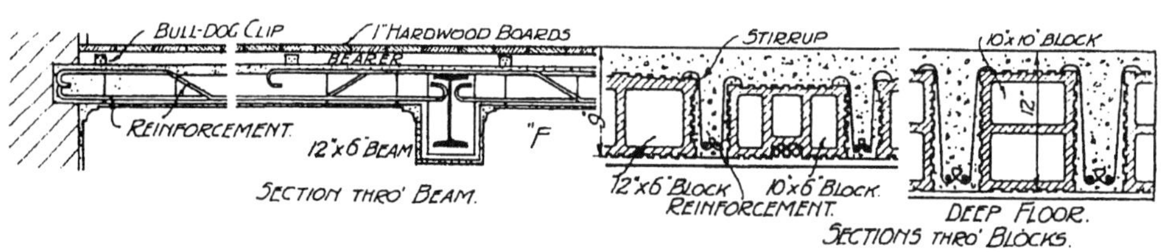

Fig. 30

Fig. 28. Cullum Floor: variant of Kleine Floor (see Fig. 39) with sound-proof raised timber floor (by courtesy of Cambridge University Press)[24]

Fig. 29. Dentile Floor: triangular clay mitre tiles interlocked to form voids within one-way or two-way spanning floor: (a) section; (b) plan on part of section; (c) single tile

Fig. 30. Diespeker Floor: variant of Kleine Floor (see Fig. 39) (by courtesy of Cambridge University Press)[24]

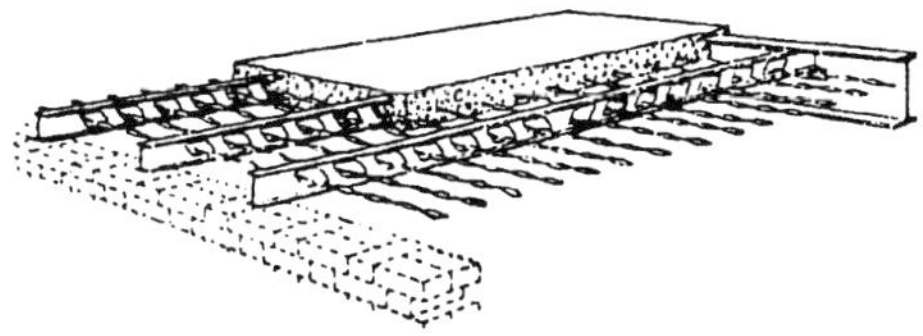

Fig. 31

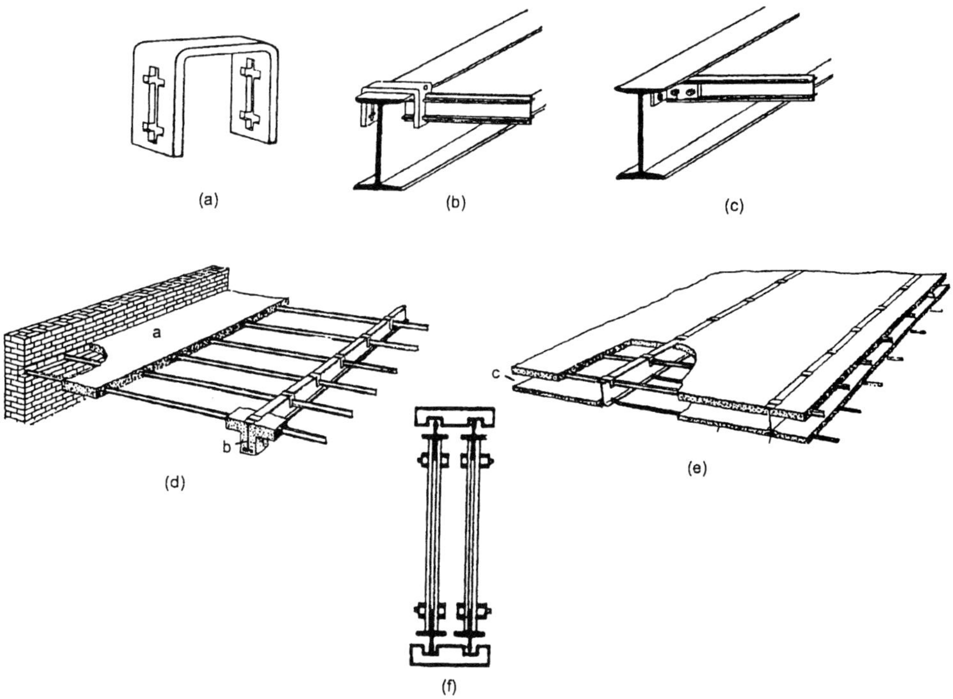

Fig. 32

Fig. 31. Fawcett (Monolithcrete) Floor: unusual floor with early perforated-web steel beams and transverse reinforcement from twisted flat bars (a form of filler joist construction)

Fig. 32. Feresco (known also as Bonna and Columbian. see also Fig. 1): distinctive cross-profiled rolled bars used as filler joists: (a) and (b) supported on hangers or (c) by bolting to beams; (d) used to form single and (e) double floors, the latter giving excellent sound-proofing and an early 'services void'; (f) bars in beams were spaced and framework could be supported by cross braces

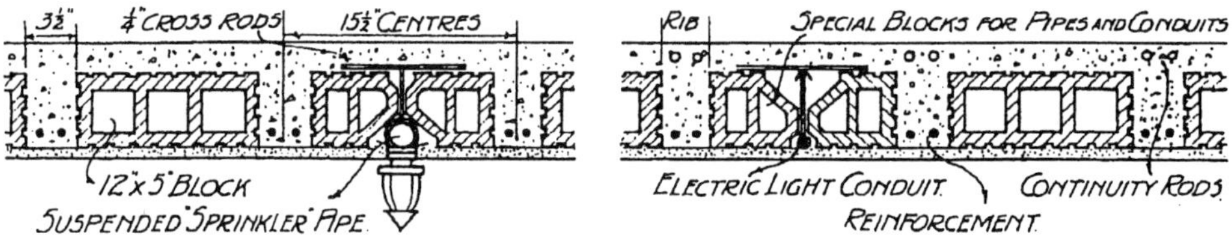

Fig. 33

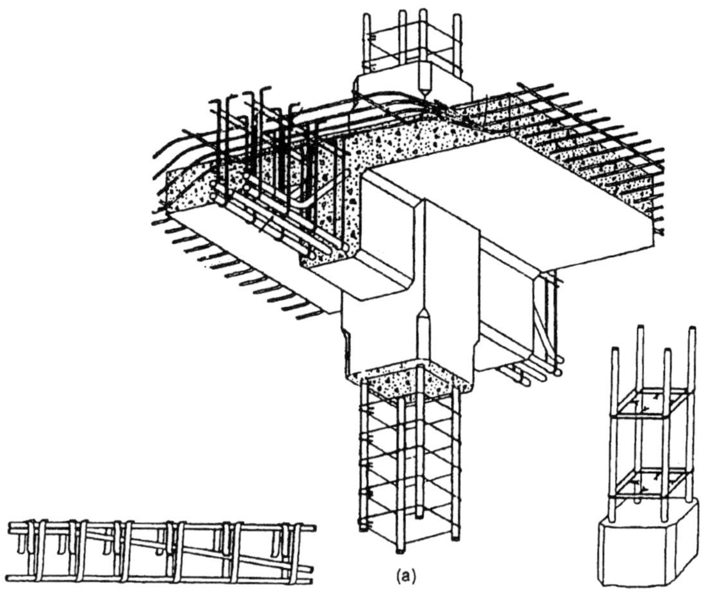

(a)

(b) (c)

Fig. 34

Fig. 33. Helicon Floor: hollow tile floors similar to Kleine (see Fig. 39) By courtesy of Cambridge University Press.[24]

Fig. 34. Hennebique: very widely used system, with distinctive deep narrow beams, often square columns and particularly noticeable chamfers on beam and column arrises: (a) bottom bars cranked up to form top steel over supports in continuous spans; (b) stirrups and links originally flat strips, later round bars; (c) columns sometimes linked by 'sausage' bars around main bars in some cases (see also Fig. 7)

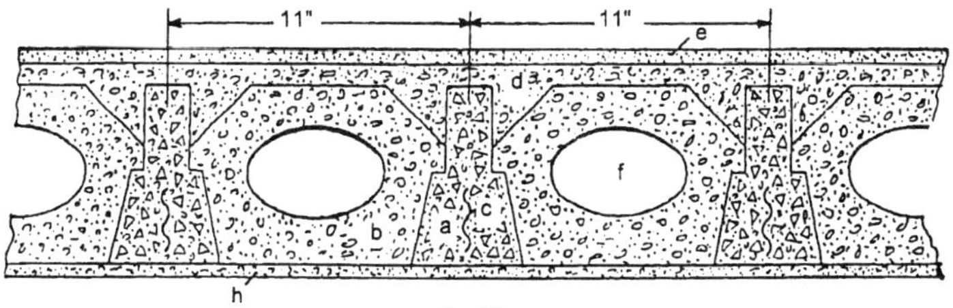

Fig. 35

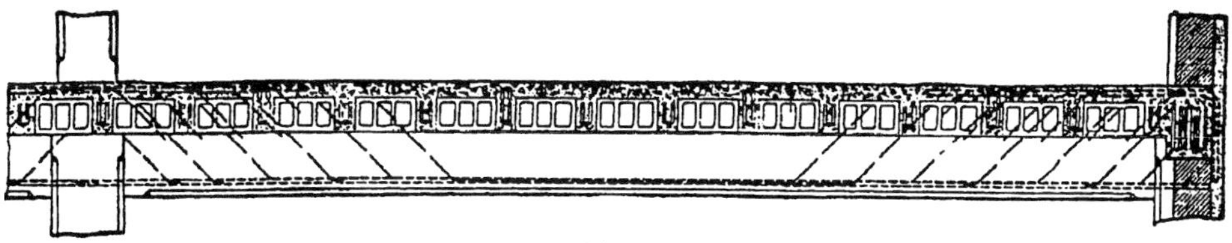

(a)

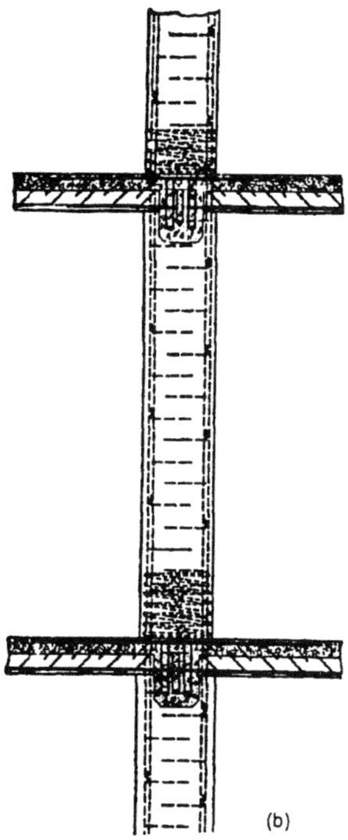

(b)

Fig. 36

(c)

Fig. 35. Herbst Tubular Floors: precast 1:3 cement: aggregate concrete ribs with corrugated strip reinforcement (typically 11 in c/cs) with 1:6 cement: breeze aggregate infill blocks

Fig. 36. Kahn (also Truscon): system employing the distinctive Kahn bars (see Fig. 13), with foldable fins that could be bent up to 90° from axis of main square bar to form: (a) beam stirrups or (b) and (c) column links

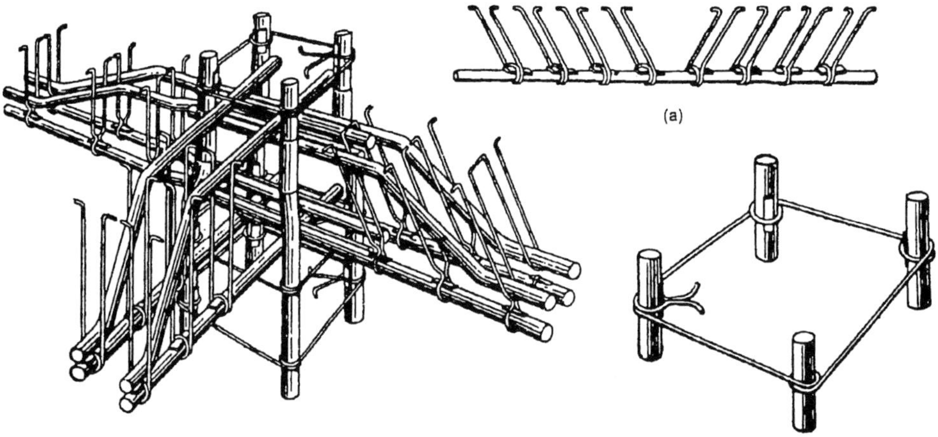

Fig. 37

Fig. 38

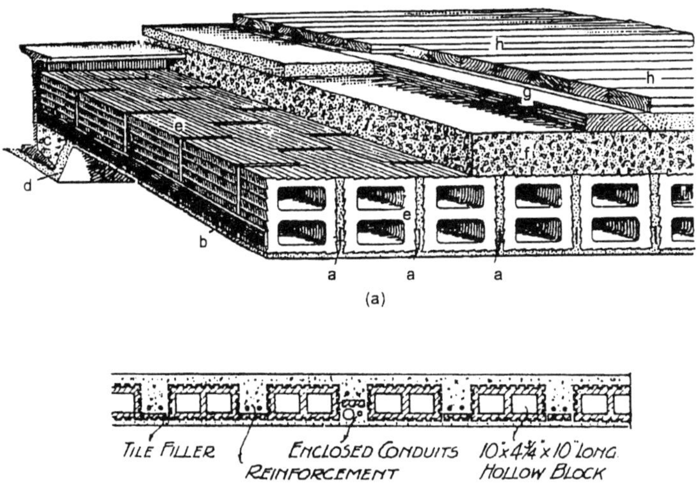

Fig. 39

Fig. 37. Keedon: system in which stirrups and links were mechanically 'keyed on' to main bars by wedges inside fully enclosing loops: (a) stirrups inclined to improve shear resistance in beams; (b) some bottom bars bent up in continuous spans; (c) column links

Fig. 38. King Floor: fireclay hollow tiles with reinforcement in ribs and concrete topping. King's Cross Floor profile is shown (By courtesy of Cambridge University Press).[24]

Fig. 39. Kleine Floors: early hollow clay tile floor, introduced to Britain in 1896, used by Higgs and Hill 1905: (a) reinforcement was flat steel on edge in cement mortar joints between blocks with coke-breeze concrete topping; (b) later, clay blocks and slip tiles were used with in situ concrete ribs and topping, usually with steel beams, often concrete-cased. (by courtesy of Cambridge University Press).[24]

Fig. 40. Monier: more commonly used in Germany where known as Wayss und Freytag; with round main bars bent up over continuous supports

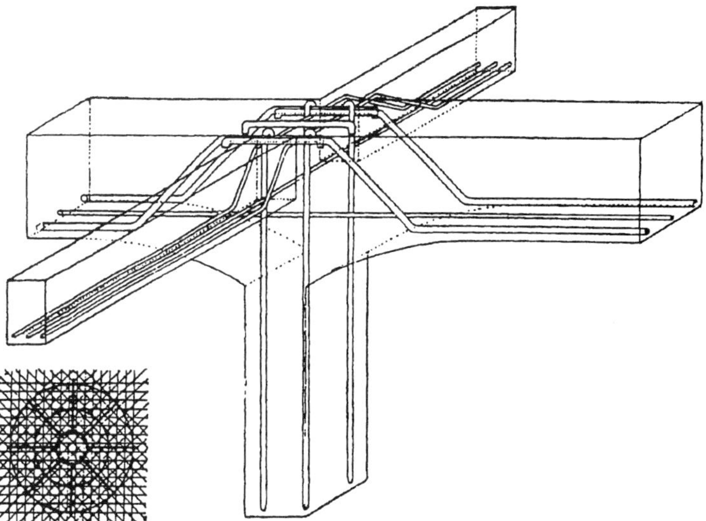

Fig. 40

(a)

Fig. 41. Mushroom: notable both for its flat slab construction and for the use of four layers of slab steel: (a) plan with broken lines suggesting 'mushroom heads' perspective; (b) indicating draping of slab reinforcement

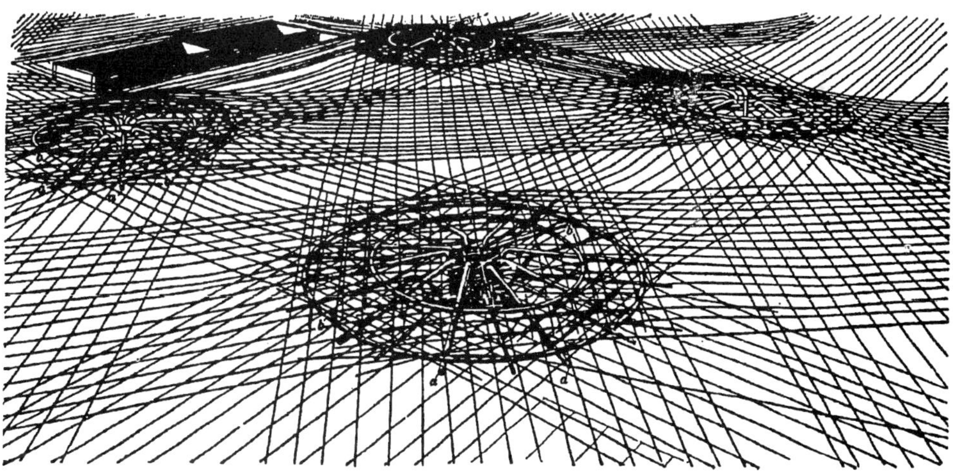

(b)

Fig. 41

22 PROPRIETARY REINFORCING SYSTEMS

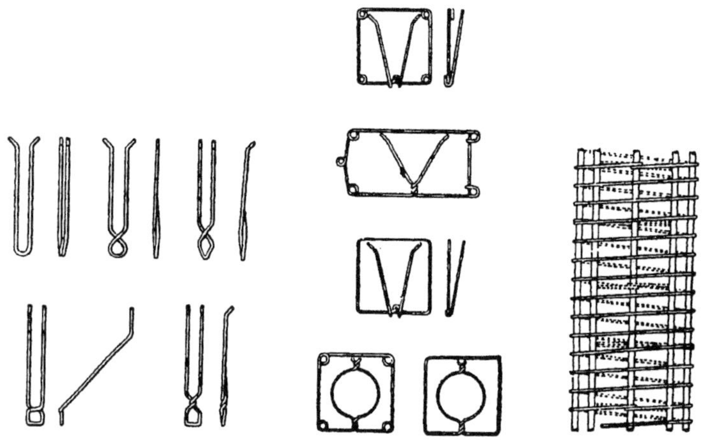

(a) (b) (c)

Fig. 42

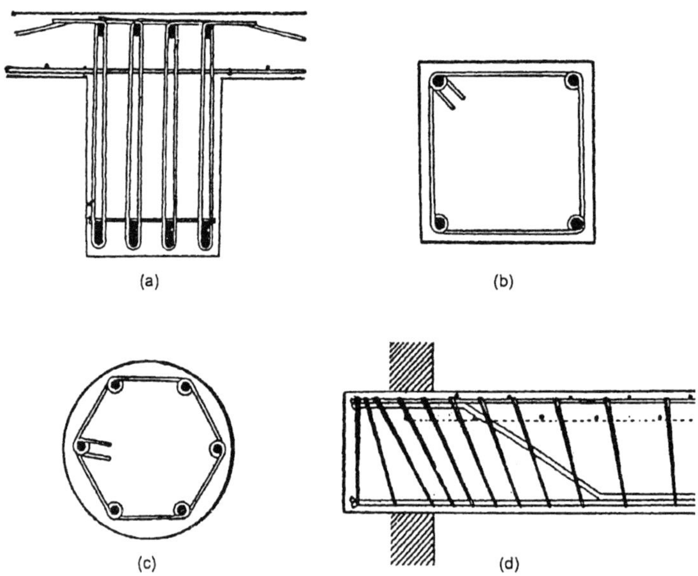

(a) (b) (c) (d)

Fig. 43

Fig. 42. Paragon: (a) and (b) distinctive stirrups and links to ensure mechanical anchorage to main bars; (c) helical column links

Fig. 43. Piketty: system with: (a) distinctive wrap-around stirrups in beams; (b) and (c) links in columns; (d) stirrups often gradually inclined near beam ends to enhance shear resistance

Fig. 44

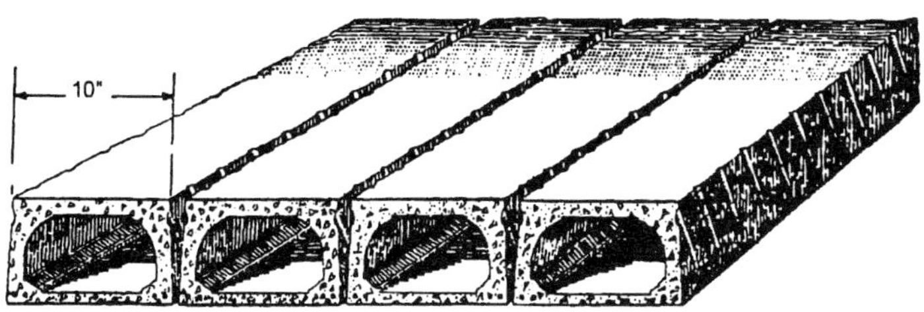

Fig. 45

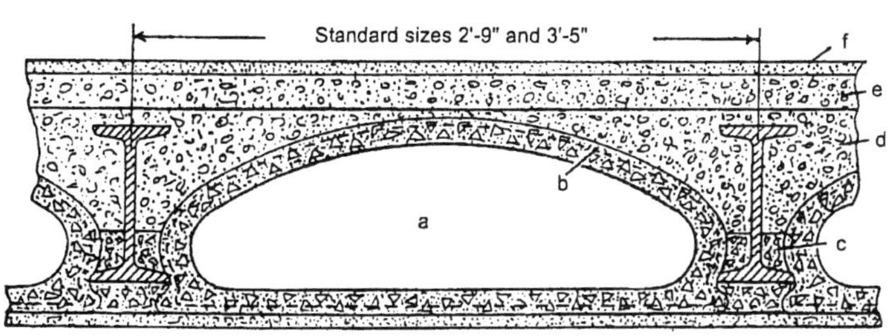

Fig. 46

Fig. 47

Fig. 44. Ransome: system using the Ransome cold-twisted square bars (see also Fig. 17), much employed in the USA. Early use of coffered slab construction for longer slab spans

Fig. 45. Siegwart Floor: hollow precast units originally reinforced with corrugated sheeting, later with round bars

Fig. 46. Somerville Floor: hollow concrete blocks infilling between steel joists at 2 ft 9 in – 3 ft 5 in c/cs, with in situ concrete topping

Fig. 47. Stolte Floor: German floors using reinforced hollow concrete blocks with flat bar embedded reinforcement

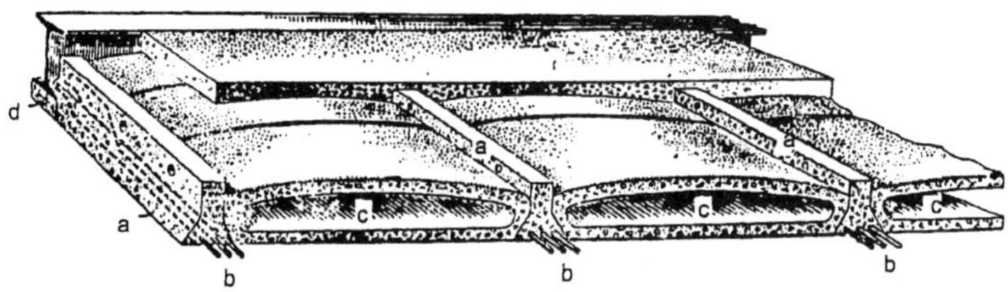

Fig. 48

Fig. 49

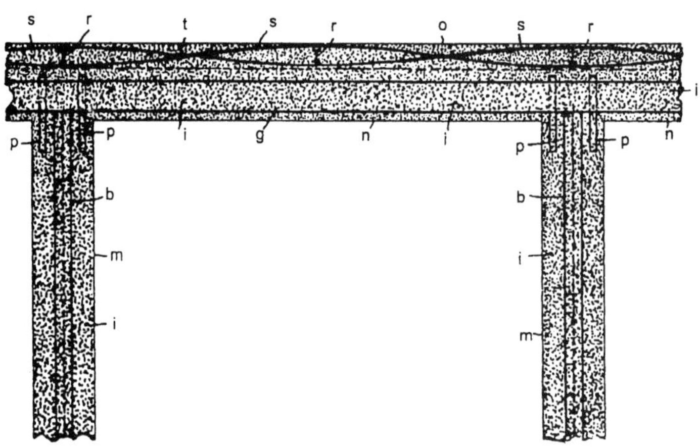

Fig. 50

Fig. 48. UK Floor: a precast concrete floor comprising ribs, at 3 ft c/cs typically, and infill blocks

Fig. 49. Unit: a prefabricated reinforcement system, originally using Corr bars (see also Fig. 3); later with plain round bars

Fig. 50. Williams system; filler joist reinforcement with draped transverse steel rods

Fig. 51. Weaver's Mill, Swansea (1897, demolished). (by courtesy of J. W. Figg)

Fig. 52. Co-operative Wholesale Society Warehouse at Quayside, Newcastle upon Tyne (1897–1900)

Fig. 53. General Post Office, King Edward VII Building, City of London (1907–1910). (by courtesy of British Architectural Library)

Fig. 54. Royal Liver Building, Pier Head, Liverpool (1909). (by courtesy of British Cement Association)

Fig. 55. Precast half-arch floor unit from the Joshua Hoyle building, Manchester (1904–1905)

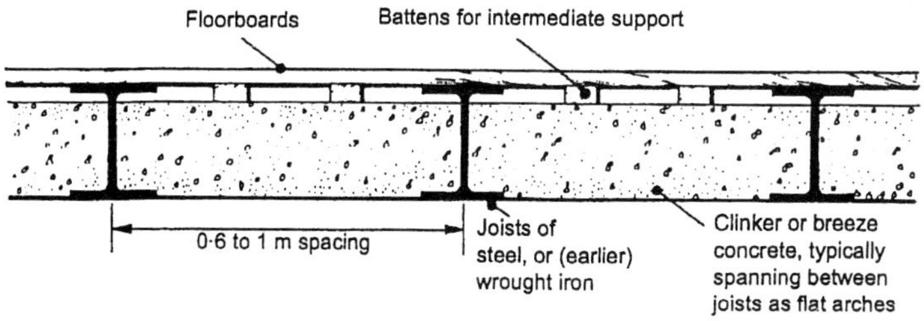

Fig. 56. Typical section through filler joist floor (by courtesy of Construction Industry Research and Information Association)[25]

The First World War and later

44. The outbreak of the First World War in 1914 brought many things to an end. For today's structural engineer, it can be said to mark the end of the 'golden age' of the proprietary systems. The 1915 publication of the LCC Reinforced Concrete Regulations put an acceptable method of designing reinforced concrete structures into the public domain.[17] Using plain mild steel bars, the engineer or numerate architect or builder could now do the calculations and design concrete structures. That is not to say that the specialists disappeared, but an element of 'free market' theory had arrived. Henceforward it would not need a specialist to design concrete structures that could be approved under regulatory by-laws. Nevertheless, companies such as Hennebique continued to design and build structures. L. G. Mouchel, for example, formed a consulting engineering practice that continues to this day. Companies making reinforcement continued to offer design services—indeed, some still do.

45. Another aspect of the First World War was, of course, the application of reinforced concrete to military structures, most commonly as protection against shelling. The defensive 'pill-box' has become for some an archaeological novelty, but it should be recalled that its design, however empirical, involved the first serious attempts to assess the effects of blast loading on reinforced concrete structures.[20]

46. Paper 5[27] continues this historical review of the development of structural concrete. It examines advances in design theory and practice from the era of proprietary systems, through early codification, towards the present day.

47. Additional general sources for the period covered by this paper are noted below.[21,22,23] Guidance for those called upon to assess or alter such structures is cited in the following paper.[27]

Acknowledgements

48. The author wishes to thank contributors to this series of papers for encouragement and helpful criticism in the drafting of this paper. The source of Table 1 was drawn to the author's attention by Mr B. N. Sharp.

References

1. Brown J. M. W. B. Wilkinson (1819–1902) and his place in the history of reinforced concrete. *Trans. Newcomen Soc.*, 1966–7, **39**, 129–142.
2. Cusack P. François Hennebique: the specialist organisation and the success of ferro-concrete: 1892–1909. *Trans. Newcomen Soc.*, 1984–5, **56**, 71–86.
3. Coignet E. and Tédesco N. de. *Du calcul des ouvrages en ciment avec ossature métallique.* La Société des Ingénieurs Civils de France, Paris, 1894.
4. Jones B. E. (ed.). *Cassell's reinforced concrete.* Waverley Book Company, London, 1920, 2nd edn.
5. Marsh C. F. *Reinforced concrete.* Constable, London, 1904.
6. L. G. Mouchel & Partners. *Standard specification for ferro-concrete.* L. G. Mouchel & Partners, London, 1917.
7. Engineering Standards Committee. *Standard specification for structural steel for bridges and general building construction.* Engineering Standards Committee, London, 1912, BS 15.
8. Anon. Ferro-concrete warehouse at Newcastle-on-Tyne. *Engineering*, 1903, April 17, 514–515.
9. Likeman A. The YMCA building, Manchester. *Concr. Constr. Engng*, 1911, **6**, 368–377 (and discussion, 501–515).
10. Joint Committee on Reinforced Concrete. Report of the Joint Committee on Reinforced Concrete, *J. R. Inst. Br. Archit.*, (3rd ser.) 1907, **14**, 513–541.
11. Hamilton S. B. *A note on the history of reinforced concrete in buildings.* HMSO, London, National Building Studies Special Report No. 24, 1956.

12. TWELVETREES W. N. *Concrete-steel buildings.* Whittaker & Co., London, 1907.
13. GRAY A. S. *Edwardian architecture: a biographical dictionary.* Wordsworth Editions, Ware, 1988, 2nd edn.
14. ANON. A Piccadilly improvement. *Manchester City News*, 1905, January 21.
15. ETCHELLS E. F. (ed.). *Modern steelwork.* Nash & Alexander, London, 1927.
16. BRITISH STANDARDS INSTITUTION. *Specification for the use of structural steel in building.* BSI, London, 1932, BS 449.
17. LONDON COUNTY COUNCIL, Reinforced concrete regulations. LCC, London, 1915.
18. MARSH C. F. and DUNN W. *Manual of reinforced concrete and concrete block construction.* Constable, London, 1908 (and various other editions).
19. TWELVETREES, W. N. *Concrete-steel.* Whittaker & Co., London, 1906.
20. MALLORY K. and OTTAR A. *Architecture of aggression: a history of military architecture in North West Europe.* Architectural Press, London, 1973.
21. DE COURCY J. W. The emergence of reinforced concrete 1750–1910. *The Struct. Engr.*, 1987, **65A**, 315–322 (and discussion of paper in *The Struct. Engr.*, 1988, **66**, No. 8, 128–130).
22. STANLEY C. C. *Highlights in the history of concrete.* Cement & Concrete Association, Slough, 1979.
23. MAIDSTONE R. J. *Developments in structural form*, Allen Lane, London, 1976.
24. JAGGARO W. R. and DRURY F. E. *Arch. Buildg. Constr.* (2nd ed.), 1937, **3**.
25. CONSTRUCTION INDUSTRY RESEARCH AND INFORMATION ASSOCIATION. Report 111, *Structural renovation of traditional buildings.*
26. HURST B. L. Concrete and the structural use of cements in England before 1890. *Proc. Instn Civ. Engrs Struct. & Build.*, 1996, **116**, 284–295.
27. BUSSELL M. N. The dsevelopment of reinforced concrete: design theory and prastice. *Proc. Instn Civ. Engrs Struct. & Build.*, 1996, **116**, 317–334.
28. WITTEN A. The Concrete Institute 1908–1923, precursor of the Institution of Structural Engineers. *Proc. Instn Civ. Engrs Struct. & Build.*, 1996, **116**, 472–481.
29. NEWBY F. The innovative uses of concrete by engineers and architects. *Proc. Instn Civ. Engrs Struct. & Build.*, 1996, **116**, 264–283.

11
Architects and the reinforced concrete specialist in Britain, 1905–1908

Patricia Cusack

The reinforced concrete specialist in Britain in the early 1900s was of recent vintage and essentially foreign (French) origin and it was as something of a 'parvenu' that he was criticized by British architects from the time that he became properly established.[1] Reinforced concrete then existed as a multitude of patented 'systems', or methods of disposing reinforcement, controlled by patentees.[2] Among these, the most prolific was François Hennebique's French-Belgian system. Hennebique's success was due not only to the system itself; it depended largely on the specialist commercial-technical organization which he evolved, and extended worldwide, to control both the design and construction of his reinforced concrete works (and which was imitated by others).[3] When Louis Gustave Mouchel became Hennebique's agent in Britain, from 1897–98, his business organization followed Hennebique's model with specially trained regional engineers to execute working drawings (after initial designs by an architect or engineer), contractors 'licensed' to execute them in return for royalties, and nationwide expansion.[4] By 1905, most reinforced concrete works and all framed buildings in Britain were in Hennebique's system. It was from about this date that growing criticism of the 'system specialist', and of Mouchel personally, was voiced there. Three institutions were initiated, which were intended both to curtail the 'specialist monopoly' of reinforced concrete in Britain and to promote the knowledge and use of the material among architects in particular. These were: the RIBA Committee on Reinforced Concrete (1905–07), a journal, *Concrete and Constructional Engineering* (1906) and the Concrete Institute (1908).

THE 'SPECIALIST MONOPOLY'

Sir Henry Tanner, Chairman of the RIBA Committee on Reinforced Concrete, maintained in 1908 that the general adoption of reinforced concrete in Britain was being hindered partly because it was practically controlled by 'specialists and patentees'.[5] The specialists' use of special systems was itself open to question. Charles Marsh, author of a substantial early text on reinforced concrete, observed before the newly-formed Concrete Institute, 'It could not reasonably be expected that reinforced concrete would remain as a multitude of various more or less rational systems, relying on more or less valid patents'.[6] Some patented features were better adapted to particular structural elements. *Concrete and Constructional Engineering* proposed that an ideal building would include: Considère and Indented Bar columns, Kahn beams and Hennebique piles.[7]

The promotion of particular systems by specialists in Britain and elsewhere was associated with a practice of trade secrecy which was also of concern to critics. Apart from preventing an interchange of knowledge among specialists, general knowledge about reinforced concrete was restricted. Indeed, caution was said to extend to withholding working drawings from a supervising architect.[8] Hennebique himself guarded his methods from other engineers and the press, although it is unlikely that he withheld information from an architect or engineer involved in one of his buildings.[9] Neither is there any evidence that Mouchel, in Britain, discouraged architects and engineers on the job from seeing working drawings or calculations. Indeed there is evidence to the contrary, since, if competent, they checked such work. However, G. P. Manning, who worked for L. G. Mouchel & Partners between 1913–15, states that at that time employees had to sign an agreement (not enforceable in law) not to work for another firm of reinforced concrete specialists within five years of leaving Mouchel's, to preserve the secrecy of their methods.[10]

Doubts were also cast on the reliability of this guarded expertise. William Dunn, who wrote part of the RIBA Report on Reinforced Concrete (1907) and had used Hennebique's system, in 1905 maintained that, 'In this country . . . the (reinforced concrete) experts, generally speaking, have very little scientific knowledge or training, and . . . few experiments are made or published here.'[11] In 1906, *Concrete and Constructional Engineering* pointed out that specialists' licensees were frequently ignorant about reinforced concrete or its constituents, and in 1907, the same journal attacked 'self-opinionated specialists whose obstinacy prevents their realising some of the early faults of this method of construction', claiming that the ' "infallibility" of a certain class of specialist' was now doubted.[12] It is true that Mouchel expressed scepticism about laboratory tests and 'theories' and that he tended to insist instead on the evident successes of Hennebique's system.[13] At the same time, Mouchel gained a considerable reputation as a civil engineer; he personally trained licensees, and the safety record for his works was excellent.

Finally, *Concrete and Constructional Engineering* referred to a belief among 'certain well-known architects' that Mouchel was trying to monopolize reinforced concrete construction in the UK and even alluded vaguely to his 'intelligence' department.[14] There were at least some grounds for this belief for Mouchel demonstrated an intolerance of other reinforced concrete systems. He publicly disparaged Considère's system (1904), and he prosecuted his own licensees who attempted to employ another system contrary to the terms of their licence (1906). Mouchel also took legal action against Coignet's representatives in Britain in 1906 ostensibly for patent infringement, although the case was unjustified and one of Mouchel's engineers claimed later that the motive had been commercial advantage.[15]

THE ARCHITECT'S ROLE IN REINFORCED CONCRETE DESIGN

Architects were frequently employed, together with specialists, for early reinforced concrete works, particularly framed buildings, but for various reasons they found their role unsatisfactory. Although usually uninformed about reinforced concrete, architects initially designed buildings for it in some detail. The specialist then adapted reinforced concrete to the architect's design, providing all structural details. However, the architect had overall charge of the supervision of the work. W. Dunn observed that while specialists (with their contractors) carried out the reinforced concrete work, the final responsibility for failure, despite the specialist's guarantees, lay with the architect.[16] Therefore, as matters stood, specialists calculated work which the architect or engineer was unqualified to check, although liable in a general way for its safety. R. T. Surtees said it was 'repugnant' to many to have to entrust work to patentees without being able to have it checked by an independent authority.[17] No doubt the combined offices of designer and contractor, which specialist firms effectively represented was also repugnant to some architects.

Architects were reluctant to give up their overall responsibility for design and supervision, however, therefore they had either to design in reinforced concrete themselves, or be able to check the specialist's work and properly supervise its construction.[18] Sir Henry Tanner wished to see architects accepting responsibility for reinforced concrete which was nominally theirs anyway.[19] Addressing the Architectural Association in 1909, he warned that architects would have to take up reinforced concrete, or be relegated to interior decoration and external design by the reinforced concrete specialists or, indeed, by the civil engineer.[20] C. A. Harrison, for example, regarded the architect's function in reinforced concrete design as embellishing the exterior or interior of buildings, while the engineer designed the framework: such construction, said Harrison, 'will necessarily add to the work of the engineer'.[21]

It was also suggested that architects should be able to design in reinforced concrete for aesthetic reasons.[22] Mouchel and his assistant T. J. Gueritte both attached importance to the architectural use of reinforced concrete and no doubt welcomed the employment of architects to give their material an artistic advantage.[23] In practice, since architects were usually ignorant of reinforced concrete, they customarily designed as if for traditional materials; hence, according to *Ferro-Concrete*, the usual imitation of stone buildings, although certain aesthetic preferences may also have been operative.[24]

There was undoubtedly interest in reinforced concrete among architects, perhaps more than among engineers, around 1900 in Britain. For example, the Northern Architectural Association had demonstrated an early interest in Expanded Metal reinforcement in 1899.[25] Two papers on reinforced concrete to the RIBA in 1904, by William Dunn and L. G. Mouchel, attracted a 'record audience', although it was augmented by a certain number of guest engineers and Portland cement manufacturers.[26] At tests (1905) of the North Eastern Railway Company's New Bridge Street Goods Station, Newcastle upon Tyne, eleven visiting architects were present, compared with seven civil engineers, excluding NER and local government officials.[27] William Bell, the architect of the station, was mentioned by T. J. Gueritte (1926) as one of the eminent British architects and engineers who 'by their open-mindedness' contributed to the spread of reinforced concrete in the first decade.[28]

WILLIAM DUNN AND THE RIBA JOINT COMMITTEE ON REINFORCED CONCRETE

William Dunn who by 1910 had become Consulting Architect to HM Office of Works to advise on designs tendered by reinforced concrete firms, apparently did not discuss the aesthetic aspects of reinforced concrete design in his many lectures on the material.

However, Dunn believed architects should themselves design in reinforced concrete and he was the main instigator of a report on reinforced concrete by the RIBA and other bodies, for architects' use, which was intended to make architects (or engineers) less dependent on specialists or on patent systems.

According to *Concrete and Constructional Engineering*, the origin of the RIBA Joint Committee on Reinforced Concrete (1905–06) was generally considered to be mainly due to the influence of Charles F. Marsh, who wanted some independent ruling in Britain for reinforced concrete, and to 'the desire of certain well-known architects (notably, William Dunn: see below) to thwart what they thought — rightly or wrongly — to be an effort on Mr Mouchel's part to make a monopoly of his system of design', by issuing independent guidance for architects and engineers.[29] E. O. Sachs said it was an, 'open secret that Mr Dunn was one of the primary movers . . . in obtaining for the architectural profession the Reinforced Concrete Committee which was formed by the RIBA . . .'.[30] In 1905, William Dunn had proposed a committee of the RIBA, possibly with the Institution of Civil Engineers, to prepare a standard specification for reinforced concrete; his object was to improve the architect's position in reinforced concrete design *vis-à-vis* the specialist, whose competence he doubted.[31] Such a committee was recommended by the RIBA Science Standing Committee, of which William Dunn was a member, and the recommendation was adopted by the RIBA Council in October 1905.[32]

The Building News in 1897 described William Dunn and his partner, Robert Watson, as among the 'leading English architects'.[33] By 1910, Dunn was also an associate member of the ICE.[34] He endeavoured to educate architects in modern construction and to inculcate, for example, that 'strength to the eye' was an inappropriate criterion.[35] He studied and advocated the use of reinforced concrete, of which he had a technical knowledge, and he lectured on the subject to the RIBA (1904), the ICE (1910) and to other bodies.[36] Indeed, Dunn was regarded as a 'leading exponent on all questions relating to reinforced concrete'.[37] He collaborated with Charles Marsh in a new edition of Marsh's *Reinforced Concrete* first published in 1904 (which ran into many more editions);[38] he was a member of the Concrete Institute at its formation and chairman of its Science Standing Committee.[39]

In 1905, Dunn said that he used reinforced concrete frequently.[40] He and Watson had employed Hennebique's system in 1903–04 in an extension to the premises of the Clerical, Medical and General Life Assurance Society, St James's Square, London, and they employed reinforced concrete for floors, for instance, in St John's Institute, Larcom Street, Walworth, London.[41] Robert Watson had assisted in the offices of (Sir) Rowand Anderson, who also employed Hennebique's system in 1910 for a chapel dome at Dunblane.[42]

In 1904 however, Dunn described Considère's spiral reinforcement for columns as 'more scientific' than systems such as Hennebique's.[43] Mouchel's somewhat unreasonable attitude towards Considère's system, and, indeed, Mouchel's dismissal of a test of a spirally reinforced column arranged and reported by Dunn, in 1903 and 1904, may well have provoked Dunn's concern to promote disinterested study of reinforced concrets in Britain.

The chairman of the RIBA Joint Committee was Sir Henry Tanner, at this time, and until 1913, the principal Architect and Surveyor at HM Office of Works. Tanner was especially interested in fireproof construction and in 1906 he personally recommended Hennebique's ferro-concrete for a large GPO extension at St Martins-le-Grand, London.[44] In 1907 Tanner said he had started 'about three' works with reinforced concrete, so he had some faith in it.[45] William Woodward commented that he had known Sir Henry Tanner for thirty to forty years and, 'no more careful man than he existed in the profession of architect'.[46] Before adopting ferro-concrete, Tanner had used coke breeze for floors for many years.

The RIBA President, when the Joint Committee was formed (1906) and its report adopted (May 1907), Thomas E. Collcutt, also had some prior interest in concrete and reinforced concrete. Collcutt had made a personal discovery of the use of concrete blocks for an old aqueduct near Algeciras, Spain,[47] and in 1904, he had observed tests of a Hennebique building in Tottenham, London.[48] Collcutt believed that 'no student (was) equipped . . . unless a builder as well as a designer of architecture', and that an architect should know the principles of modern construction, although he himself was ignorant technically of reinforced concrete.[49] During the discussion of the RIBA report, J. J. Burnet's nervousness about adopting it as a 'RIBA textbook' could possibly have afforded a lead not to do so, had not the president moved its adoption, even though the discussion revealed some doubts about the practicability of reinforced concrete.[50]

The RIBA Science Standing Committee had previously shown an interest in concrete; in its Annual Report for 1901–02 the committee said it was considering 'the subject of cement-concrete in flats and floors and are awaiting the result of some further experiments to . . . report on the behaviour of concrete'.[51] By March 1906 this committee had instigated — and provided a basis for — a joint committee of RIBA members and other interested bodies to draw up rules for architects' guidance in the use of reinforced concrete. The committee included representatives from the District Surveyors' Association, the Institute of Builders, the Incorporated Association of Municipal and County Engineers, the War Office, the Admiralty, and three individual members, Charles F. Marsh, Professor W. C. Unwin and Colonel F. Winn.[52]

The ICE was invited to join the committee, but refused.[53] *Concrete and Constructional Engineering* observed that for many years the ICE had treated reinforced concrete with 'indifference or hostility' and only later was it persuaded to take up the subject.[54] If the ICE Council was intransigent at this time with regard to the study of reinforced concrete, there was some interest in the subject earlier among individual engineers, apart from C. F. Marsh. In 1900, A. T. Walmisley claimed that tensile reinforcement in concrete was engaging engineers' attention.[55] W. Armstrong had adopted Hennebique's system for various buildings for the Great Western Railway Co., from the late 1890s. However, H. E. Steinberg, who joined Considère's company in Britain in 1909, recalled antagonism towards reinforced concrete among consulting engineers.[56] Lack of initiative in the use of reinforced concrete among engineers in the United Kingdom was not typical of other countries.[57]

A leading article in *The Builders' Journal* in 1907 commented that the committee members lacked a thorough practical knowledge of reinforced concrete, unlike the members of the French Government's Commission on Reinforced Concrete who included François Hennebique.[58] On the RIBA Committee there was no official representation of — nor consultation with — specialist designers, although they were consulted unofficially by some committee members. However, the view held of their expertise, apart from that of William Dunn, is perhaps indicated by a comment in the RIBA Journal with regard to the proposed report, that so far no responsible body in Britain 'no one man, even, whose name carries great weight' had advocated reinforced concrete — thus ignoring Mouchel's work in Britain and his advocacy of Hennebique's system, for example, in a lecture to the RIBA in 1904.[59]

The report was intended to provide an independent and unbiased working guide to the use of reinforced concrete; as such, not only did the committee decide to exclude any representatives of commercial systems, but its proposals avoided any preference for 'patent' features.[60] Efforts to be unbiased were regarded as precluding from consideration all 'special forms of bar', only referring to plain, round bars.

At the Joint Committee's first meeting (April 1906), the proposed scope of the report included as well as technical questions such as fire-resistance and the question of rust, how far building regulations should be altered to permit reinforced concrete walls, the contracting and carrying out of reinforced concrete work and the architect's or engineer's responsibilities *vis-à-vis* the specialist firms.[61] By December 1906 three sub-committees were appointed to deal with fire-resistance, materials and formulae, and the chief headings of the report were Materials and Methods of Calculation.[62] Although notes upon fire-resistance and on the need for altering by-laws to accommodate reinforced concrete were included, the respective roles and responsibilites of architects and specialists in reinforced concrete design were not. A possible limitation of scope may have resulted from the committee's intention to produce a report by the following spring. This it accomplished, as the report was issued on 27 May 1907. There was also an unexpected quantity of work involved in translating and studying 'the great mass of literature and records of tests upon reinforced concrete'.[63]

The RIBA Committee decided to collate information about reinforced concrete instead of carrying out additional testing, relying largely on foreign reports and experiments and a few tests on beams given to the committee by Cubitt & Co.[64] William Dunn, who contributed three of the five Technical Appendices to the report was also something of a linguist, which enabled him to translate foreign regulations for the Joint Committee.

The RIBA Report was the first 'official' pronouncement upon reinforced concrete in Britain, preceding the first ICE Report on Reinforced Concrete of 1910 by several years. *Concrete and Constructional Engineering* called it a 'momentous statement' which sanctioned a novel method of construction.[65] The report considered various technical and practical aspects of reinforced concrete construction, often without conclusions, because of the difficulty of prescribing general rules.[66] Col. Winn said that it was deliberately simple, without advanced mathematics.[67] The need for skilled workmanship was emphasized, no doubt in contradistinction to Mouchel's expressed view that 'no skill and no mental effort is required' in reinforced concrete construction. There is no aesthetic comment, but this was never part of the committee's 'brief'.

Advance copies of the report were sent in April 1907 to all RIBA members and Major E. M. Paul maintained that 'any one with this Report in his hands would be able to strike out confidently in the design of reinforced structures without having recourse to any particular patent system'.[68] The report was said to be widely used, but doubts were expressed whether it achieved its objects, because it was at once insufficient, and difficult for architects.[69] *Building Industries* thought it contained safe commonplaces together with abstruse matter timorously advanced, and that it represented a 'darkening of counsel', but *Concrete and Constructional Engineering* concluded it was the best guide available.[70]

EDWIN SACHS AND *Concrete and Constructional Engineering*

While the RIBA's Joint Committee was being constituted, Edwin O. Sachs started a new journal, *Concrete and Constructional Engineering* (March 1906), which he founded, managed and edited, with financial help from sympathizers as the journal could not pay its way on the basis of the popularity of the subject.[71]

Edwin Sachs was born in 1870 and educated in London and at the University of Berlin. He travelled extensively before commencing practice as an architect in London in 1892, where he was architect to the Royal Opera House, Covent Garden, reconstructed between 1898 and 1903.[72] Sachs had a particular interest in fire-prevention (especially in theatres) and at the age of 27 founded and chaired the British Fire Prevention Committee in 1897, with a broad-based membership, including many eminent architects.[73] This committee undertook a vigorous programme of tests of materials, including reinforced concrete, and testing stations were established abroad, for instance in St Petersburg, on the London model. Sachs published numerous papers and books on the subject of fire-prevention, especially in public buildings. His interest in reinforced concrete no doubt resulted from, and was certainly associated with, this interest; for example, a resolution proposed by Sachs at the International Congress of Architects in 1906 drew attention to conditions for the fire-resistance of reinforced concrete.[74]

In 1906 Sachs said that, as an architect, he was 'astounded at the ugliness' of examples of reinforced concrete buildings (not necessarily British). He believed that the design of reinforced concrete buildings should be kept in architects' hands although he also considered that 'concrete engineering works' had been aesthetically successful.[75] Later, he stated, in apparent contradiction, that concrete architecture was 'likely to find its salvation in the hands of the civil engineer'.[76]

Sachs described the current organization of reinforced concrete work in Britain as a 'harmful' situation of '*laisser-faire*'.[77] This was inaccurate, since Mouchel's safety record for buildings, for example, was good. None the less, from its foundation in 1906, *Concrete and Constructional Engineering* maintained an editorial line opposing the commercial-specialist 'monopoly' of design and execution, which was largely Mouchel's. Between 1906 and 1908 the journal had clear objectives, evident in its editorials. These were, the independent propagation of reinforced concrete, with unbiased reports, independent design by engineers or architects, selecting from systems as required, and lastly, the amendment of building laws to encourage reinforced concrete construction, by extending the periods for Local Government Board loans, which were restricted for reinforced concrete.[78] This may have been of particular interest to Sachs, because such loans influenced the use of reinforced concrete for public buildings. In accordance with the first objective, it was the policy of *Concrete and Constructional Engineering* not to gloss over failures of reinforced concrete and to urge independent investigation. Its coverage of 'interesting failures in reinforced concrete construction whilst strongly advocating that method' received more comment than any other point in the journal and the policy was generally approved, except by 'a few all too optimistic . . . engineers and . . . representatives of the "specialist systems"'.[79]

The journal favoured the designation 'reinforced concrete', which was not apparently associated with a particular system, to names which were used by particular specialists — ferro-concrete (Hennebique's system), armoured concrete and concrete-steel — which it dismissed as 'fancy descriptions'.[80] Advertisements in *Concrete and Constructional Engineering* promoted only 'independent' contractors, such as D. G. Somerville & Co., who expressed their willingness to construct in any system and used Kahn's, for instance, which did not require a licence.[81] An editorial commented that 'the most difficult problem . . . has been to retain absolute independence in respect to the conflicting interests of the various 'specialist' concrete systems, contractors and traders'. However, the journal gave less attention to the Hennebique system than the extent of its use justified, and it favoured, for example, the Patent Indented Steel Bar Company.[82]

Concrete and Constructional Engineering combined an internationalism of content and comment with a somewhat jingoistic attitude to foreign specialists in Britain such as Mouchel. Unsigned articles referred to 'pretentious foreigners (who) attempted to monopolise reinforced concrete', and asserted, vaguely, that the ways of foreigners working 'so-called systems . . . are anything but in accordance with British traditions'.[83] This was perhaps a reference to the method of licensing contractors instead of allowing 'free competition',[84] although in 1907, it was stated, again without elucidation, 'we . . . do not always agree with Mr Mouchel's methods of propagation'.[85] The journal's claims that Mouchel had had 'linguistic difficulties' and that 'he only took up civil engineering at a somewhat late period of his life' both appear to be unjustified.[86] The RIBA Report was welcomed as representing, 'independent data compiled by British colleagues'.[87]

THE CONCRETE INSTITUTE

The Concrete Institute was again largely the creation of E. O. Sachs. According to S. B. Hamilton, it was formed to pre-empt a combination of specialists which was proposed in a circular from a specialist firm to others in 1907, and to provide instead a more broadly based permanent association.[88] Therefore, like the RIBA Committee on Reinforced Concrete, the Concrete Institute was to some extent a 'defensive' response to specialist initiative; it was also instigated in default of interest in reinforced concrete by the major engineers' institutions, although *The Builders Journal* (1908) maintained the call for a Concrete Institute (to parallel the Iron and Steel Institute) was 'pressing and widespread'.[89]

The Concrete Institute was being formed in January 1908, and by February had its 100 founders. At a lunch held by Sachs at the (steel-framed) Ritz Hotel, in July 1908, prepared regulations were agreed and the Earl of Plymouth was elected first President, a titular head to give prestige to a small, new society.[90] The objects of the Concrete Institute were twofold: to advance the knowledge of concrete and reinforced concrete and, like *Concrete and Constructional Engineering*, to refrain from indulging in 'unnecessary laudation', and, to provide a means of communication between the different professions and trades involved with reinforced concrete.[91]

From the beginning, a beneficial meeting of different interests was anticipated and a better cooperation was hoped to result.[92] The Concrete Institute is interesting as including a mixed membership with diverse interests in reinforced concrete: it was not only a civil engineers' institution, as sometimes implied, although *Concrete and Constructional Engineering* claimed that only civil engineers were specifically invited to join, originally.[93] The British Fire Prevention Committee, of which Sachs was also chairman, and some of whose work the Concrete Institute anticipated taking over, itself comprised a broad, professional membership. Institute members could be: those professionally or practically engaged in the application of concrete/reinforced concrete or the production of their constituents, persons of relevant scientific, technical or literary attainments, or 'special subscribers' such as public authorities. Unlike the RIBA Committee, the Concrete Institute was not intended to exclude specialists or commercial interests. Sachs, Chairman of the Executive, viewed the Institute as combining different, often divergent, interests: engineers, architects, government officials, manufacturers and specialist firms.[94]

The vice-presidents represented architecture (Sir Henry Tanner), civil engineering (Sir William Preece F.R.S.) and industrial interests (Sir William Mather). Only British subjects could be council members, which would have debarred Mouchel, who remained a French citizen, had he lived. Council members in 1908, numbering about twenty, included the architects and the engineer most active in the RIBA Joint Committee, notably, William Dunn and Charles Marsh. The council also included heads of specialist firms, among them, J. S. E. de Vesian for L. G. Mouchel & Partners, G. C. Workman of E. Coignet Ltd, and L. Serraillier of the Patent Indented Steel Bar Company. Four main sections of membership were identified (1909): architects, engineers, specialists and cement manufacturers, with engineers and architects in the majority.[95]

The first president, the Earl of Plymouth, urged members to consider the aesthetic — as well as the practical — aspects of reinforced concrete. Beresford Pite proposed an Art Committee of the Concrete Institute and Charles Marsh suggested a role for the Institute in 'adapting this material to architectural uses'.[96] By 1911, with membership weighted towards London, there were 582 engineers, 91 architects and surveyors, 31 specialists, 45 chemists and cement manufacturers and 28 contractors, a considerably increased proportion of engineers, perhaps due to the technical bias of papers. *Concrete and Constructional Engineering* commented that with the somewhat diverse membership, '*esprit de corps* (was) sadly lacking' and meetings were not well attended; membership declined and in 1911 the president, Sir Henry Tanner, proposed a 'broader' scope as an Institution of Structural Engineers. *Concrete and Constructional Engineering* considered such a title would entail loss of prestige, since 'structural engineer' was not a professional term, being applied to steel contractors or to failed civil engineers.[97]

Perhaps Sachs's personal influence was significant in the momentum of the Concrete Institute; he had ceased to manage its affairs in 1909, but he was also absent altogether subsequently for a long period through illness.[98] Perhaps concrete and reinforced concrete had less need of a forum, partly because the Institute's educational function was being provided for by new classes in reinforced concrete. Tanner said that the Institute was unable to compete with the LCC Schools or London or other universities, in such courses.[99] The diversity of its members and interests was evidently a problem rather than a bonus. Finally, the ICE by now had 'a little concrete institute of its own'.[100]

No doubt prompted by the formation of the Concrete Institute, which had so far drawn about two hundred of its members, in December, 1908 the ICE appointed a committee to study reinforced concrete, none of whose members belonged to the Concrete Institute, and only one of whom (Professor W. C. Unwin) was a member of the RIBA Joint Committee. The ICE thus began work on reinforced concrete, but 'in "splendid" isolation' and *Concrete and Constructional Engineering* pointed to their committee's inexperience of reinforced concrete practice, though Sir Henry Tanner, later, became a member.[101]

Unlike the RIBA Report, which avoided mentioning special systems, the ICE Report catalogued them although with significant omissions. It referred to the experiences of a small sample of engineers with reinforced concrete, including their relation to specialists, and to examples of work without giving reinforcement details.[102] Although the committee recognized Mouchel's important role in introducing reinforced concrete in the United Kingdom it did not refer to the work of the British Fire Prevention Committee, nor to that of the RIBA Joint Committee. Furthermore, *Ferro-Concrete* (1910) described part of the Report as out-of-date.[103] Unlike the RIBA Report, whose purpose to provide a practical guide for architects was clear, if not wholly achieved, the ICE's preliminary report does not appear to have had a defined object beyond an outline survey of current systems and contracting procedures.

CONCLUSION

Although, by means of his 'specialist organisation', L. G. Mouchel successfully established reinforced concrete construction in Britain, some architects and engineers, from about 1904–05, were dissatisfied with the conditions of reinforced concrete practice. Although architects or engineers had overall responsibility for reinforced concrete buildings, they relied for their structural designs upon specialists, whose methods were guarded from public knowledge, and on the specialist's contractor for the construction. This arrangement also entailed the use of special 'systems', one of which was selected for an entire job. There was also an impression that Mouchel was attempting to monopolize reinforced concrete construction in Britain.

There was therefore felt to be a need to encourage 'independent' design by architects and engineers, drawing on different systems as required with reference to some objective standards or rules for reinforced concrete, and with contractors free from obligations to particular specialists. Although the arguments for 'independent' instead of 'specialist' design applied equally to engineers, the initial impetus towards this end was mainly provided by architects, who wished to retain responsibility for designing reinforced concrete buildings. Furthermore, some architects believed that for aesthetic reasons architects themselves should be capable of designing in reinforced concrete.

This situation resulted in the formation of three organizations in Britain directed towards objective studies of reinforced concrete and towards educating architects and engineers in reinforced concrete design: the RIBA Joint Committee on Reinforced Concrete (1905), *Concrete and Constructional Engineering* (1906) and The Concrete Institute (1908). These organizations were largely responses to specialist initiative, and to some extent in default of interest in reinforced concrete by the major engineers' societies. They are all interesting examples of 'institutions' (which, effectively, endured) created by a very small number of active and influential individuals, the most prominent of whom were three architects: William Dunn, Edwin O. Sachs, Sir Henry Tanner, and an engineer, Charles F. Marsh.

While the RIBA Joint Committee maintained independence from specialists, *Concrete and Constructional Engineering* actively opposed the specialist system and 'pretentious foreigners', notably L. G. Mouchel. The Concrete Institute, on the other hand, encouraged a broad membership, including architects, engineers and reinforced concrete specialists 'to resolve differences'; by the time this was being constituted, L. G. Mouchel himself (who died in May 1908) was no longer a dominant influence or a supposed threat to British reinforced concrete practice.

The achievement of the RIBA Report or the Concrete Institute in educating architects or engineers in reinforced concrete design, may not have been significant: the usefulness of the RIBA Report for architects in this respect was questioned, and Sir Henry Tanner in his first presidential address to the Concrete Institute, commented that the Institute was not equipped to compete with other educational institutions for lack of finance. However, both, together with *Concrete and Constructional Engineering* and other journals, no doubt encouraged such bodies to include the subject in their syllabuses for both architects and engineers. William Dunn, in 1910, was one of the first lecturers on reinforced concrete at the University of London.[104] At the end of the first decade, however, the system specialist was still well-established in Britain.[105]

NOTES

Abbreviations

C&CE *Concrete and Constructional Engineering.*
CI The Concrete Institute.

1 This article is based on a chapter from Patricia Cusack, 'Reinforced Concrete in Britain: 1897–1908' (unpublished PhD thesis, University of Edinburgh, 1981), 351–94.

2 By 1904 there were over fifty different systems or methods of reinforcement available. Charles F. Marsh, *Reinforced Concrete* (London, 1904), 2.

3 Patricia Cusack, 'François Hennebique: The Specialist Organisation and the Success of Ferro-Concrete 1892–1909', *Transactions of the Newcomen Society*, 56 (1984–85), 1986.

4 Mouchel and his agency in Britain are discussed in Cusack, thesis, 291–349.

5 Paper to Royal Sanitary Institute, Cardiff, in *C&CE*, 3, no. 4 (September 1908), 263–64.

6 Charles F. Marsh to CI, in *C&CE*, 3, no. 6 (January 1909), 474.

7 'Editorial Notes', *C&CE*, 2, no. 1 (March 1907), 4.
8 'Chronicle', *JRIBA*, 13 (28 April 1906), 339; 'The British Attitude to Ferro-Concrete Construction', *Building Industries*, 17 (15 June 1906), 34.
9 F. Hennebique, 'Note sur les brevets et travaux de la maison Hennebique', *Le Béton Armé* (September 1900), 4; Paul Christophe, 'Le béton armé et ses applications', *Le Béton Armé* (April 1900), 1.
10 G. P. Manning, M.ENG., FICE, letter to P. Cusack, 9 October 1980; G. P. Manning, 'A Hennebique Bridge', letter to editor, *Concrete*, 11, no. 10 (October 1977), 20.
11 William Dunn, FRIBA, 'Reinforced Concrete Floors', letter to editor, *JRIBA*, 12 (8 April 1905), 373.
12 Editor's note, *C&CE*, 1, no. 4 (September 1906), 275; 'The Progress of Reinforced Concrete in Great Britain during 1906', *C&CE*, 2, no. 1 (March 1907), 8, 11.
13 L. G. Mouchel, 'Discussion on Reinforced Concrete', *JRIBA*, 12 (10 December 1904), 85.
14 'Report of the Committee on Reinforced Concrete', *C&CE*, 2, no. 3 (July 1907), 176. 'Editorial Notes: The Late Mr. L. G. Mouchel', *C&CE*, 3, no. 3 (July 1908), 180.
15 See: Cusack, thesis, 322, 335, 337.
16 Dunn, op. cit., 373.
17 R. T. Surtees, 'A Few Experiments and Tests with Reinforced Concrete', *C&CE*, 2, no. 1 (March 1907), 43.
18 For example see A. H. Alban Scott, MSA, discussion at CI, *C&CE*, 4, no. 1 (March 1909), 54; Dunn, op. cit., 373.
19 Sir H. Tanner, 'Discussion on the Report of the Joint Reinforced Concrete Committee', *JRIBA*, 14 (1 June 1907), 503.
20 Sir H. Tanner, addressing the Architectural Association, *C&CE*, 4, no. 6 (November 1909), 377–78.
21 C. A. Harrison, 'Reinforced Concrete Railway Structures', *C&CE*, 2, no. 3 (July 1907), 233.
22 Cusack, thesis, 196–97.
23 L. G. Mouchel, 'Monolithic Constructions in Hennebique's Ferro-Concrete', *JRIBA*, 12 (26 November 1904), 57, 59; T. J. Gueritte, 'Ferro-Concrete Construction', paper to Edinburgh Architectural Association, *Transactions*, 5 (1910), 27–28.
24 'Ferro-Concrete in Architecture', *Ferro-Concrete*, 2, no. 1 (July 1910), 2; Cusack, thesis, 193.
25 'The Use of Expanded Metal in Concrete', *The Builder* (15 September 1900), 231.
26 Harold H. D. Anderson, 'Discussion on Reinforced Concrete', *JRIBA*, 12 (10 December 1904), 87–94; 'Chronicle', *JRIBA*, 12 (26 November 1904), 62.
27 L. G. Mouchel, *The Hennebique Ferro-Concrete System* (London, 1905), 174.
28 T. J. Gueritte, 'The First Decade of Reinforced Concrete in the UK (1897–1906)', *C&CE*, 21 (1926), 92.
29 'Report of the Committee on Reinforced Concrete', *C&CE*, 2, no. 3 (July 1907), 176.
30 E. O. Sachs at CI, *C&CE*, 3, no. 6 (January 1909), 479.
31 Dunn, op. cit., 373; idem, 'Discussion on the Report of the Joint Reinforced Concrete Committee', *JRIBA*, 14 (1 June 1907), 502.
32 'Chronicle', *JRIBA*, 12 (10 June 1905), 516; Sir H. Tanner, 'Prefatory Remarks, Report of the Joint Committee on Reinforced Concrete', *JRIBA*, 14 (15 June 1907), 513.
33 *The Building News*, 73 (24 September 1897), 436.
34 J. Gibson Fleming, 'Review of Lectures on Reinforced Concrete by William Dunn, FRIBA, Assoc. ICE', *JRIBA*, 18 (30 June 1911), 588–89.
35 William Dunn, 'The Principles of Dome Construction — I', *The Architectural Review*, 23 (1908), 73.
36 W. Dunn, 'The Bending Stresses in Flat Rectangular Concrete Floors', *JRIBA*, 7, 369–73; idem, 'Construction and Strength of Reinforced Concrete Columns', paper to RIBA, *JRIBA*, 12, 21–46; 'New Books: Lectures on Reinforced Concrete delivered at ICE, November 1910', *C&CE*, 6, no. 9 (September 1911), 716–17; 'University of London: Mr. Wm. Dunn's Lectures', *JRIBA*, 17 (22 October 1910), 798.
37 Sachs, op. cit., 479.
38 Charles F. Marsh and William Dunn, *Manual of Reinforced Concrete* (London, 4th edn 1922).
39 'Chronicle', *JRIBA*, 15 (9 May 1908), 412; Sachs, op. cit., 479.
40 Dunn, op. cit., 373.
41 *Mouchel-Hennebique Ferro-Concrete. List of Works* (London, 1920); W. Dunn and R. Watson, letter to A. D. Besant, 4 September 1903; 'Current Architecture', *The Architectural Review*, 18 (July–December 1905), 184–85. On a visit to the CMGLA Society's buildings at St James's Square (July 1977) little evidence was found of reinforced concrete, which may have been used in some floors and roofing.
42 William Dunn, 'Robert Watson, 1865–1916', *JRIBA*, 23 (19 February 1916), 142; Cusack, thesis, 540.
43 Dunn, op. cit., 44.
44 Sir H. Tanner, *Evening Standard*, 25 February 1914, cited in: *Mouchel-Hennebique Ferro-Concrete* (London, 1921), 88; 'Editorial Notes', *C&CE*, 1, no. 6 (January 1907), 403; cf. *The Durability of Reinforced Concrete in Buildings*, NBS Special Report no. 25 (London, 1956), 3.
45 Tanner, op. cit., 502.

46 William Woodward, *C&CE*, 6, no. 4 (April 1911), 304.
47 T. E. Collcutt, *JRIBA*, 14 (22 December 1906), 120.
48 B. Hannen, 'Discussion on Reinforced Concrete', *JRIBA*, 12 (10 December 1904), 91; *Mouchel-Hennebique Ferro-Concrete. List of Works* (London, 1920).
49 T. E. Collcutt, *JRIBA*, 9 (1901–02) 415; idem, 'President's Address to Students', *JRIBA*, 14 (9 February 1907), 216–17; idem, 'Discussion on the Report of the Joint Reinforced Concrete Committee', *JRIBA*, 14 (1 June 1907), 503.
50 J. J. Burnet, ARSA, 'Discussion on the Report of the Joint Reinforced Concrete Committee', *JRIBA*, 14 (1 June 1907), 502. F. T. Reade questioned the sanity of any engineer using reinforced concrete for girders or columns because of the heterogeneity of concrete and steel, ibid., 503.
51 'Report of the Council', *JRIBA*, 9 (10 May 1902), 335.
52 'Chronicle', *JRIBA*, 13 (24 March 1906), 271; 'Report of the Joint Committee on Reinforced Concrete', *JRIBA*, 14 (15 June 1907), 513.
53 William Dunn, 'American Committee's Report on Reinforced Concrete', *JRIBA*, 17 (20 November 1909), 87.
54 'Editorial Notes', *C&CE*, 5, no. 10 (October 1910), 703; 'The Inst. C. E. Report on Reinforced Concrete', *Ferro-Concrete*, 2, no. 5 (November 1910), 136.
55 'The Use of Expanded Metal in Concrete', *The Builder*, 79 (15 September 1900), 231.
56 H. E. Steinberg, *C&CE*, 51, no. 1 (January 1956), 31.
57 Dunn, op. cit., 87. The American Committee on Reinforced Concrete was initiated (1903) by the American Society of Civil Engineers; see also Professor Schüle, cited in 'Editorial Notes', *C&CE*, 3, no. 5 (November 1908), 349.
58 'The RIBA and Reinforced Concrete', *The Builders' Journal*, 25 (8 May 1907), 225; W. Noble Twelvetrees, 'François Hennebique: A Biographical Memoir,' *Ferro-Concrete*, 13, no. 5 (November 1921), 140.
59 'Chronicle', *JRIBA*, 13 (28 April 1906), 338; Mouchel, op. cit., 47–61.
60 'Report of the Joint Committee', *JRIBA*, 14 (15 June 1907), 514.
61 'Chronicle', *JRIBA*, 13 (28 April 1906), 338–40.
62 'Chronicle', *JRIBA*, 14 (8 December 1906), 96–97; 'Report of the Joint Committee', *JRIBA*, 14 (15 June 1907), 513 ff.
63 'Chronicle', *JRIBA*, 14 (8 December 1906), 97; Dunn, op. cit., 87; 'Report of the Joint Committee', *JRIBA*, 14 (15 June 1907), 513.
64 Tanner, 'Report of the Joint Committee', *JRIBA*, 14 (15 June 1907), 514. W. B. Wilkinson's pioneering work (1854–55) was acknowledged.
65 'Editorial Notes', *C&CE*, 6, no. 8 (August 1911), 563; see also: 'Editorial Notes', *C&CE*, 6, no. 9 (September 1911), 645; compare 'The RIBA and Reinforced Concrete', *The Builders' Journal*, 25 (29 May 1907), 259.
66 For example, 'Report of the Joint Committee', *JRIBA*, 14 (15 June 1907), 518.
67 Lt-Col J. Winn, 'The "Advance" in the Concrete Age', *C&CE*, 3, no. 1 (March 1908), 6.
68 Major E. M. Paul, RE, 'Discussion on the Report', *JRIBA*, 14 (1 June 1907), 500.
69 Dunn, op. cit., 86; *Ferro-Concrete*, 1, no. 2 (August 1909), 28; 'The RIBA and Reinforced Concrete', *The Builders' Journal*, 25 (29 May 1907), 259; John Slater, 'Discussion on the Report', *JRIBA*, 14 (1 June 1907), 498.
70 'Joint Committee on Reinforced Concrete', *Building Industries*, 18 (15 June 1907), 33–34; Editor's note, *C&CE*, 3, no. 3 (July 1908), 181.
71 *C&CE*, 51, no. 1 (January 1956), 1. *The Builder* and *The Builders' Journal* had already taken up the subject, for example, 'The Student's Column. Concrete-Steel', *The Builder*, 85 (4 July 1903), 19; *The Builders' Journal, Concrete and Steel Supplement*, 23 (January–June 1906).
72 *Who's Who in Architecture* (Westminster, 1914), 193; see also Max Clarke, *JRIBA*, 4 (26 August 1897), 444; *C&CE*, 51, no. 1 (January 1956), 2.
73 *C&CE*, 51, no. 1 (January 1956), 3; compare S. B. Hamilton, *A Short History of the Structural Fire Protection of Buildings Particularly in England*, NBS Special Report no. 27 (London, 1958), 31; *The Builders' Journal*, 7 (February–August 1898), 272.
74 'Summary of Proceedings', Congress Number, *JRIBA*, 13 (1906), xxxii.
75 E. O. Sachs, *Building Industries*, 17 (16 May 1906), 17; 'Reinforced Concrete at the ICA', *C&CE*, 1, no. 4 (September 1906), 291; *Transactions*, RIBA (London, 1908), 246.
76 E.O. Sachs, discussion at CI, *C&CE*, 6, no. 7 (July 1911), 498.
77 E.O. Sachs, cited in 'Editorial Notes' *C&CE*, 3, no. 4 (September 1908), 262.
78 For example, 'Characteristics of the Chief Systems of Reinforced Concrete Applied to Buildings in Great Britain', *C&CE*, 2, no. 6 (January 1908), 428; 'Editorial Notes', *C&CE*, 2, no. 1 (March 1907), 3–4.
79 E. O. Sachs, 'Reinforced Concrete at the ICA', *C&CE*, 1, no. 4 (September 1906), 293–94; 'Editorial Notes', *C&CE*, 2, no. 1 (March 1907), 1.
80 'Memoranda', *C&CE*, 2, no. 4 (September 1907), 320. For a brief discussion of the term 'reinforced concrete' see Cusack, thesis, 128. 'Editorial Notes', *C&CE*, 2, no. 5 (November 1907), 344.

81 D. G. Somerville & Co., advertisement, *C&CE*, 2, no. 6 (January 1908), 510; D. G. Somerville & Co., advertisement, *C&CE*, 3, no. 4 (September 1908), 344.
82 'Editorial Notes', *C&CE*, 2, no. 1 (March 1907), 1; 'Memoranda', *C&CE*, 1, no. 3 (July 1906), 214.
83 'The Progress of Reinforced Concrete in Great Britain During 1906', *C&CE*, 2, no. 1 (March 1907), 11.
84 Lucien Serraillier, 'The Reinforced Concrete Specialist', *C&CE*, 7 (1912), 94.
85 'Report of the Committee on Reinforced Concrete', *C&CE*, 2, no. 3 (July 1907), 176.
86 'Editorial Notes: The Late Mr. L. G. Mouchel', *C&CE*, 3, no. 3 (July 1908), 180.
87 'Report of the Committee on Reinforced Concrete', *C&CE*, 2, no. 3 (July 1907), 175.
88 S. B. Hamilton, *A Note on the History of Reinforced Concrete in Buildings*, NBS Special Report no. 24 (London, 1956), 15.
89 'Editorial Notes', *C&CE*, 3, no. 3 (July 1908), 177–78; 'The Concrete Institute', *The Builders' Journal, Concrete and Steel Supplement*, 27 (20 May 1908), 436; compare 'Chronicle', *JRIBA*, 15 (9 May 1908), 412.
90 'A Concrete Institute', *The Builders' Journal, Concrete and Steel Supplement*, 27 (29 January 1908), 95; 'The Concrete Institute', *The Builders' Journal*, 27 (26 February 1908), 197; Sir H. Tanner, *C&CE*, 6, no. 12 (December 1911), 930; E. O. Sachs, *C&CE*, 6, no. 7, 494.
91 'Objects of the Institute', *CI: Transactions and Notes*, 1, part 2 (April 1909); *C&CE*, 3, no. 6 (January 1909), 471.
92 'A Concrete Institute', *The Builders' Journal, Concrete and Steel Supplement*, 27 (29 January 1908), 95; 'Editorial Notes', *C&CE*, 3, no. 1 (March 1908), 1–2.
93 See Hamilton (1956), 16; M. Bowley, *Innovations in Building Materials* (London, 1960), 88–89; P. Collins, *Concrete*, (London, 1959), 134; 'Editorial Notes', *C&CE*, 6, no. 12 (December 1911), 890.
94 'The Concrete Institute', *The Engineer*, 105 (8 May 1908), 490; E. O. Sachs, 'Editorial Notes', *C&CE*, 3, no. 4, 262.
95 'Editorial Notes', *C&CE*, 3, no. 2 (May 1908), 89–90; 'Memoranda', *C&CE*, 3, no. 4 (September 1908), 341; F. A. White, *C&CE*, 4, no. 4 (September 1909), 290; 'Editorial Notes', *C&CE*, 6, no. 7, 483–84; 'Chronicle', *JRIBA*, 16 (8 May 1909), 483.
96 The Earl of Plymouth, cited in 'Editorial Notes', *C&CE*, 3, no. 4, 262–63. B. Pite, and C. F. Marsh, *C&CE*, 3, no. 6 (January 1909), 474, 482.
97 Sir H. Tanner, President's Address to CI, 9 November 1911, *The Builders' Journal*, 34 (22 November 1911), 548; 'Editorial Notes', *C&CE*, 6, no. 7 (July 1911), 483, 485; H. Tanner, *C&CE*, 6, no. 12, (December 1911), 930–31, 935; 'Editorial Notes', *C&CE*, 6, no. 12, 892–93. The Institution of Structural Engineers was formed in 1922.
98 E. O. Sachs, *C&CE*, 6, no. 12 (December 1911), 936; idem, *CI: Transactions and Notes*, 3 (November 1911), 3.
99 H. Tanner, *C&CE*, 6, no. 12, 933.
100 Alexander Siemens, President, ICE, at CI, *C&CE*, 6, no. 7, 492.
101 'Editorial Notes', *C&CE*, 4, no. 1 (March 1909), 2; see also E. O. Sachs, *C&CE*, 6, no. 7 (July 1911), 494; 'Editorial Notes', *C&CE*, 4, no. 2 (May 1909), 78; 'The ICE and Reinforced Concrete', *C&CE*, 5, no. 10 (October 1910), 707; E. O. Sachs, *C&CE*, 6, no. 12 (December 1911), 935.
102 'The ICE and Reinforced Concrete', *C&CE*, 5, no. 10 (October 1910), 711–13, 715.
103 'The Inst. C. E. Report on Reinforced Concrete — II', *Ferro-Concrete*, 2, no. 6 (December 1910), 163; compare 'The ICE and Reinforced Concrete', *C&CE*, 5, no. 10 (October 1910), 711; *The Builders' Journal, Concrete and Steel Supplement*, 32 (9 November 1910), 487.
104 'University of London: Mr. William Dunn's Lectures', *JRIBA*, 17 (22 October 1910), 798. The first detailed course of instruction on reinforced concrete in Britain was introduced in 1911, at the LCC School of Building, Brixton, 'Editorial Notes', *C&CE*, 6, no. 9 (September 1911), 651.
105 For example, 'A Correspondent: The Choice of Reinforcement', *C&CE*, 5, no. 8 (August 1910), 552. Contrast Collins, op. cit., 81.

12
Genesis of reinforced concrete construction

W.K. Hatt

Prefatory Notes.

It is the custom in some institutions of learning that some one of the officials shall give, each year, an address entitled "In Praise of the Founders." It has occurred to the present speaker that it would be appropriate on this occasion to discuss, in a very incomplete manner it must be confessed, the contributions of the early workers in the field of reinforced concrete.

This service could only be completely rendered after a long and difficult research. Examples of constructions are given in books without the date of their design. The history of these early periods is scattered through sources that cannot be attained in any one library.

The speaker has consulted all the resources at his disposal in the three languages of French, German and English, endeavoring to go back in every case to the original sources. Where this was not possible the excellent treatise of Christophe has been relied upon. He has endeavored to avoid any questions of a controversial nature.

If this address will impress upon our present-day engineers the extent of the achievements of the early builders and investigators and will stir up an interest on the part of someone to carry forward an historical account of the development of reinforced concrete, its purpose will have been fulfilled

Early Ideas in Reinforced Concrete.

As in the history of other arts, we may find early disclosures of the principles of action of various later constructions.

The early date at which reinforcement was considered is instanced in the following quotation from *Bridge Engineering*, by H. G. Tyrell, 1911, page 407, Chapter 17, on "Reinforced Concrete Bridges."

"In the construction of masonry arches, it has long been observed that the arches settle at the crown when the temporary centers are removed, and the extrados joints tend to open at the haunches, these points being known as points of rupture. To prevent these joints from opening, rods and iron bands have often been used in the extrados of the arch rings extending from the piers and abutments up to or beyond the points of rupture. Brunel, when experimenting with arches, built a semi-arch of brick 60 ft. long with hoop iron bond which supported itself by cantilever action." (About 1835.)

"Sir Shafto Adair built a concrete bridge in 1871 over Waveney at Homersfield, England, with model reinforcing frames from designs by H. M.

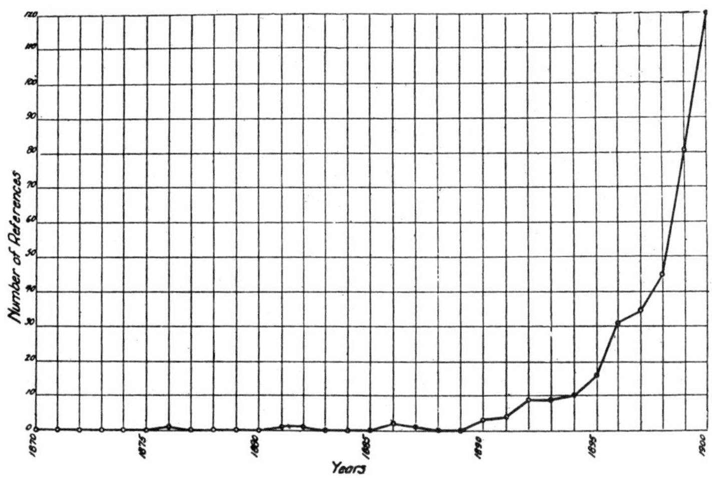

FIG. 1.—CHART SHOWING NUMBER OF REFERENCES.

Eyton of Ipswich. The arch had a span of 50 ft., a rise of 5 ft. 3 in. and the skeleton iron frames were imbedded in Portland cement concrete, over 100 tons of concrete being used."

We know that as early as 1838 the action of iron ties to give flexural strength to brick masonry was well understood. The two materials were shown to act as one composite material. [See Proc. Inst. C. E., Vol. 1, 1838, pp. 16 and 20.] It would seem to be a natural step to use such iron ties in concrete.

Many will be surprised to learn of the extent of the remarkable development of reinforced concrete construction in Europe during a period when the art was in a crude state in this country. In Europe, cheap labor and high cost of materials, together with a reliance upon theory and the certainty of good workmanship led to constricted sections. There the art was developed by commercial companies and had little publication of actual designs, or little recognition by officials during its early rapid growth.

To determine a date back of which this literature might be examined is somewhat an arbitrary selection. I, however, have consulted the very complete bibliography found in Chapter 10 of the French treatise by R. Feret, entitled "Étude Expérimentale du Ciment Armé." The references in this bibliography cover the principal countries of Europe and the United States, and apparently no publication of even slight importance has escaped. Plotting the number of references of succeeding years, as shown in Fig. 1, it is apparent that the years prior to 1900 may fairly be considered as early.

The French Exposition at Paris in the year 1900 contained such advance applications of reinforced concrete as to mark an epoch. Just before this date, we find appearing in the *Annales des Travaux Publique de Belgique* a series of articles by P. Christophe, afterwards published in an enlarged form in a well-known volume by Christophe entitled "Le Beton Arme." This series of articles was marked by the publication of large plates of drawings showing the actual reinforcement of various structures, with dimensions. Previous to this time there had been photographs of structures and skeleton drawings, but it would appear that the actual designs were considered to be trade secrets.

For these three reasons, namely, the display of the work at the French Exposition, the appearance of the Christophe articles, and the state of the literature as shown by the bibliography, it seemed to the writer that he might most profitably discuss the literature mainly prior to 1895.

LITERATURE PRIOR TO 1900.

The writer must confess that this paper cannot be complete, because certain of the early publications have not come to his hands in their entire form, especially the publication "Le Ciment" and "Le Beton Armé."

The original book of Coignet was published in 1861. In this he has described the advantages of the use of iron in masonry or in concrete. In construction of floors, Coignet has indicated the method of construction in iron and concrete as follows:

"Below this network of iron, a wooden falsework is placed; on this falsework is turned the concrete in thin successive layers and is tamped vigorously. It raises itself little by little up to the beams and envelopes them entirely and finally covers them with a layer of from 5 to 6 centimeters.

"At the end of some days, the concrete having attained the hardness of rock, the falsework is taken away and there remains a true slab of concrete forming a ceiling underneath and a floor above.

"In this system of flooring the iron work is completely enclosed in a slab of hard rock. One sees then that an iron work so placed in stone cannot be bent without bending the stone itself."

The increase of resistance to tension which comes to the cement by the introduction of iron is brought to view very clearly in the following paragraph:

"Concrete offers great resources—not only over and above a mass of ordinary masonry, but it possesses a resistance to tearing apart of 10 kg. per sq. cm. It is also possible to introduce, during its manufacture at the time of tamping, ties or iron chains of such a kind that by this means the strength of the concrete, already so large, is increased by these materials that are introduced."

Coignet's patent of 1869 is for improvements in artificial stone monolithic structures. It "consists in the introduction in the body of the structure of double-headed nails, double T-pieces, clamps, hoops, scraps of twisted or irregular shaped irons for the purpose of strengthening same and giving it greater cohesive strength. The irons to be thus introduced may be arranged in such a manner as to interlace each other, so that by the combination of this metallic skeleton, and of agglomerated artificial-stone paste, the thickness of the walls or size of articles may be considerably reduced and yet great strength be attained."

He mentions pipes, troughs or waterways. In the latter, "angular bent angle or L-shaped pieces may with good effect be introduced in the body of the material, to give greater strength to the angles and prevent the trough from spreading asunder at those points."

He claims no novelty in use of iron clamps or framework of metal in ordinary masonry or brickwork for strengthening same.

His claims are for:

(1) The combination of agglomerated artificial stone paste with iron straps of irregular shape, such as nails, double-headed nails or bolts, rings, hooks, clamps, wire, substantially in the manner and for the purpose set forth.

(2) The introduction in the body of artificial stones, or in the body of artificial stone monolithic structures made of agglomerated artificial stone paste, of skeletons or metallic framework linked or arranged so as to strengthen the same, substantially as specified.

(3) The application of agglomerated artificial stone paste to the protection and isolating of telegraphic wires.

Monier's descriptive memoirs of August, 1873, with drawings, shows his system of reinforced concrete arch rings. Round rods are used, constituting an intersecting grillage which is in one, two or three layers, depending upon the span and load. These are thus round bars in extrados and intrados not connected with metal through the thickness of the arch ring.

Monier's Patent Specification of September 15, 1890, Patent No. 208,250, applies to cement constructions with iron skeleton adapted for reservoirs, silos, tanks, cisterns, etc. In erection of reservoirs, etc., he prepares the foundation frame, and then puts in the first heavy irons necessary to give them their shape and desired dimensions as well as great resistance, as shown in his drawings. He describes in detail the method of constructing a reservoir resting on poor soil and having a diameter of more than 20 m. He describes the use of concrete columns with spiral and longitudinal reinforcements, and counterforts to reservoir walls or pipes of similar construction. His concrete is of mortar, of lime or cement, with sand or gravel slag.

In addition to the work of Coignet and Monier, we may especially mention certain early publications. The book printed by Thaddeus Hyatt

(Fig. 2) in 1877 for private circulation, whose title page is shown in Fig. 3, discloses the improvements in fireproof construction, describing certain tests of reinforced concrete beams and the principles of action of these.

According to an article by Professor Spofford, Hyatt was a lawyer by education, born in New Jersey in 1816, and lived most of his life in New York and London. He died in 1901.

Hyatt compares the first successful application of cast iron beams in buildings in 1801 with the method of construction shown in Fairburn's book

FIG. 2.—PORTRAIT OF THADDEUS HYATT.

of 1854, and asserts that the only difference between the early 1801 and later type of structure, 1877, was the form of the beam.

Hodgekinson's experiments in 1827–30 on metal beams had shown the economy of the I-over the T-form. In 1849–50, through the experiments to determine the construction of Stephenson's tubular bridges, wrought iron was substituted for cast-iron beams.

He calls attention to the fact that iron was believed to be, but is not, a fireproof material; that the bottom of the iron beams, against which the floor arches abut, will soften by heat, and the construction will collapse. The system adopted by Hyatt proceeds on the assumption that iron is not fireproof, and that in a floor it must be protected with a fireproof coating on all sides, the amount of which he has determined from experiment. His constructions show girders protected on all sides with concrete and light iron joists carrying a slab between the girders. And between these joists run vertical bars connected by wires to hold the ceiling.

He shows a construction in which floor joists are dispensed with, "flat tie-irons being substituted for the joists, the concrete in this case becoming the compressive member of the beam or slab."

He shows why it is a waste of material to use iron in beam form in concrete. His diagram shows a slab first with I-beams embedded, and second with only the lower flange of the I-beam left, a section of 2 x $\frac{1}{4}$ in. spaced 12 in. c.c. in a slab of 10 in. depth. Hyatt takes the axis at half the depth of the beam, as did Koenen in 1887. He then has 60 sq. in. of concrete in compression as the fulcrum for the tie to act against in lieu of the $\frac{1}{2}$ in. of metal in the top flange of the rolled joists. The question is, can those 60 in. of concrete when brought into compression be made to do the duty of the $\frac{1}{2}$ in. of metal at the top of an iron joist? He computes the compressive resistance of the concrete at 2000 lb. per sq. in. and has a mean of 1000 lb. acting at $2\frac{1}{2}$ in. above the neutral axis, or 60 x 1000 x $2\frac{1}{2}$ = 150,000 in. lb. moment in compression. The moment of the metal in tension is 60,000 x 2 x $\frac{1}{2}$ x 4 = 120,000 in. lb.

He therefore concludes that the compressive surface of the concrete is in excess of the demands of the tie metal.

Then he refers to his experiments to show the possibility of uniting metal to concrete as a bottom flange is held to its web in a rolled or riveted beam and that the two metals will act in concert. His experiments include beams as follows:

One of cement, one of concrete, four with firebrick tiles, and the remainder of concrete with various forms of reinforcement, round rods, iron plates of various width placed both vertically and flatwise in the beam. In certain beams some of the round rods are turned up from the bottom to the top of the beam. Rods ending in the top of the beam have heel plates. The plate ties when placed vertical are connected by rods running transversely in the beam. Plate ties placed horizontally are anchored to the top of the beam by bolts.

The beams varied in cross-section, 8 x 12, 9 x 8, 12 x 12 in., and were tested on a 5-ft. or an 8-ft. span. Of course, we now know that a 12-in. beam on a 5-ft. span will give rise to large shearing and bond stresses. His reinforcement was from 1.37 per cent.

His tests proved to him that the two materials worked in harmony.

Hyatt had evidently an analytical faculty. He sees that other matters besides strength need investigation. The fireproof qualities of Portland cement; the ratio of its expansion and contraction as compared with iron under like conditions; the effect of the two in combination when heated;

the heat-conducting powers of concrete; laws of its conduction; the compressibility and extensibility of cement as compared with iron.

He tests fireproof qualities and conduction. He found the expansion of concrete to be 0.00137 for 180° F., as compared with 0.00140 for wrought iron, and he finds also that no differential action exists in actual reinforced

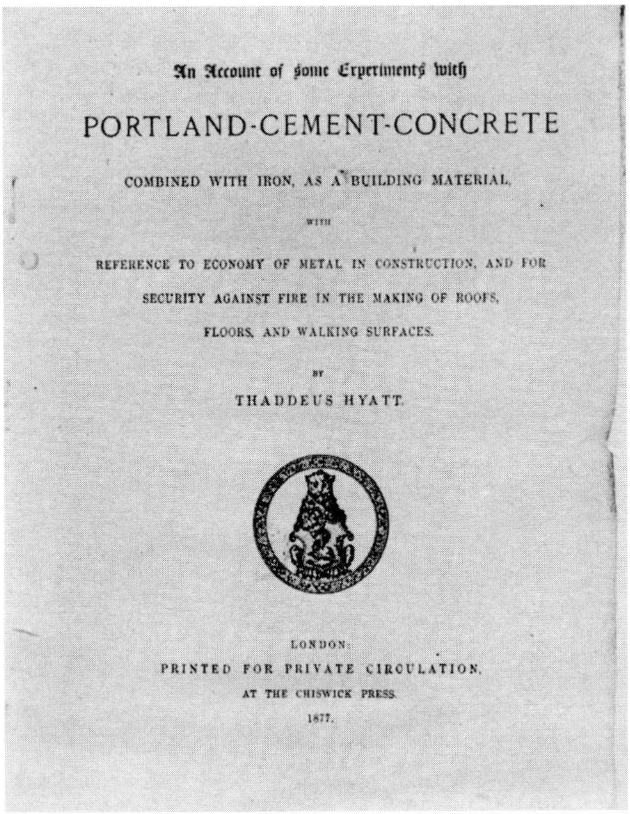

An Account of some Experiments with

PORTLAND-CEMENT-CONCRETE

COMBINED WITH IRON, AS A BUILDING MATERIAL,

WITH

REFERENCE TO ECONOMY OF METAL IN CONSTRUCTION, AND FOR

SECURITY AGAINST FIRE IN THE MAKING OF ROOFS,

FLOORS, AND WALKING SURFACES.

BY

THADDEUS HYATT.

LONDON:

PRINTED FOR PRIVATE CIRCULATION,

AT THE CHISWICK PRESS.

1877.

FIG. 3.—TITLE PAGE OF HYATT'S BOOK.

concrete bars when heated in the fire. His tests for the modulus of elasticity in compression show 0.00096 for 1000 lb. per sq. in., or $E = 1{,}000{,}000$ lb. per sq. in. approximately.

In tension the extension was 42/1,000,000 part of its length at a breaking stress of 100 lb. per sq. in. Here $E = 2{,}240{,}000$. Hyatt reasons that it was not possible to obtain the tensile modulus with the same accuracy as the compressive modulus. In addition to his laboratory tests, he applies the test of fire and water and a load of 200 lb. per sq. ft. to a real floor section 6 ft. long by 2 ft. wide by $7\frac{1}{2}$ in. thick, containing tie metal in the gridiron form in the middle of its thickness.

The fire was incandescent, bed 6 in. thick, 12 in. under the slab, kept

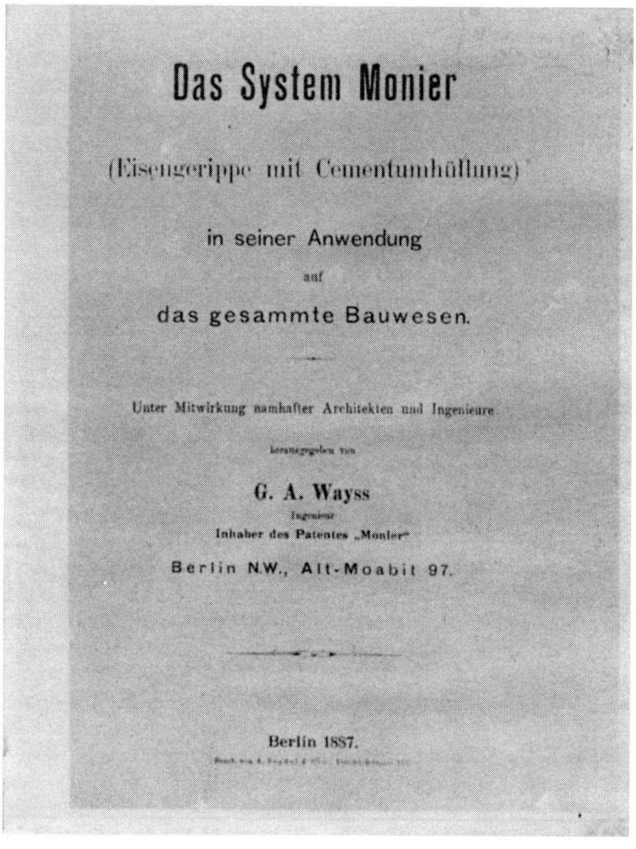

Das System Monier

(Eisengerippe mit Cementumhüllung)

in seiner Anwendung

auf

das gesammte Bauwesen.

Unter Mitwirkung namhafter Architekten und Ingenieure

herausgegeben von

G. A. Wayss

Ingenieur

Inhaber des Patentes „Monier"

Berlin N.W., Alt-Moabit 97.

Berlin 1887.

FIG. 4.—TITLE PAGE OF DAS SYSTEM MONIER.

for 10 hours, during which the deflection increased from $\frac{1}{32}$ in. to $\frac{3}{8}$ in. Then a stream of cold water was thrown on the bottom of the slab for 15 to 20 minutes. The under side of the slab was found to be unimpaired and when cold had returned to its original shape without deflection. A second test was made on the same slab. He also experimented with plaster of

Paris, concretes more or less porous and also with air spaces, and found that the best material to protect the metal against heat was that which was strongest in compression, viz, Portland cement concrete of the best quality, no advantage for any purpose being found from fibers of any kind, not even asbestos.

On speaking of certain of his test beams in which flat ties were connected by bolts with the upper part of the beam, he says that no contrivance of this kind is necessary. The tie requires no attachment to it other than one which will prevent it from sliding upon the web. Hyatt's analyses shows him to be in possession of the mechanics of beam action. He prefers the flat tie on account of its greater holding surface. We know now that

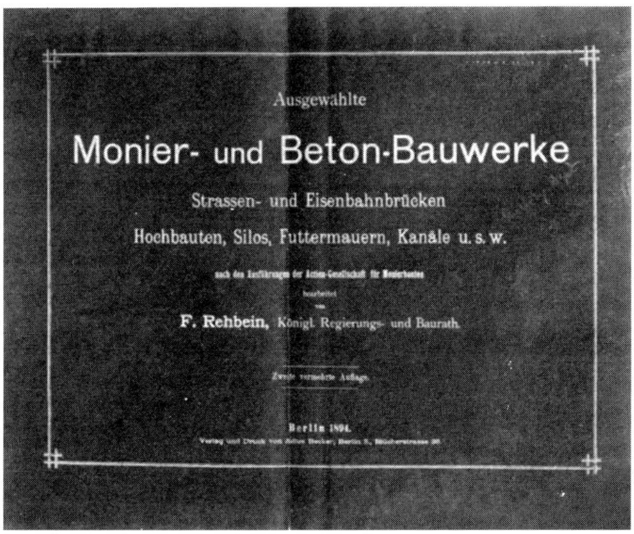
Ausgewählte

Monier- und Beton-Bauwerke

Strassen- und Eisenbahnbrücken

Hochbauten, Silos, Futtermauern, Kanäle u. s. w.

F. Rehbein, Königl. Regierungs- und Baurath.

Berlin 1894.

FIG. 5.—TITLE PAGE OF REHBEIN.

this flat tie is not efficient in bond. He puts the flat tie edgewise in the beam and connects the ties with horizontal transverse wires that serve to prevent sliding. He states that hollow floors may be made by using the ties for the top flange as well as for the bottom flange. Floors of 40 ft. span he says need to be 2 ft. thick, with bars 2 x $\frac{1}{4}$ in. spaced 2 in. apart for a load of 223 lb. per sq. ft.

Hyatt's "stone light" was well known, a combination of his gridiron slabs with glass inserts. He says he is prepared to undertake the construction of domes of any span from 2 ft. to 200 ft. Arch roofs with lights are recommended for picture galleries. He outlines the possibility of the use of his construction for chimneys and lighthouses, stairways, bridges, etc.

He acknowledges his debt to Mr. Thomas Beckett of Birmingham, an engineer, for assistance in working out his problems.

In this book of date November, 1877, we have the first recorded broad and definite treatment of reinforced concrete, and an attempt to establish an experimental basis for design.

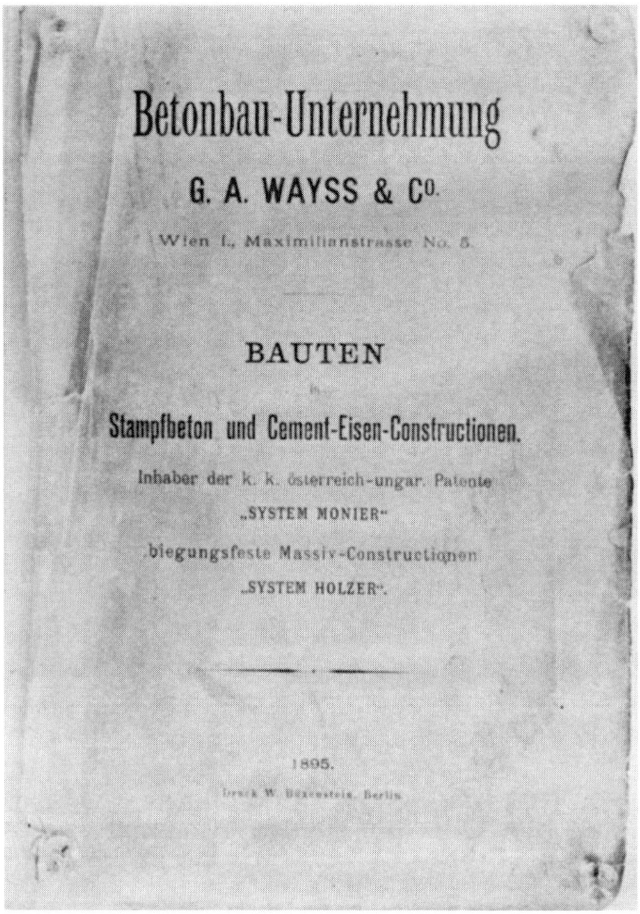

FIG. 6.—TITLE PAGE OF BETONBAU-UNTERNEHMUNG.

The publication in 1877, entitled *Das System Monier*, describes the advantages of reinforced concrete, makes the record of certain tests and explains the method of calculations derived from the tests of the firm of Wayss and by Engineer Koenen. The title page of the book is shown in Fig. 4.

A further description of the work of the firm of Wayss & Co. is found in the publication by Rehbein in 1894 entitled "Monier und Beton Bauwerke." A number of interesting bridges are shown in this publication (Fig. 5).

The publication entitled "Betonbau-Unternehmung" of Wayss, in 1895, shows some excellent photographs of the structures to that date (Fig. 6).

Certain publications describing the doings of special companies, for instance in France, *Le Ciment*, appearing monthly after 1896. In Germany a journal, *Tonindustrie Zeitung*, contained many articles on cement and concrete. The publication, *Zement und Beton*, from 1902 is especially devoted to the concrete industry. In France, *Le Béton Armé* was the organ of the agents of the Hennebique system and appeared monthly after June, 1898. A very important journal, *Beton und Eisen*, published at Vienna by von Emperger, was created in 1902, and published monthly after 1905. This may be considered an international journal on reinforced concrete.

The book by P. Christophe, entitled "Béton Armé," is probably the best complete work on reinforced concrete from the descriptive and the historical side. He is not content with illustrating the works, but has given dates of the appearance of a large part of the constructions of our illustrations. The book is not only valuable matter, but is written in excellent style.

The work of Berger and Guillerme, published in 1902 by Dunod, Paris, is especially valuable in the description of the various systems of construction with an account of the works and methods of calculations by these systems. It is weak in theoretical treatment and account of experiments. Curiously enough, in a work which has for its main field historical records, the authors have in very few cases given dates.

A very excellent work from the standpoint of discussion of experiments and theories and regulations is the book by Professor E. Mörsch of Zurich, entitled "Concrete-Steel Construction." This has been translated by Mr. E. P. Goodrich and published by the *Engineering News* Publishing Co. in 1910. The work was originally published by Wayss and Freytag in 1902 and then re-published by Mörsch in 1905.

A work in the English language from the press of Van Nostrand under the authorship of Marsh is a clear and readable account of the European work as disclosed in the books by Christophe and Berger.

METHODS OF CALCULATION.

Methods of calculating the strength of these structures were defined by the promoters of the various systems of construction as early as 1876.

The various Monier constructions built throughout Germany and Austria by the firm of Wayss & Company were based upon the method of design developed for this firm by Koenen. Koenen assumed a rectilinear variation of compressive stress; the tensile stresses were neglected. His equations are as follows:

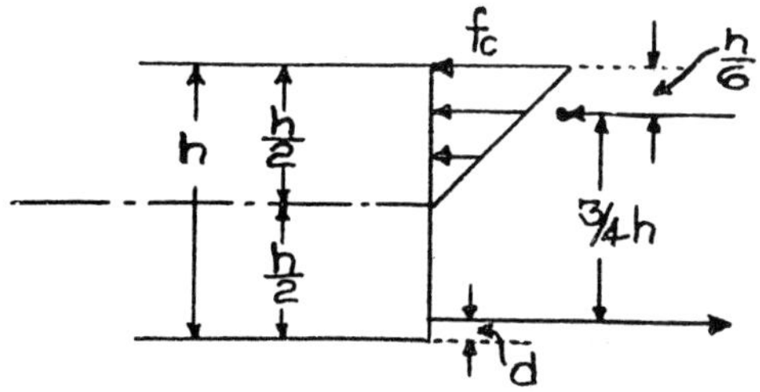

b = width of beam

In practice $d = \frac{h}{12}$

$$\frac{f_c h}{4} = f_s\ A_s$$

$$M = \frac{f_c h}{4} b \frac{3}{4} h = \frac{3}{16} f_c b h^2$$

$A_s = \frac{1}{4}\frac{f_c}{f_s} hb$ and

$$h = 2.31\sqrt{\frac{M}{f_c b}}$$

For fixed ends $M = \frac{1}{24\ w\ l^2}$ at center.

Christophe says that the error in locating the neutral axis at one-half the depth has this effect, that for 1 per cent reinforcement the compressive stresses are really about 50 per cent greater than Koenen fixed them, and the tensile steel stress is but 8 per cent less.

The method used by Hennebique is evidently very faulty; he assumed the compressive stress to be uniform above the neutral axis. He omits the tensile stress on the concrete. He supposed that the moment of the compressive stresses equaled the moment of the force of the steel. That is to say that either one of these equaled one-half of the external moment. Actually, as we now know, the concrete supplies about one-fourth and the steel and concrete in tension altogether three-quarters of the resisting moment. In working calculations, the height of the beam was supposed to be known. Hennebique's equations follow below.

$$\text{Moment of compression} = \frac{f_c a^2 b}{2} \quad \ldots\ldots\ldots\ldots\ldots\ldots (1)$$

$$\text{Moment of tension} = A_s f_s (h\text{-}a) \quad \ldots\ldots\ldots\ldots\ldots\ldots (2)$$

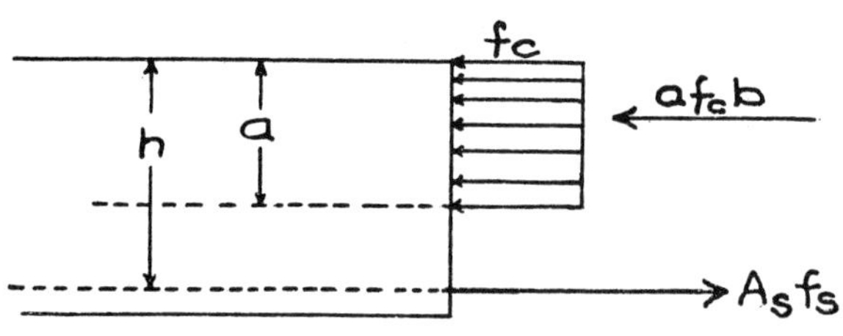

(1) = (2) and one-half of the external moment equals either (1) or (2). When b and h are known, h and A are calculated.

Generally h is assumed and the other quantities calculated. Hennebique used 25 kg. per sq. cm. for the concrete and 10 kg. per sq. mm. for iron.

In construction $h = \frac{3}{2}a$, for small h, to $h = 5/2a$ for large h.

Christophe shows that Hennebique's formulas yield a factor of safcty that depends upon the height of the piece. For shallow slabs where Hennebique calculates 25 kg. per sq. cm. in compression and 10 kg. per sq. mm. in tension, he really has 60 kg. and 7.8 kg. respectively. For thick slabs he really has 38 kg. and 12.86 kg. respectively.

For fixed beams $M = \frac{1}{10} WL$. For two-way slabs $M = \frac{1}{36} WL$.

Melan, in 1896, proposed a method involving a fictitious cross-section in which the dimensions of steel reinforcing were increased in proportion to the ratio between coefficients of elasticity of steel and concrete. Indeed, this method was clearly shown by the late J. B. Johnson in the *Engineering News* of January, 1895.

The recognition of the different properties of the two materials and the location of the neutral axis appeared as early as 1894, when the paper by Coignet and Tedesco, published in *Proceedings* of the Soc. des Ingénieurs Civils, approximated nearly all modern theories of flexural strength. The differing moduli of the two materials were taken account of and a quadratic equation fixed the location of the neutral axis.

An article in *Le Ciment*, in 1896, refers to P. Planat as one of the first writers upon the theory of reinforced concrete. P. Planat was the editor of the journal of architectural construction entitled *Le Construction Moderne.* In the journal after December, 1893, Planat published a series of articles of the theory of reinforced concrete which afterwards appeared in book form without date under the title of "Recherches sur la Théorie des Ciments Armés," Aulanier & Co., Paris. Planat also has another book published without date entitled "Théorie des Poutrées Droites en Fer et Ciment." By a complicated process of analysis Planat arrived at the proportionate depth of the cross-section available for compression. In this he assumes the compressive stress to vary by a rectilinear law. He omits the tensile stresses. His moment is the steel force times the leverage to the centroid of compression. The depth of the compression area he determines to be from $\frac{1}{5}$ to $\frac{1}{3}$ of the depth of the section, depending on the nature of the loading of the beam. Planat appears to have been a competent writer, and his calculations of many forms of construction such as beams, roofs, steeples, reservoirs, retaining walls of cantilever type, were plain and suggestive. He appears to have had extensive knowledge of the physical side of the subject.

Mention should be made of the work of Sanders of Amsterdam in 1898, who assumed a curvilinear elevation between stress and strain on the concrete. Spitzer, of the firm of Wayss & Co., proposed a parabola for this relation in 1897.

It cannot be said, however, that the principles of action had been properly defined and limited prior to 1900. The recognition of the part played by tensile strength of the concrete and the calculation of bond, and of shear reinforcement had yet to be determined upon the basis of laboratory examination. The early constructors, however, were not prevented from applying their systems in successful form throughout Europe, and the researcher into the literature of these early constructions cannot but be impressed with the apparent completeness with which designers had met the constructive necessities in an economic and skillful manner. Practical constructors like Coignet, Hennebique, Cottancin, Bordenave, Bonna, Matrai, Melan, Wayss, Ransom, Thatcher, built bridges, floors, foundations, aqueducts, factories, silos and retaining walls that showed a general knowledge of the strains to be resisted and a mechanical skill in meeting these strains with economical design that cannot but excite the admiration of present-day constructors. They must have had the elements of faith and imagination to project these structures and also a reliance upon the application of such theoretical knowledge as they possessed.

These early practitioners were, no doubt, without extensive knowledge of the durability of the construction and effect of variations of temperature, but they went ahead and their work was accompanied by test. In fact, if all matters were to be settled before action, very little would be done in this world. Indeed, our own regulations of today, as in case of flat slabs for instance, have been fixed in the absence of a complete knowledge of the manner of functioning of the construction.

It is necessary to call attention to these facts, because those interested mainly in theory, and accustomed to demand a rational process, are likely to undervalue the intuition and mechanical sense of the constructor controlled also by experience and by failures. While the engineers of official bodies doubted and learned persons calculated, these early practitioners, who were in many cases also theorists and calculators, applied their ideas and perfected them; their experiences led them to new departures.

The development of this art on its constructive side was associated with the activities of builders of patented systems. Some of the earliest records exist in the trade publications of these companies.

Monier's simple desire to reduce the thickness of the walls of his pots of concrete led him to the introduction of a metallic network. Monier's first French patent was of date of 1876. This was followed by other patents on reservoirs, floors straight and arches beams. Morsch says that "in his patent drawings are already disclosed all the elements which are today employed in the various construction details of the various systems." This was the point of departure of many applications. Later, in 1884, the rights of the Monier patents were purchased by Freitag and Heidschuh in Neustadt-on-the-Haardt and of Martenstein and Josseaux in Offenbach-on-the-Main. The firm of Wayss & Co., of Berlin, obtained later the rights of the Monier system for the whole of Germany. Experiments and study of the methods of calculations were published by Wayss in the brochure, "Das System Monier," 1887. The firm of Wayss & Co. built many structures throughout Germany and Austria on the Monier system. Monier has often been associated with all forms of reinforced concrete construction. Christophe says that the countries of German language made the fortune of reinforced concrete where it was introduced into construction of buildings and applied to public works.

The sequence of dates below will serve to outline the development. (See Table I.)

TABLE I.—REFERENCES PRIOR TO 1900.

(*a*) UNITED STATES.

1874–84. Ransome building various structures, wire rope and hoop iron.
1875. Concrete house slab and beams, tank tests. Built by Ward.
1878. Hyatt patent, structures, deformed bar.
1884. Ransome twisted bar.
1886. P. H. Jackson's patent.
1888. Ransome, wine cellar atSt. Helena, Academy of Science.
1889. Reinforced arch, Golden Gate Park.
1889. Ransome, Borax works at Alameda (ribbed floor).
1891. St. Louis Expanded Metal Co's. slabs.
1892. Leland Stanford University Museum.
1892. Monier U. S. Patent.
1893. Monier Arch, Pennypack Creek, Philadelphia.
1893. Northwestern Expanded Metal Co.
1894. Haymarket Exchange, Boston.
1894. Melan Arch, Rock Rapids, Iowa.
1895. Eden Park, Bridge, Cincinnati.
1898. Hennebique Patent.
1899. Thatcher patent.

(*b*) EUROPE.

1854. Wilkinson Patent.
1855. Boat exhibited.
1861. Coignet book.
1867. Monier patent.
1868–73. Reservoirs by Monier.
1869. Coignet patent.
1871. Reinforced concrete arch, Ipswich. 50-ft. span, skeleton frame.
1877. Hyatt's book, theory, tests.
1879. Hennebique building floors.
1883. Hennebique, reservoir.
1884. Monier patents purchased by Freitag and others for South Germany.
1885. Monier patents, relinquished to Wayss.
1887. Wayss' book, Das System Monier preceded by tests and theory of Koenen.
1887. Bordenave construction.
1888. E. Coignet, theory.
1889. Cottancin patents.
1890. Wildegg Bridge, 37.2 m. span, rise 3.5 m.
1892. Wunsch System, Budapesth.
1892. Hennebique patent.
1892. Melan bridges.
1893. Hennebique girder bridge at Don, France.
1893. Water conduit at Huesden, Holland, Nolthenius tests, retaining wall.
1894. Coignet and Tedesco, theory.

1894. Planat, theory and calculations of retaining walls.
1894. Moller, of Brunswick, bridges and retaining walls.
1894. Coignet's project for Alex III bridge.
1895. Austrian tests on arches.
1892–99. Hennebique, 3000 constructions.
1896. Hennebique piles.
1896. Foundations of church at Brebieves.
1896. Maryborough bridge, Australia.
1898. 3-hinged arch at Steyer, Austria.
1899. Chattellerault Bridge.
1887–99. 320 Monier bridges by Wayss.
1899. Considere and Ritter, theories.

In 1892 the important names of Hennebique and E. Coignet appeared in the history of reinforced concrete. Their designs of beams and bridges extended reinforced concrete to new construction of a monolithic character. Especially is the name of Hennebique prominent in the old applications, such as bridges, retaining walls, stairways, reservoirs, piles, sheet piling. Christophe says that after having worked in France and Germany, Hennebique established branches of his plans in all countries of Europe. In a space of eight years, between 1892 and 1899, Hennebique applied reinforced concrete to 3000 constructions. When one considers the elegance of Hennebique's constructions from the engineering standpoint, he is not so ready to admire the crude efforts of others who received recognition at a later date.

EXAMPLES OF SELECTED EARLY STRUCTURES IN REINFORCED CONCRETE.

1852.—Roof by Fr. Coignet (64 years ago).* (See Fig. 7.)

At St. Denis from three-story brick building. Roof about 12 in. thick, resting on brick walls. Span of slab, 20 ft. Mixture, ½ cement, 1 hydraulic lime, 5 gravel. Reinforcement, H-bars, about 3 in. deep on under side of slab. Both steel and concrete in 1910 were excellent condition. Construction cracks apparently filled up with cement. Roof watertight.

1877.—Thaddeus Hyatt, a lawyer by education, born in New Jersey in 1816, lived most of life in New York and London. The manufacture of his illuminated sidewalk grating was profitable and left him able to devote himself to research. His book in 1877. He died in 1901. His U. S. Patent No. 206112, July 16, 1878. During years preceding Civil War he was active in the antislavery side. Imprisoned for refusing to testify before U. S. Senate in relation to Harper's Ferry incident. Assisted in raising $1,000,000 for relief in Kansas.

1887.—Gas tank.† Height, 5.0 m.; diameter, 12.6 m.; thickness of walls, 13 cm.

Bauschinger's test of Monier Bridge at Munich, 1887. Load, 3824 kg.

* From *Concrete*, June, 1910. London, Vol. V, No. 6.
† From "Das System Monier," 1887.

per sq. m. Rupture, 4116 on one-half of span and 3252 kg. per 39 m. on the other. Span, 10 m.; rise, 1 m.; thickness, 9.9 to 12.3 cm.; width, 1 m. One mesh at introdos. Total cross-section of resisting bars, 14.9 (cm.)². (Fig. 8.)

Cupola of the Mausoleum of Frederick III at Potsdam. Built by Bonna. Covers, 25 x 37 m. Cantilever brackets project 0.5 m. and carry rolling load of 27,000 kg.

1890.—Foot bridge at Wildegg, Switzerland. Span, 37.22 m.; rise, 3.50 m.; width, 3.9 m.; crown thickness, 0.17 m. at crown; 0.25 at spring. Built by Wayss. Skew, 45 deg. Two meshes of reinforcing. Load, 500 kg. per sq. m.

1890.—Monier Bridge at North German Exposition at Bremen. Span, 40 m.; crown thickness, 0.25 m.; 0.55 m. at spring; rise, 4.5 m.; width,

FIG. 7.—BUILT BY FRANÇOIS COIGNET IN 1852.
ROOF OF HOUSE AT ST. DENIS.

3.0 m. at crown; 8.00 m. at spring. Built in 36 hours. Test load, 1000 kg. per sq. m.

1891.—Bridge near Graeffenhagen. Shown in Rehbein, 1894. Under-crossing for heavy wagon bridges. Span, 2.0 m.; height, 1.5 m. Founded in gravel 1 m. deep, foundation designed with cantilever toe. Reinforced counterfort retaining wall of counterfort type.

1894.—Canal bridge at Draulitten (Canal d'Oberland). Elliptic arch, 26.30 m. span; rise, 6.4 m.; crown thickness, 0.4 m.; at spring, 0.80 m. Reinforced with Monier net, double at spring and single at crown. Design load, 400 kg. per sq. m. Test load, 800 kg. per sq. m. Nine-ton roller.

1896.—Building at Lille. One of the first applications of Hennebique system. Span of girders, 6.80 m.; beams, 4.68 m.; slabs, 1.70; load, 450 to 800 kg. per sq. m.

1896.—Court house at Verviers. Built by Hennebique. Girder, beams and slabs; continuity; stirrups; 500 kg. per sq. m.; girders, 8.9 m. span; beams, 5.10 m.

1897.—Skodsburg Bridge. 21.85 m. span; 2.57 m. rise. Rail reinforcements about 2 per cent at the crown.

1897.—Bridge on Lausanne, Geneva line. Built by Hennebique on skew of 65 deg. Span on skew, 4.25 m.

Bridge at La Rue Valette. (See Marsh, p. 486.) Shows cantilever brackets and slab walls. Steel from barrel into spandrel wall. Retaining

FIG. 8.—TEST OF AN ARCH BY BAUSCHINGER.

wall with relieving arches. Span, 15.0 m.; rise, 1.25 m.; width, 32.8 ft.; load, 28-ton roller.

1899.—Chateaus d'Eua. A remarkable work, built by Ed. Coignet for the Paris Exposition of 1900. It consists of a half cylinder surmounted by a half dome, 144 ft. high. From this niche issues a cascade flowing over stepped slabs supported on beams and radiating arches. Below are passageways and galleries. These are side galleries. In this structure may be found applications of the art in beams, arches, slabs, retaining walls, etc. The face arch has a span of 23.8 m. and is composed in section of a hollow rectangle 2.8 m. deep and 2.0 m. wide with six square bars in the thickening corners. Back of this are two other single arches on the line of the dome. Radiating ribs connect these arches. This admirable work will repay study.

1899.—Chatellerault Bridge. Built by Hennebique in 1899; has a total length of 135 m. There are two arches of 40 m. span and 4 m. rise and a central arch of 50 m. span and 4.8 m. rise. Sidewalks were cantilevered. The deck, 8 m. wide, is carried by posts spaced 2 m. apart on four arch ribs spaced 2 m. and 0.5 m. wide. These ribs vary in depth from 0.45 m. at the crown to the rock foundation. Piers and abutments are cellular. The design load was 600 kg. per sq. m. of roadway and 400 m. per sidewalk; also a 11-ton truck. The bridge was tested to 800 kg. on the road and 600 kg. on the sidewalk. The deflection was 1/7300 part of the span in end arches and 1/5000 in middle arch. It returned to place without cracks. It was further tested with wagon traffic, marching soldiers, etc., with excellent results. There were no expansion joints. Considere says that the concrete floor, the longitudinal chords and the shortest vertical posts of the Chatellerault bridge are streaked wholly or partly by cracks and fissures, much wider fissures than generally seen in armored constructions.

Hennebique bridges: 1893, 1 girder bridge; 1894, 5 bridges; 1895, 3 bridges; 1896, 10 bridges (girder, 42 ft. span); 1897, 13 bridges (girder, 49 ft.); 1898, 29 bridges (52 ft. span girder); 1899, 42 bridges.

1901.—Bridge at Altare, Italy. Built by Maciachini on Walser-Gerard system. Segmental arch, 59 ft. span and 6.9 ft. rise, 24.93 ft. wide; depth at crown 11.8 in. and 19.7 in. at spring. Cellular spandrel. Reinforced in extrodos and introdos with 0.63 in. rods spaced 7.9 in., tied together. Rods run from ring to foundation of counterfort abutment.

1903.—Pont d'Ivry Bridge. A test span of 65.6 ft. (one-third of the span of the proposed bridge). Two parabolic bowstring reinforced trusses. Girders connected by wind bracing, and a decking supported on beams. Under design load of 60 tons applied for 12 hours, the deflection at center was 1/2000 span. At a load of 180 tons no sign of failure. Failure then began with tension cracks in lower chord. At a load of 220–240 tons the concrete outside the spiral began to scale at a core stress of 5972 lb. per sq. in. A load of 241 tons produced failure.

1894.—Moller retaining walls with arch anchors.

1897.—Cantilever wharf at Chantenay sur Loire. Projects 6 m.; anchored by ties to reinforced earth anchor. Retaining wall, 4.60 m. high between brackets with counterforts, supported on concrete piles.

1898.—Quay wall at Southampton. Hennebique, builder for London and Southampton Railway.

1899.—Retaining wall, Quai Debilly, Paris. Length, 256 m.; height, 5.48 m. to 0. Counterforts spaced, 1.50 m. Projecting platforms at port height with special counterforts. Spaced, 3 m. Well developed toe.

13

The Ward House: a pioneer structure of reinforced concrete

Ellen W. Kramer and Aly A. Raafat

In recent years there has been a growing interest in the contributions of the later nineteenth century to modern developments in planning and construction. The origins of reinforced concrete, whose possibilities are being so dramatically exploited in contemporary architecture, also lie in the middle of the last century. Early European contributions to the development of reinforced concrete are well known; to date, however, there has been no adequate coverage of the American story. Indeed, it comes as something of a surprise that so little is known about its beginnings in this country. In an effort to fill a part of this gap in our knowledge, we present here the Ward House, Comly Avenue, Port Chester, New York (fig. 1). Picturesquely situated on spacious grounds overlooking the Byram River, the house is still occupied by the Ward family. Built from 1873 to 1876, it is to the best of our knowledge the first reinforced concrete structure in the United States.

The development of reinforced concrete was a direct result of the search by engineers for a fireproof material which would give full protection against the threat to life and property caused by the disastrous fires of the period. Cast iron, in which many had such implicit faith, was completely discredited in this country after the fires of the early 1870s in Boston and Chicago. A few decades earlier attempts had been made to fireproof cast iron by enveloping the beams in concrete. The designs of Fox and Fairbairn in England and of Hyatt in the United States were the result of efforts to render the material fireproof. While these were not the ultimate development in this direction, they led to the next step in the search for fireproof structural materials.

Fig. 1. Ward House, Port Chester, N. Y. (courtesy Atlas Portland Cement Co.).

* This article developed as the result of a chance visit to the Ward House in 1951 by Mrs. Kramer who knew of it through descriptions in the *American Architect and Building News* I (1876), 112, and II (1877), 266–267. Correspondence with the late Warren L. Ward, a grandson of the builder, ensued. A few years later Mrs. Kramer mentioned the house to Mr. Raafat, who was at work on his doctoral dissertation at Columbia; the building is discussed briefly in his *Reinforced Concrete in Architecture* (New York, 1958), pp. 24–25, 27. For a description of the house by its architect, Robert Mook, Mrs. Kramer is indebted to the architect's daughter, Mrs. Jessie M. Hebert of Hartford, Conn. William E. Ward's obituary in the *New York Times* (4 March 1900), p. 7, contains interesting biographical information. Ward's son, 'Boss Ward', achieved fame as Republican Chairman of Westchester County, a post he held for almost half a century: see Agnes E. Meyer, *Out of these Roots. The Autobiography of an American Woman* (Boston, 1953), who describes 'Ward's Castle' as a forbidding stone [*sic*] structure. Additional material, including architectural drawings, was made available to the authors by Mr. Evans Ward, another grandson, upon the occasion of their visit to Port Chester in June 1957. The article was completed at this time, prior to Dr. Raafat's return to Egypt.

The next stage came with the realization that concrete surrounding cast-iron or steel joists could perform another function besides that of fireproofing. Concrete and steel would work together in a monolithic section. When such a section is subjected to bending stresses, enough bond exists between the two materials to enable concrete to provide compressive strength and for steel to resist the tensile stresses. The placement of the steel in the section is governed by the position of the tensile stresses. The independent realization of this phenomenon by men like Wilkinson in England, Coignet and Monier in France, and Hyatt and Ward in the United States, resulted in the invention of reinforced concrete. A period of experimentation followed in which the basis of the new invention was widened.

Among these early pioneers in reinforced concrete was William E. Ward of Port Chester, New York, who actually constructed his own residence with the new material. Ward (1821–1900), the founder of the Russell, Burdsall & Ward Bolt & Screw Company, which is still very much in business today, was a mechanical engineer with a keen interest in science and an inventive turn of mind. According to his own statement, he was completely unaware of the technical work preceding and paralleling his own experiments. In a paper read in 1883 before the American Society of Mechanical Engineers, of which he was a charter member, Ward stated:

> . . . the incident which led the writer to the invention of iron with béton occurred in England in 1867, when his attention was called to the difficulties of some laborers on a quay trying to remove cement from their tools. The adhesion of the cement to the iron was so firm that the cleavage generally appeared in the cement rather than between the cement and the iron.[1]

Ward then began a series of experiments from which he concluded:

> . . . the utility of both iron and béton could be greatly increased for building purposes through a properly adjusted combination of their special physical properties, and very much greater efficiency be reached through their combination than could possibly be realized by the exclusive use of either material separately, in the same or in equal quantity.[2]

The Ward House was a technical tour de force unique in its day. In contrast to all previous experiments, this was the first architectural realization of the possibilities of reinforced concrete. Not only were the exterior and interior walls, cornices, and towers constructed of concrete, but all the beams, floors, roofs, closets, stairs, balconies, and porticoes were exclusively of concrete reinforced with light iron beams and rods. Wood was used only for doors and window frames. The concrete floors, smoothly sanded, were supported on I-beams placed six feet apart; Ward put them near the bottom of the concrete beams in order 'to utilize the tensile quality for resisting the strain below the neutral axis.'[3] He suggested that 'for future construction . . . an inverted T-beam would furnish a more preferable distribution of iron in the composite beam than the I-beams which were used.'[4] As for the floor slabs, rods of iron five-sixths of an inch in diameter were placed longitudinally and laterally at a uniform distance of eight inches apart over a thin course of concrete. A top layer of concrete two inches thick was carefully tamped on. Finally, a half-inch coat of cement and fine beach sand, applied after the hardening of the slab, brought its total thickness to three-and-one-half inches. As in the case of the beams, the iron rods were placed near the bottom of the slabs. Over 13,000 square feet of flooring and roofing constructed in this manner was used for the house. The paneled ceilings of the first and second stories were constructed by arranging very light I-beams in a complicated network forming the desired decorative pattern. Concrete was poured around all the parts of this framework; the slabs in between were constructed of concrete and iron rods. The ceilings were later plastered and ornamented directly on the concrete (fig. 2).

Fig. 2. Detail of ceiling, Ward House (photo: Aly Raafat).

An interesting feature of the house is a balcony weighing nearly four tons, boldly cantilevered four feet from the face of the building. In a period when architects prided themselves on the thickness of walls used in the better types of houses, Ward dared to suggest that a two-and-one-half inch thick reinforced concrete partition was equal in resistance to an eight-inch brick wall. He even suggested that a system of thin reinforced cavity walls six to ten inches apart with connections every two to three feet, could be built up to a height of forty feet at a cost not

1. William E. Ward, 'Béton in Combination with Iron as a Building Material', *American Society of Mechanical Engineers, Transactions* IV (1883), 388.
2. Ward, 'Béton . . .', pp. 388–389.
3. Ward, 'Béton . . .', p. 391.
4. Ward, 'Béton . . .', pp. 391–392.

36

Fig. 3. Detail of tower, Ward House (photo: Aly Raafat).

exceeding that of brick walls. He pointed out the advantages of such a system with regard to protection against stormy weather, dampness, and fire. This cavity, moreover, was put to clever use in heating the house. Warm air from the hot air furnace in the basement was forced through heating flues arranged inside the walls and floors —a precursor of the modern system of slab heating.

A spacious veranda surrounds nearly half of the house. Here extremely large slabs were used, in some cases measuring thirty feet in length by twelve in width. The simply treated Doric columns, likewise of concrete, are hollow, the center spaces serving as leaders to carry off the rain water. The remander cylinders are of concrete reinforced with circles of three-eighths-inch rods in the centers. These are among the earliest known examples of reinforced concrete columns. The house had its own water supply system with two storage tanks in the tower. Spring water was piped into the upper tank; rain water was collected in the basement and pumped up to the lower tank.

According to contemporary accounts in the *American Architect and Building News* of 1876 and 1877, the house took four years to build. Mr. Evans Ward, a grandson of William Ward, suggests that this may have been due in part to the lack of knowledge on how to cure concrete in the winter. Another contributing factor was undoubtedly the care with which William Ward undertook numerous field tests of his new system from which acceptable results for deflection and strength were obtained. Before constructing any of the floors, he subjected a sample beam to a weight far beyond its normal carrying load. After the parlor floor had been laid for one year, he piled a weight of twenty-six tons between the two central beams, leaving it there over the winter; at the end of the testing period the amount of deflection was only one hundredth of an inch. Ward was an eminently practical man who believed in putting his theories to the most rigid tests. He also carried out experiments on flat slabs of concrete supported on all four sides.

This remarkable house was the result of the collaboration of architect and engineer. The design of the house was furnished by Robert Mook (1830–1917), a New York architect who enjoyed a large practice during the 1860s and 1870s, but whose name is forgotten today.[5] According to P. B. Wight's account of 1909, Ward requested that Mook furnish him designs 'for a large, first-class, comfortable house, with walls such as would be required if they were of brick with a hollow space, and floors of the usual thickness required for construction with timber, furred off on the underside, paneled and elaborately ornamented with plaster'.[6] These designs no longer exist, though Mr. Evans Ward still has in his possession four rough working drawings executed after the house was already under construction. Wight's statement is extremely interesting: it reveals the profound schism between technology and architecture which characterized the history of building in the later nineteenth century. The design possibilities of the new material were overlooked; concrete, as had been the case previously with cast iron, was treated in a traditional and entirely conventional manner. Stylistically, the building, with its rather curious combination of Hudson River Gothic and French Second Empire features, was not at all advanced for its date; if anything, it was *retardataire* (fig. 4).

Yet, in spite of the fact that no attempt was made to recognize the special qualities of the material in the design, reinforced concrete did exert an influence on the final executed design. This is particularly evident if we compare the Ward House with Mook's design for a row of houses built a few years earlier on Fifth Avenue, his famous 'Marble Row', demolished in 1929.[7] In the Ward House the simple, rather crude execution of what was intended to have been fine stone and brick detailing reflects

5. For information on Mook, see Ellen Kramer, 'The Domestic Architecture of Detlef Lienau, a Conservative Victorian' (unpublished dissertation, N. Y. U., 1958), p. 168, n. 5.

6. Peter B. Wight, 'The Pioneer Concrete Residence of America', *Architectural Record* xxv (1909), 359–363.

7. See Kramer, 'The Domestic Architecture of Lienau . . .', p. 166, fig. 63; also Moses King, *King's Handbook of New York City* . . ., 2d ed. (Boston, [ca. 1893]), p. 747.

Fig. 4. Side view, Ward House (photo: Ellen Kramer).

the difficulties of transposing a design from one material to another. This is particularly apparent in the treatment of columns, window architraves, the arches over the dormer windows, parapets, cornices, and pediments. The corner quoining, structurally uncalled for and difficult to execute, was evidently a feature considered indispensable to dignified architectural appearance. The extreme simplification dictated by the nature of the material results, nonetheless, in a certain picturesque boldness which is noticeable even in a photograph (fig. 3). Moreover, there was no attempt to conceal the monolithic quality of the house by dividing the surface into ashlar courses, nor was it concealed under a coating of a more noble material, a practice which is still popular at the present time, when reinforced concrete is often faced with brick. The only concession to exterior finish is seen in the use of a thin coat of plaster. A one to two mixture of Portland cement and sand was used, then rubbed down with stone, sand, and water.

This house aroused considerable interest at the time of its construction, when it was popularly referred to as 'Ward's Folly'. The correspondent of the *American Architect and Building News*, P. B. Wight, described it in a two-page spread in 1877 in which he pointed to its revolutionary use of materials and to other new features introduced here. In 1883, when Ward read his paper, 'Béton in Combination with Iron as a Building Material', to his fellow engineers, an animated discussion followed in which interest centered on the fireproof qualities of the new type construction and the heating system. In his paper Ward expressed unlimited confidence in his new invention, indicating his desire to extend the use of reinforced concrete to contemporary projects such as the Hudson River Tunnel and the Statue of Liberty, then in the course of erection on Bedloe's Island. He himself went ahead and used it with excellent results in heavy foundations for stationary machinery and for lining a reservoir of 96,000 gallons capacity. Every part of his factory at Port Chester exposed to fire risk was constructed of reinforced concrete. P. B. Wight, himself a great authority on fireproofing, returned to a discussion of the house in an article published in the *Architectural Record* of 1909, asserting that in constructing this house Ward had 'invented' reinforced concrete. While William Ward cannot claim to be the sole inventor of reinforced concrete, he appears to have been the first to give the new material architectural expression.

The dichotomy between structure and design, between the engineer Ward and the architect Mook, which this house so aptly expresses, has continued down to the present day. The structural capabilities of the new material, which had been realized by Ward in his two-dimensional reinforcement for his slabs, was developed in later decades by both European and American engineers. Only now are the unlimited possibilities resulting from a successful pooling of the talents of the engineer and the architect gradually achieving recognition.

14

The first reinforced-concrete skyscraper: the Ingalls Building in Cincinnati and its place in structural history

Carl W. Condit

Among the major pioneer works of reinforced-concrete construction, the Ingalls Building (now the Transit Building) in Cincinnati possesses a special importance. A skyscraper according to the standards of the time, its design involved peculiar problems, the successful solution of which provided valuable lessons for the subsequent evolution of the structural arts. To understand the meaning of this achievement, however, requires that we place the building in its full historical context, first describing the previous innovations which with suitable modifications were incorporated in the finished work, and finally examining briefly certain separate but parallel developments that soon converged into a unified body of techniques available to the whole range of concrete construction.

I. *General Background of Construction in Reinforced Concrete*

The history of reinforced-concrete construction is sometimes regarded as originating in the seventeenth-century invention of strengthening masonry by the introduction of iron tie rods and armatures into the body of the masonry work.[1] The essential factors in the progress of the art, however, are the revival of concrete as a structural material and the practice of reinforcing it with metal in order to secure a property thought to be lacking in the material itself. The precise beginning, then, may be said to have occurred in 1848, when Josef Lambot of France built a rowboat of poured concrete reinforced with a grid of

[1] See, e.g., S. B. Hamilton, "Building and Civil Engineering Construction," in Charles Singer, E. J. Holmyard, A. R. Hall, and Trevor I. Williams (eds.), *A History of Technology* (5 vols.; Oxford: Clarendon Press, 1954–58), IV (1958), 474–75, 480.

2 *Carl W. Condit*

thin iron rods to increase the cohesion of the concrete and hence its total strength. There is no evidence that Lambot, or anyone else for more than twenty years, understood that the role of the metal is mainly to impart a workable tensile strength to a material which, like stone, possesses this property only to a negligible degree. From 1848 to the end of the century there flowed a steadily broadening stream of patents, experiments, and structures embodying various techniques for reinforcing mortars, plasters, and concretes.[2] The decisive innovations came in the decade of the 1870's, and for the next thirty years the typical evolution of a pioneer stage of technical development took place, one in which a kind of natural selection operated on the diversity of inventions through the medium of increasing scientific insight into the properties of this novel composite material.

A small group of European and American inventors were the chief figures in the early phase. The French took an early lead, but the United States quickly caught up, in spite of the scientific backwardness that marked its technological progress in the nineteenth century. François Coignet obtained a patent in 1855 for a two-way grid of iron rods imbedded in concrete floor slabs and securely fixed to the walls, but his rather ambiguous description suggests that his intention was as much to prevent the overturning of the walls in monolithic construction as it was to increase the strength of the concrete. Josef Monier was destined to play a more important role when he was granted a patent in 1867 for the reinforcing of flowerpots and garden tubs by means of a heavy wire mesh imbedded in the mortar shell. Ten years later he received another patent for a considerably more advanced technique, namely, the reinforcing of concrete columns and girders with a grid of iron rods. He recommended this technique for the construction of rail and highway bridges, but it is doubtful whether Monier ever understood its true function, believing that its purpose

[2] There are several accounts in English of this development, all of them superficial and none perfectly reliable in its chronology. Chief among them are Peter Collins, *Concrete: The Vision of a New Architecture* (London, 1959), Part I; Carl W. Condit, *American Building Art: The 19th Century* (New York, 1960), chap. viii; S. B. Hamilton, "Building Materials and Techniques," in Singer *et al.* (eds.), Vol. V (1958), Part VIII; and Aly Ahmed Raafat, *Reinforced Concrete in Architecture* (New York, 1958), chap. i. A more detailed work is Gustav Haegermann, *Von Cæmentum zum Spannbeton* (Berlin, 1964-65). In addition to these there is a valuable paper by A. W. Skempton, "Portland Cements, 1843–1887," *Transactions of the Newcomen Society*, XXXV (1962–63), 117–52, but it is concerned mainly with the physical properties of Portland-cement concrete and the testing methods by which they were determined. All these works include bibliographies of varying lengths, those of Collins and Raafat being the most thorough.

was to increase general inner cohesion rather than to provide a tensile strength. The transformation of Monier's invention into a valuable scientific technique was largely the work of German builders and engineers.

Before this next phase of European development occurred, however, two extremely important contributions came, respectively, from the United States and England. In 1871–76 William E. Ward built a house for himself on Comly Avenue in Port Chester, New York, that proved to be the first complete work of reinforced-concrete construction. The architect was Robert Mook, but the structural designer was Ward himself. His account of the genesis of the idea, while illuminating in one respect, makes little reference to the role of the iron reinforcing, although it is clear that Ward understood this very well. "The incident which led the writer to the invention of iron with béton occurred in England in 1867, when his attention was called to the difficulties of some laborers on a quay trying to remove cement from their tools. The adhesion of the cement to the iron was so firm that the cleavage generally appeared in the cement rather than between the cement and the iron."[3] Ward's own experiments, conducted during the early phase of construction, led him to an understanding of the statical action of the two materials. In his own words, "The utility of both iron and béton could be greatly increased for building purposes through a properly adjusted combination of their special physical properties, and very much greater efficiency be reached through their combination than could possibly be realized in the exclusive use of either material separately, in the same or in equal quantity."

The structural system of the house is redundant for the load requirements of a residence, as Ward's exacting tests revealed. The floor slabs are reinforced with a two-way grid of iron rods and supported on concrete beams reinforced with wrought iron I-beams. The iron members are correctly located in the lower portion of the beam, where the tensile stress would be concentrated under a positive bending moment. The columns were poured as hollows cylinders reinforced with hoops. A heavy balcony at the rear of the house is cantilevered 4 feet from the wall. Ward tested the parlor floor under a weight of 26 tons located at the center of the 18-foot span. After leaving it

[3] William E. Ward, "Béton in Combination with Iron as a Building Material," *Transactions of the American Society of Mechanical Engineers,* IV (1883), 388–89; quoted in Ellen W. Kramer and Aly A. Raafat, "The Ward House: A Pioneer Structure of Reinforced Concrete," *Journal of the Society of Architectural Historians,* XX, No. 1 (March 1961), 35. For a more analytical treatment of the structure of the house, see Condit, pp. 233–34, 337–38.

throughout the winter, he measured a resulting deflection of only 0.01 inch. The house was undoubtedly influential in the United States and France, even though Ward himself did no further work in building.[4]

The major event in the evolution of concrete construction as a scientific technology came in 1877, when the Amercan inventor Thaddeus Hyatt performed his experiments on the behavior of reinforced concrete. Although he was a New Yorker, he carried on these experiments in London in collaboration with David Kirkaldy, a pioneer in the development of industrial testing machines. Without Hyatt's investigation, progress in concrete construction would have had to depend in large measure on primitive and dangerous trial-and-error methods. In addition to understanding the proper forms of reinforcing to counteract tension in beams, slabs, and columns, he made two valuable discoveries in the physical properties of the material, namely, that the coefficients of thermal expansion of iron and concrete are nearly identical and that the elongation of the two under load is virtually the same for the two materials. The full title of Hyatt's report of his experiments, published at London in 1877, nicely summarizes his aims—*An Account of Some Experiments with Portland-Cement-Concrete Combined with Iron as a Building Material, with Reference to Economy of Metal in Construction and for Security against Fire in the Making of Roofs, Floors, and Walking Surfaces.*

Shortly after the publication of Hyatt's book the scene of vital activity shifted to Germany.[5] In 1879 the German builder G. A. Wayss bought the German rights to Monier's patents. In the following year Rudolph Schuster bought the Austrian rights, and he and Wayss formed a construction company to exploit the French inventions. The first commercial use of reinforced concrete, however, does not appear to have come until 1884, when the contracting firms of Freitag and

[4] The Ward house was publicized in the *American Architect* (1877), *Transactions of the American Society of Mechanical Engineers* (1883), and the *Gazette des architectes,* though the precise influence is difficult to estimate (Collins, p. 58). Although the technique of I-beam reinforcing was primitive, it survived into the early twentieth century in the method of reinforcing arch ribs developed for concrete bridge construction by Josef Melan of Vienna.

[5] In addition to the accounts given in Collins, Haegermann, and Raafat (n. 2 above), a further summary of events in Germany may be found in O. Kohlmorgen, "Evolution of Reinforced Concrete in Germany," *American Architect and Building News,* XC, No. 1616 (December 15, 1906), 188–90. This account was originally published in the British periodical *Concrete and Constructional Engineering* (November 1906). There are minor discrepancies between the chronologies given by Collins and Kohlmorgen, but Collins appears to be the more reliable of the two.

Heidschuch, and Martenstein and Josseaux, acquired still other rights to the Monier patents and began to use them in practical building. The two firms carried on independent activity for only a year, both having been acquired by Wayss in 1885.

In collaboration with the Freitag and Martenstein engineers, Wayss began experiments to determine the capacity and behavior of reinforced concrete under load, the resistance of concrete to fire, and the corrosion resistance of the iron reinforcing. The results of these experiments were published at Berlin in 1887 under the title *Das System Monier*. At the same time that Wayss undertook his experiments, the Prussian state architect, K. Koenen, developed a theory for calculating the strength of Monier-reinforced floors, vaults, and free-standing cylindrical water tanks, publishing it as a series of articles in the 1886 volume of the *Centralblatt der Bauverwaltung*. By the time Monier's primitive system of reinforcing left the hands of Wayss, Koenen, and their collaborators, it had been transformed into a well-developed scientific technology. The irony in all this is that, although Monier was increasingly neglected, the Wayss-Koenen system was known as *Monierbau* in Germany and, after it crossed the Atlantic, as *Monier* construction in the United States.[6] By 1890 the techniques and theories developed by the German investigators were being applied to all structural components of buildings and to pipes, tanks, reservoirs, and arch bridges. The form of the reinforcing retained Monier's original simplicity, that of a two-way grid of rods, but the location of this grid in the tension zones of beams, slabs, and arch barrels indicated that the Germans understood very much better than Monier the fundamental purpose of reinforcing.

While the work of Wayss and Koenen was receiving wide public attention, the French builder François Hennebique was privately carrying on experiments that were to lead to equally valuable innovations. Hennebique must have understood the necessity for reinforcing in concrete about the time that Monier received his patent in 1877. Unlike Monier, however, he launched out immediately into practical use of the new technique: his first work of concrete construction, a house completed in 1879, included concrete beams reinforced in the tensile zone with iron rods. In the following year he made his greatest contribution when he discovered the existence and nature of shearing stress in a concrete member subject to deflection. Since the material has a negligible shearing strength, Hennebique first introduced small

[6] For the importation and development of the Monier system in the United States, see below, pp. 10–11.

vertical plates and later U-shaped stirrups set at intervals along the length of the beam to bind the reinforcing in the lower or tension zone to the upper or compression zone of the beam.[7] For the next twelve years Hennebique carried out secret experiments on the most efficient forms of reinforcing for columns, beams, and slabs. He was granted his first patents in 1892 for the most scientific system so far devised: in addition to the tension bars and stirrups, it included such original elements as reinforcing for continuity and the bending up of the ends of tension bars to resist diagonal shear at the ends of the beams. Before the close of the century several French engineers were proposing alternatives, chief among them Armand Considère, who was the leading exponent of helical reinforcing for the compression members of building frames and bridge trusses.

II. *American Development, 1880–1900*

With the exception of the Ward house, which remained an isolated phenomenon, the history of reinforced concrete in the United States until the end of the century was chiefly bound up with the work of Ernest L. Ransome of San Francisco. His system of reinforcing, whether for buildings or bridges, was the only one up to 1899 to involve the use of bars rather than wire mesh, expanded metal, or I-section ribs.[8] As an American builder of English origin, Ransome was undoubtedly familiar with the published descriptions of the Ward house and with Hyatt's report of his experiments, but beyond these his work was largely a matter of independent development.

Ransome's first commission for a building to include a mature form of reinforced-concrete construction was the Bourn and Wise Wine Cellar at St. Helena, California (1888). Most of this structure was built of masonry and plain concrete (with an unusual aggregate in the form of crushed basalt), but the second floor was of reinforced concrete de-

[7] An account of Hennebique's experiments and reinforcing techniques has yet to be written. For descriptions of the external appearance and main structural elements of his buildings, see Collins, pp. 64–75; for a generalized sketch of the final form of reinforcing in a standard column-and-girder frame, see Raafat, p. 28.

[8] The best account of Ransome's early work is George W. Percy's "Concrete Construction," *Engineering Record,* XXIX, No. 17 (March 24, 1894), 272–73. Percy was a San Francisco architect who designed a number of buildings constructed by Ransome's contracting firm. For later accounts, see Collins, pp. 61–64; Condit, pp. 234–40; and E. L. Ransome and Alexis Saurbrey, *Reinforced Concrete Buildings* (New York, 1912). For a brief discussion of Ransome's tests on the elasticity of reinforced-concrete beams, see Edwin Thacher, "Concrete-Steel Bridge Construction," *Engineering News,* XLII, No. 12 (September 21, 1899), 179.

signed for a load of 220 pounds per square foot. The floor was poured with its undersurface in the form of a series of narrow vaults—a technique that was nearly universal at the time for hollow-tile floor construction—and the reinforcing consisted simply of two longitudinal bars located near the bottom of each beam.

The success of the St. Helena building led Ransome to break completely with traditional floor construction in his structural design of the California Academy of Science, erected in San Francisco in 1889. Here the modern technique appeared for the first time in American practice: the floor slab spanned between beams of rectangular section, slab and beams together cast as a monolith. Again the reinforcing consisted of two bars set low in the tension zone of the beam. Two other innovations in the building extended the range of concrete techniques: one was the method of constructing balconies around a central light well by cantilevering the floor slabs inward from the two inner rows of columns; the other was the introduction of square-twisted bars in the upper (tension) zones of the cantilevers. Ransome secured a patent on twisted bars, and they became one of the distinguishing features of his work. The twisted shape provided a stronger bond between the steel and the concrete, while the process of cold twisting increased the tensile strength of the metal.

During the next decade Ransome extended and refined these rather simple techniques and acquired patents on the innovations as they appeared. In the factory of the Pacific Coast Borax Company at Alameda, California (1889), he introduced concrete columns and a two-way floor-framing system in which closely spaced beams spanned between the girders that rested on the columns. Reinforcing consisted of a single bar near the bottom of the beam and a pair of bars in the girder. A 10 × 20-foot section of the floor in the Borax factory was tested in February 1893 under loads ranging from 234 to 551 pounds per square foot, with a deflection of only ¼ inch under the maximum load. The Leland Stanford, Jr., Museum at Palo Alto, California (1892), was a complete work of reinforced-concrete construction, "probably the largest and most important building in the world constructed entirely of concrete," as George W. Percy described it.[9] The chief additions to the vocabulary of reinforcing techniques were the overlapping roof tiles and the central dome, in which a two-way system of ribs carried a skylight of glass inserts.

With the essential features of tension reinforcing established, Ransome's next major achievement was the introduction into American

[9] Percy, p. 272.

building of stirrup reinforcing to take shearing stresses. The first work to embody this practice was the eastern factory of the Pacific Coast Borax Company, erected at Bayonne, New Jersey, in 1898, with Ransome's firm acting as designer and contractor.[10] The entire structure of this four-story building was a homogeneous concrete monolith: the floor slab was poured integrally with a frame of heavy girders, beams, and closely spaced joists; the floor and roof frames rested on solid concrete columns and on piers set in hollow concrete walls. The main girders, with an over-all depth of 28 inches, were designed to carry a total load of 6,545 pounds per lineal foot over a span of 11 feet. The tension reinforcing was more elaborate than in previous Ransome designs: single bars were located at both the top and bottom of the beams and joists and paired bars at the top and bottom of the girders. The presence of the bars at the tops of these members suggests that Ransome thought of the floor frame as made up of continuous elements. The stirrups were confined to the joists, where they were used to unite the upper and lower longitudinal bars. Since the Bayonne factory was built six years after Hennebique received his first patents, it is likely that Ransome arrived at the idea of reinforcing against shear on the basis of the French builder's precedent rather than through his own independent discovery. The reinforcing elements in the factory structure were square-twisted bars throughout.

Ransome's next major work might seem retrogressive in some respects, but it embodied structural details essential to the progress of the art. The factory of the Varley Duplex Magnet Company, erected in Jersey City, New Jersey, in 1901, was the joint work of the architect L. H. Broome and Ransome's engineering and construction company.[11] Although the building was much smaller than the Borax works, it involved ingenious solutions to two special problems of construction. The poor bearing quality of the soil required that footing pressure be kept below 2,000 pounds per square foot, a limitation that necessitated spread footings of considerable area. At the same time, one wall of the building fell so close to the lot line that a footing of the required area would have extended under the neighboring building. Ransome's solution was to cantilever the first floor and the wall along the side of inadequate clearance outward three feet from the outer face of

[10] "A Large Monolithic Factory Building," *Engineering Record*, XXXVIII, No. 9 (July 30, 1898), 188–90, and "Constructing a Large Monolithic Concrete Building," *Engineering Record*, XXXVIII, No. 12 (August 20, 1898), 254–56.

[11] "A Jersey City Concrete Factory Building," *Engineering Record*, XLV, No. 12 (March 22, 1902), 270–71.

the foundation wall. This required a series of underfloor cantilever and anchor beams massive enough to support a wall load of 24,000 pounds per lineal foot. The reinforcing consisted entirely of a single layer of bars set close to the upper or tension surface of these beams. The whole system probably represents the first use of concrete cantilevers large enough to support a superimposed wall load. The reinforcing in the footing was a simple two-way system probably designed to withstand a certain amount of shear as well as tension induced by bending under the concentrated load of the foundation wall.

Ransome was still the leading American builder in reinforced concrete at the turn of the century, but shortly before 1900 others offered alternative techniques to challenge his long-held supremacy. Around 1893 William Orr, the superintendent of mills of the New Jersey Wire Cloth Company, a subsidiary of the John A. Roebling's Sons Corporation, invented a system of floor construction in which the familiar arches springing between the floor beams were made of concrete reinforced with woven wire netting strengthened by being tightly bound to parallel iron rods. Orr's invention was known as the Roebling system of reinforcing.[12] An installation in an office building at Broadway and Broome Street in New York was tested in July 1895 under loads as high as 1,200 pounds per square foot without failure. At the same time the floor slabs of the Hotel Savoy in Boston were subjected to severe impact tests during construction to satisfy the building inspector.[13] Another form of light-weight reinforcing was the system originally patented in France in 1898 by Alphonse de Man and introduced into the United States shortly thereafter. This was a composite construction of steel columns and beams and concrete floor slabs reinforced with twisted steel strap.[14]

Although the Roebling and De Man systems played a lively role in building construction around the turn of the century, they suffered from serious limitations and were soon eclipsed. The physical flexibility of the Roebling net made it useful for floor slabs and driveways, where it could easily be unrolled into sheets and could provide the necessary

[12] John A. Roebling's Sons Corp. was established at Trenton in 1849 by the celebrated bridge-builder. By 1900 the Roebling system of reinforcing was only one of a number of patented forms of wire mesh and expanded-metal reinforcing.

[13] "The Roebling Wire and Concrete Construction for Floors and Walls," *Engineering News*, XXXIV, No. 3 (July 18, 1895), 45–46. For a variation on the system in a large industrial building, see "A Steel-Concrete Factory Building," *Engineering Record*, XLII, No. 8 (August 25, 1900), 179–80.

[14] American Fireproofing & Cement Construction Co., *The De Man System of Fireproof Construction* (New York, 1901).

reinforcing for the small bending forces involved. On the other hand, it was extremely inflexible in the range of its uses, since it could not take the place of bar reinforcing in beams, columns, and heavy slabs. The De Man system was simply a light-weight reinforcing that had to be supplemented by the usual bar forms for the framing members of multistory buildings. Finally, mesh and expanded-metal forms belonged to a prescientific building technique because the regular grid of wires or narrow straps did not follow the lines of principal stress.

The most important event outside Ransome's work was the importation into the United States of the Monier-Wayss system. The first American recognition of it was the article "Tests of Monier Construction" in the *Engineering Record* of May 12, 1894, which summarized the results of tests of reinforced-concrete bridges in Europe. The chief source of knowledge for American builders in this pioneer phase was a paper read by E. Lee Heidenreich before a meeting of the Western Society of Engineers at Chicago on June 6, 1900, and subsequently published in the society's journal.[15] Offices had been established in the United States about 1896 to find markets for the patents of Wayss and Koenen, and Heidenreich was in charge of the one at Chicago. He first proposed the European system in 1896 for a grain elevator in Minneapolis, but it was not built until 1899–1900 because the owner would not trust Heidenreich's plan until he had sent a member of his firm to Europe to study examples of similar forms in active use. The first executed work of Monier construction was a storage tank built in Chicago in 1899 by the Illinois Steel Company after Heidenreich's design. The cylindrical concrete shell was reinforced with a two-way system of annular and vertical bars. After 1900 Monier concrete structures began to multiply in the Midwest and the Mississippi Valley, in part because of the impetus given to the new techniques by the Illinois Central Railroad.

Heidenreich described a variety of structural forms—culverts, tanks, arched floor construction with arches springing between steel beams, flat slabs on steel I-beams, and slabs cast monolithically with concrete beams—but the reinforcing in all cases represented variations on a simple basic principle. The theory, as he summarized it, was the strengthening of concrete with steel rods so located as to take the principal compressive as well as tensile stresses, but in ordinary practice the use of metal to carry compressive forces was limited to vertical

[15] E. Lee Heidenreich, "Monier Constructions," *Journal of the Western Society of Engineers*, V, No. 3 (May–June 1900), 208–37, and V, No. 4 (September–October 1900), 329–39.

forms such as storage tanks and columns. In all cases the reinforcing consisted of a two-way sytem of rods bound tightly at intersections by means of wires, the rods in one direction being designated as distributing rods and those in the other as carrying rods. The first group functioned to transfer the stress in the concrete to the carriers, which acted to take the tensile stress itself. Thus the carrying rods ran continuously between supports, but the distributing rods could consist of short lapped lengths. In the reinforcing of slabs, the rods were located near the bottom if the slab was poured with joints at the I-beam supports (i.e., as a simple member), but they were placed near the top in the region of the beam and near the bottom between the beams if the slab was poured as a continuous member over a series of supports —in short, correct tensile reinforcing for continuity.

Arched undersurfaces of floors and concrete slabs on steel beams constituted the major form of concrete floor construction when Heidenreich's paper appeared, although they are little used today, whereas the less prominent slab-and-girder system is now nearly universal.[16] And yet Heidenreich dealt with the latter in a remarkably informal way, as though it were an afterthought that might suggest something useful when all else failed. Speaking of monolithic systems, he said, "I will call your attention to [these constructions], showing how to do away with I-beams in a pinch if the mill is behind time and you cannot wait. Simply use a Monier girder with a rod or two at the top for compression, and a rod or two at the bottom for tension."[17] His accompanying illustration shows two rods close to the vertical axis of the beam at the top and two larger rods somewhat more widely spaced at the bottom. This informality stands in strong contrast to the scientific spirit exhibited in the rest of the paper. As far as building construction is concerned, however, there is nothing in it that had not appeared in Ransome's work before 1898 (this is not true, however, of bridges, tanks, and elevators). What the Monier system so far lacked was reinforcing for shearing stresses, but these were to come in the next two years. Meanwhile, Ransome was refining his own techniques in the same direction, so that it is difficult to assess the relative roles of the Ransome and Monier systems. Whatever the case, the stage was set for an exploitation of the new techniques on a scale comparable to steel-

[16] The old technique of placing concrete floor slabs on top of iron or steel beams and columns must be distinguished from the recently developed technique known as composite design, in which a floor or deck slab functions as the extended top flange of steel girders.

[17] Heidenreich, p. 218.

framed construction, and there were several builders eager to try their hands. A powerful impetus to the acceptance of their art was the disastrous Baltimore fire of 1904, which paralleled the fires of the 1830's and '40's in their stimulus to cast-iron construction.

III. *The First Concrete Skyscraper*

In 1903 the leading American firms of designers and builders in reinforced concrete were Ransome and Smith of San Francisco, the Baltimore Ferro-Concrete Company, the Ferro-Concrete Construction Company of Cincinnati, and the Trussed Concrete-Steel Company of Detroit.[18] By 1905 they had been or were to be involved in the design and construction of three building triumphs that may be said to have brought the new structural techniques to maturity in the United States and to have established them as potentially competitive with steel construction. All three survive today in active use—the Ingalls Building in Cincinnati (1902–3), Terminal Station in Atlanta, Georgia (1903–4), and the Marlborough Hotel in Atlantic City, New Jersey (1905–6)—and among them the Ingalls was the most daring and influential. By the time it was placed under construction in the fall of 1902 the main streams of development in reinforcing technique had come together in the design of large buildings, and all are revealed in the structural system of the Cincinnati project. The basic techniques were the following: (1) Ransome's heavy monolithic beam-and-slab construction with tension reinforcing; (2) the two-way reinforcing systems of Monier and Wayss; (3) the bent bars and stirrups of Hennebique; (4) the hoops and continuous helixes for compression members, the former originally proposed by Hyatt and the latter by Considère.

The Ingalls was planned as a first-class high-rise office building to be constructed at the northeast corner of Fourth and Vine streets, in the heart of the commercial and banking center of Cincinnati. In addition to the department stores and financial institutions, the immediate area included two major hotels and a number of railroad ticket offices. The building was thus bound to attract the most desirable tenants and to

[18] Ransome's firm was long established; the Baltimore and Cincinnati companies were founded in 1901, and the Detroit organization was established in 1903 by Julius Kahn. The last was the most successful of the four, its activities quickly expanding to Europe as well as to major cities in the United States; yet, ironically enough, the reinforcing techniques developed by Kahn have been entirely superseded by forms essentially like those of the Ransome and Monier systems. Soon to compete with the leading four in volume of large-scale construction was the Turner Construction Co., founded in 1902 by Henry C. Turner, formerly a member of Ransome's engineering staff.

represent a highly remunerative long-term investment, characteristics that it retains to the present time. A number of talents took a hand in designing and erecting the structure. The architects were Elzner and Anderson, and the structural engineer was Henry N. Hooper, the head of the engineering staff of the Ferro-Concrete Construction Company, which acted as the contractor for the concrete structure. The reinforcing system designed by Hooper was based on the Ransome patents. The general contractor was the William H. Ellis Company, a Cincinnati firm, as were the architects and engineers.[19]

In its general functional character the Ingalls is a typical commercial building, with a bank, stores, and former ticket offices located in the two-story base and the general business offices above (Pl. 1). The overall dimensions in plan are very nearly 50 × 100 feet; the sixteen stories and one basement rise 210 feet above grade, or 235 feet above the undersurface of the foundations. A number of steel-framed buildings in New York and Chicago were considerably higher at the time, but no concrete frame exceeded half the height of the Ingalls. Concrete was chosen as the structural material chiefly on the ground of economy: it offered the equivalent of a steel frame in load-bearing capacity and other physical properties, while the structural cost of $400,000 (exclusive of mechanical equipment) was somewhat lower than that of steel construction.

The structural frame of the building is a virtual monolith of solid columns, footings, foundation walls, girders, beams, floor and roof slabs, and spandrel panels, the last of which functioned as part of the load-bearing system above the level of the third floor. Monolithic action was secured as nearly as possible by carefully bonding freshly poured concrete to partly set concrete at the joints left from successive daily operations. The reinforcing throughout all framing members and all foundations consists of Ransome's square-twisted steel bars, so located as to take all tensile and shearing stresses, thus allowing the concrete to develop its full compressive stress. In the case of the columns, however, the compressive action of the concrete is supplemented by groups of heavy round rods, four to a column.[20] The framing system was designed for the following loading factors: floor loads of 200

[19] For the design of the Ingalls Building, see "A Tall Concrete-Steel Office Building," *Engineering Record*, XLVII, No. 21 (May 23, 1903), 540–43; "Sixteen-Story Concrete-Steel Office Building at Cincinnati, O.," *Engineering News*, L, No. 5 (July 30, 1903), 90–95; and A. O. Elzner, "The First Concrete Skyscraper," *Architectural Record*, XV, No. 6 (June 1904), 531–44.

[20] For details of the column reinforcing, see below, pp. 18–19.

pounds per square foot for the first floor, 80 pounds for the second, and 60 pounds for all floors above the second; roof load of 40 pounds per square foot; direct wind load of 30 pounds per square foot of exterior wall area above a line 50 feet above the street level, with no pressure assumed to act below that line. This assumption is not strictly in accord with the facts, the empirical basis probably being that the building was protected by surrounding structures on narrow streets from any wind action other than minor turbulence. The dimensioning of framing elements was predicated on a typical range of physical properties of the materials then available.[21]

All interior vertical loads from floors and roof are divided among three parallel rows of columns, two of which stand in the walls of the long elevations and the remainder in a line located at approximately the one-third point across the short dimension of the building (Fig. 1). Floor and roof loads are for the most part transmitted to the columns by means of a simple floor frame of massive girders carried by the columns and by means of smaller beams spanning between both the girders and the columns (Pl. 2). Peripheral floor and roof loads—that is, those coming from the outer halves of the peripheral bays—are transmitted to the columns through load-bearing spandrel panels in the exterior walls (the construction photograph, Pl. 3, shows the exterior form of these panels most clearly). The bending and shearing forces of wind loads are sustained by the diaphragm action of the floor slabs and the wall panels and by the rigid joints of the monolithic floor-framing system. All this is perfectly familiar to the modern builder, since it represents no more than the action of any straightforward column-and-girder frame constructed since the builders whose work

[21] A summary of physical properties of concrete and steel used in the Ingalls Building follows: Concrete mix, 1:2:4, regarded as including enough mortar to provide a volume = 1.15 × volume of all voids among pieces of aggregate, such pieces selected to pass through a 1-inch screen. Concrete, compressive strength: 2,000 pounds per square inch. Concrete, working stresses: 600 pounds per square inch for floor slabs; 750 pounds per square inch for columns up to the tenth floor; 500 pounds per square inch for columns above the tenth floor; no tensile stress assumed in concrete. Concrete, sample blocks tested for tensile strength with following results: 175 pounds per square inch at one day; 500 pounds per square inch at seven days; 600 pounds per square inch at thirty days. Steel type: mild steel used throughout. Reinforcing bars, untwisted, stresses: working stress, 16,000 pounds per square inch; ultimate strength, 64,000 pounds per square inch. Reinforcing bars, compression in columns, stresses: same as untwisted bars. Reinforcing bars, square twisted, stresses: working stress, 20,000 pounds per square inch; ultimate strength, 80,000 pounds per square inch (greater strength assumed to be developed by process of cold twisting). Adhesion between steel and concrete: working stress, 100 pounds per square inch; ultimate strength, 500 pounds per square inch.

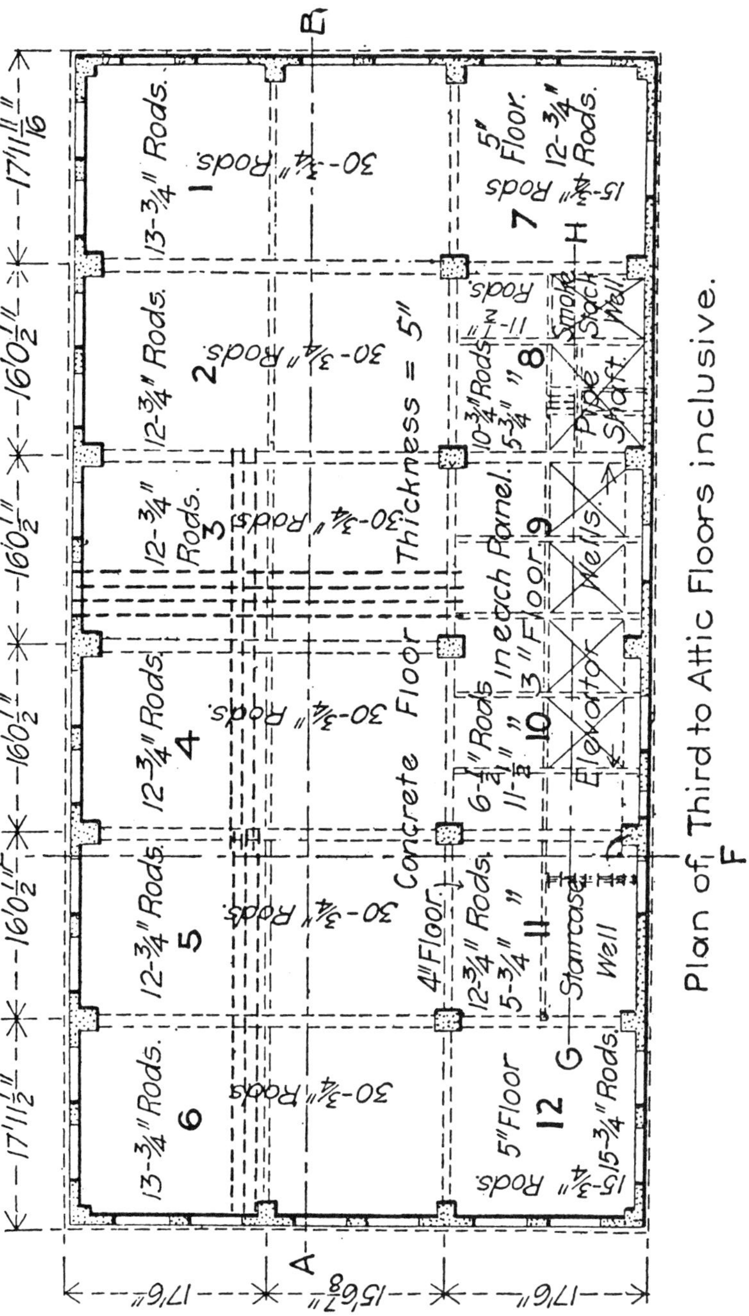

Fig. 1.–Ingalls Building. Typical floor plan. The more closely spaced columns were paired along the transverse lines on common footings. (*Engineering Record.*)

16 *Carl W. Condit*

we are describing here first showed the way. The difference between the structural system of the Ingalls and that of a contemporary building is that load-bearing wall panels are seldom thought to be necessary as supplements to the skeleton of columns and beams. The recent revival of load-bearing wall construction, however, has brought about an accompanying revival of analogous components.

The soil of the Cincinnati region varies in character with the distance from the valley floors of the main streams. The many hills are composed of relatively thin strata of Ordovician limestone interspersed with beds of soft shale. In the so-called bottom land along the Ohio River and its tributaries, however, this rock is overlain by varying depths of sand, clay, and gravel, of which the sand and gravel are sufficiently compacted to have a high bearing capacity. All columns and wall footings of the Ingalls Building are of reinforced concrete resting on a sand-and-gravel mixture with a maximum bearing capacity of 20,000 pounds per square foot, well above the actual footing load of 10,000 pounds per square foot.[22] The shapes and dimensions of the footings vary considerably, depending on whether they carry party walls, single columns, or two columns paired along the transverse column lines.

The footings for the party walls (along the elevations facing away from the streets) are continuous slabs 7 feet 6 inches wide, with the wall falling close to the outer edge of the slab. The upper surface of the inward portion slopes downward from a thickness of 2 feet at the earth side to 6 inches at the inner edge, a shape which indicates that the designers thought of the footing as an inverted cantilever extending on either side of the party wall and subject to maximum bending force immediately under the wall. The concentration of lateral reinforcing near the bottom of the footing further confirms this view. There are two types of column footings, the shape and size determined by whether the footing carries a single column or pairs of the more closely spaced columns along the transverse lines. The former is a truncated pyramid, square in plan, the upper horizontal face of which carries the cast-iron pedestal that receives the lower end of the column. Some of the single-column footings are large enough to be regarded as the type known in Chicago as floating-raft footings, since they measure 12 feet

[22] The finer surface material at Cincinnati is alluvial, while the coarser gravels are glacial in origin. The load-bearing capacity of the gravel beds may be judged from the fact that the towers of Roebling's suspension bridge at Cincinnati (1856–67) rest on this glacial gravel.

9 inches on a side.[23] The reinforcing consists of a two-way grid of longitudinal and transverse bars. The two-column footing is an elongated truncated pyramid with a rectangular plan, the upper face of which is developed into a massive rib designed to receive the column pedestals. These footings are more elaborately reinforced than the single-column forms: the grid of transverse and longitudinal bars is located near the upper as well as the lower face, while several planes of vertical bars strengthen the concrete against crushing and shearing forces. The maximum footing depth is 3 feet 4 inches.

Above the foundation and party walls—that is, above grade level—the walls of the building are divided between an open articulated base, two stories high, consisting essentially of the columns and girders of the frame, and a system of columns and spandrel wall panels extending from the third-floor level to the roof. In the base the windows occupy somewhat more than half the area of the street elevations, the glass filling the width of the bay; above the second floor, however, they are grouped in pairs in the bays. The mullion separating the pair is a thin post of reinforced concrete. The spandrels above the third floor are solid panels cast monolithically with the floor slabs and the peripheral columns. These panels are 8 inches thick and are reinforced with vertical bars set near both the inner and outer surfaces, indicating that they were designed to absorb the bending forces of wind as well as some part of the vertical wall loads. For the first three stories the walls are faced with a veneer of Vermont marble carried by a projecting concrete belt course at each floor level and fixed to the vertical concrete surfaces by means of wrought-iron anchors. Above the third floor the wall facing is gray enameled brick supported on similar horizontal courses and fixed to the panels and column surfaces by means of wire anchors (details of the spandrel panels and facing are shown in the drawings on the left side of Fig. 2).

Among the framing members the columns embody the most complex system of reinforcing (Fig. 2). The columns up to the tenth floor are rectangular in section, a typical one measuring $30\frac{3}{4} \times 34$ inches, although the dimensions vary from 30 to 38 inches on a side. Above the tenth floor all columns are square in section and 12 inches on a side. The reduction in cross-sectional area, of course, was dictated by the smaller

[23] The so-called floating raft, a widely spread concrete footing reinforced with grillages of steel rails, was developed by John Wellborn Root initially for the Montauk Building (1881–82) in Chicago. See Carl W. Condit, *The Chicago School of Architecture* (Chicago, 1964), pp. 49, 54–55.

gravity loads and the reduced bending moments arising from wind action, but a more rational indication of this decrease in column loads would be a repeated uniform reduction in area as the column rises above the grade level. The most unusual feature of the column construction is the presence of compression reinforcing in the form of four heavy, round bars set at the corners of a square centered on the axis of the column, the four bound together by hoops and by diagonal ties hooked around the rods. Designated as bearing rods in the building plans, they are not subject to tension and hence are of smooth, round shape without embossing. There can be little question that the idea of compression rods bound by hoops came from Considère, who was the leading exponent of this technique, but whether the Ingalls Building represents its

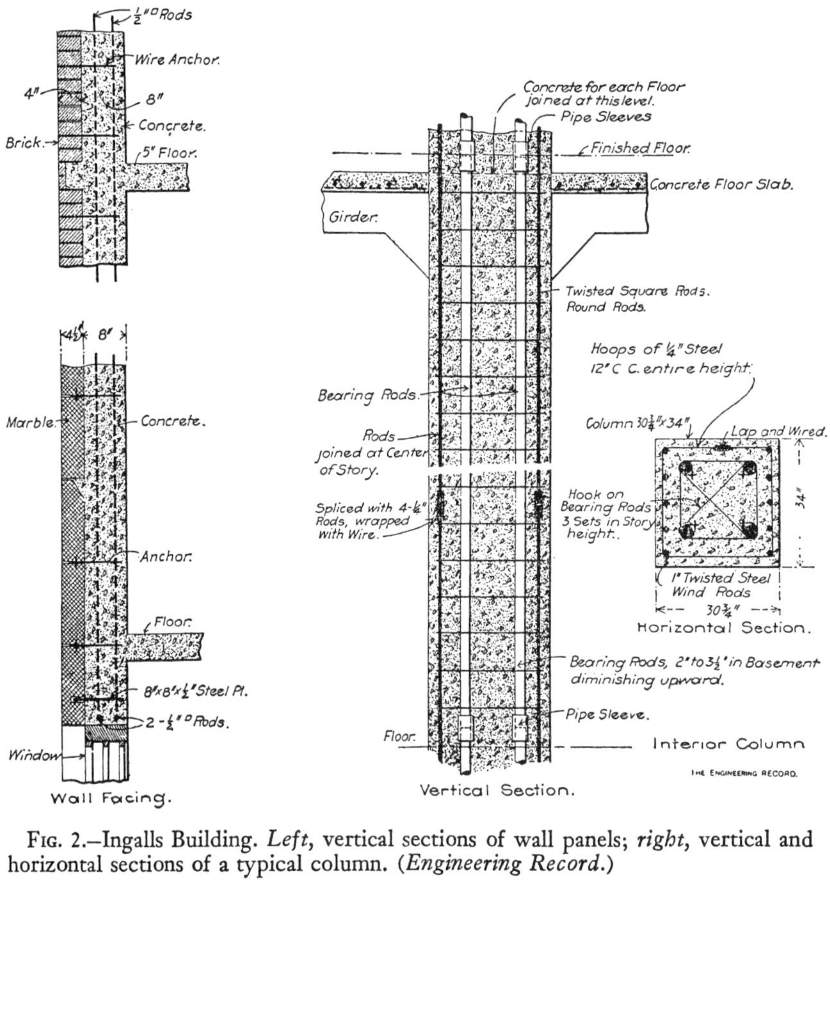

FIG. 2.–Ingalls Building. *Left,* vertical sections of wall panels; *right,* vertical and horizontal sections of a typical column. (*Engineering Record.*)

first American use is still a matter of question.[24] The remainder of the column reinforcing consists of ten square-twisted bars set five on each of the long sides and bound by hoops. Designated as wind rods by the architects, they perform the double function of absorbing the tensile stresses induced by the bending forces of the wind and resisting the tendency of the column to spread under buckling action.[25]

The columns stand in three rows parallel to the long dimension of the building, two rows respectively in the wall planes and the intermediate row at approximately the one-third point of the transverse dimension. The width of the building is thus divided into two unequal bays, respectively 17 feet 6 inches and 33 feet in span, measured from the center line of the intermediate column row to the outer faces of the exterior walls (see the dimensions given in Fig. 1). The narrower space is occupied mainly by elevators, stairways, mechanical utilities, and the main access corridors, an unusual arrangement that places the offices along one side of the building rather than symmetrically on either side of a central corridor. The floor slabs are supported by five sets of transverse girders, the spandrel panels of the two end walls, and by longitudinal beams located along the line of columns and along the center points of the long transverse girders.[26]

The columns, girders, beams, and floor slab were cast as a rigid monolith which in itself offers considerable resistance to the bending and shearing forces induced by wind. The depth of the girder is increased by means of fillets or haunches at the places where the girder and its supporting columns are joined (Fig. 3). These fillets were introduced to provide additional strength for the high shearing stresses at the ends

[24] In his description of the Ingalls Building, Elzner refers to the experiments of Considère as promising great possibilities for concrete construction (Elzner, n. 19 above, p. 543). Since Elzner was one of the architects, it is possible that the designers of the Ingalls were directly influenced by the work of the celebrated French investigator. The comprehensive report of Considère's experiments was translated into English and published in the United States in 1901 under the title *Experimental Researches in Reinforced Concrete.*

[25] The bearing rods vary in diameter from 3½ inches in the basement to 1 inch at the top floor. The successive lengths are joined by means of butt joints in pipe sleeves filled with a grout of mortar. The wind bars are 1 inch square, the 27-foot lengths lapped 21 inches and bound together by wire. The outer hoops are spaced 12 inches center to center and the inner, 6 inches.

[26] The transverse girders above the basement are 16–20 inches in width by 27 inches in depth, including the thickness of the floor slab. The basement girders measure 20 × 36 inches in section, and the longitudinal girders 9–12 inches in width by 18 inches in depth.

of the horizontal members and to increase the general rigidity of the frame.[27] Most of the resistance to shearing and tensile stresses is provided by reinforcing in the girder, which consists of horizontal tension bars located near the top and bottom of the girder, by U-shaped stirrups set in the vertical planes and spaced at decreasing intervals between the center and the ends of the girder, and by separate shear rods set at 45° to extend from the girder up and down into the column (Fig. 3).

The floor slabs vary in thickness from 3 to 5 inches, depending on the unit load, and are reinforced near the undersurface with a two-way grid of square-twisted ¾-inch bars. They are set 12 inches center to center on the transverse line and 15 inches on centers in the longitudinal direction, the shorter spacing being used for the somewhat greater bay span along the transverse line (Pl. 4). Each stairway flight and its associated parapet were poured integrally with the adjacent wall panel and the floor slab. The flight is reinforced with longitudinal bars, vertical bars in the low parapet along the outside edge of the stairs, and transverse bars for the tension arising from the cantilevering of the full width of the flight out from the wall. Another use of cantilevers occurs at the roof level, where reinforced-concrete cantilever beams project 5 feet outward from the wall to carry the cornice, which is built up as a hollow shell of terra cotta. The total weight of reinforcing steel in stairways, floor and roof slabs, and framing members in the building is 400 tons, while the total volume of concrete is 4,000 cubic yards.

Since the Ingalls Building was the first reinforced-concrete skyscraper, the process of construction was itself something of an experiment, offering valuable lessons for the subsequent development of the methods of erecting a large urban building.[28] The Cincinnati project quickly demonstrated one great advantage of concrete over steel construction: the absence of heavy structural steel members freed the site from the traffic of large vehicles and from the associated problems of unloading, storing, and placing the awkward lengths of columns and girders. As a consequence, it was unnecessary for the contractor to provide derricks, erecting towers or travelers, and heavy-duty hoisting machinery. These advantages not only meant some saving in the cost of construction but also freed the surrounding streets from the nuisance

[27] At the present time haunches are rare in column-and-beam frames, but their equivalent in mushroom capitals is still common in flat-slab framing, which was patented by Claude A. P. Turner of Minneapolis five years after the completion of the Ingalls Building.

[28] For the construction process, see "The Erection of a Tall Concrete Office Building," *Engineering Record*, XLVII, No. 3 (July 18, 1903), 64–67.

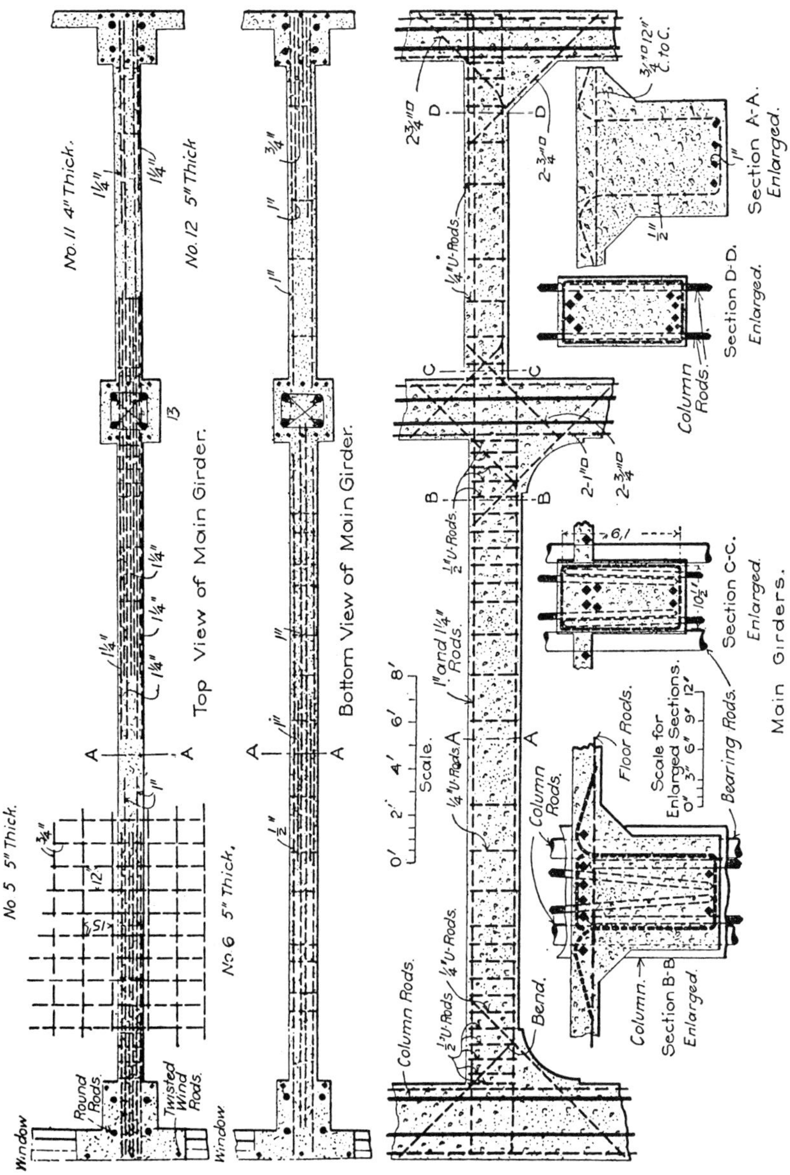

Fig. 3.–Ingalls Building. Horizontal and vertical sections of girders. (*Engineering Record.*)

and danger of handling large structural members in a densely built urban area. These savings were (and continue to be) partly offset by the high labor costs of constructing the wooden formwork necessary for the pouring of concrete frames and slabs. Formwork, like the tedious job of placing and joining reinforcing, is still largely done by expensive and time-consuming hand operations. In the case of the forms for the Ingalls Building, the contractors could take advantage of the various patents which had grown out of Ransome's twenty-year experience as a builder of concrete structures. Over the years he had developed many ingenious techniques for securing the close alignment of boards, tight joints, the rigidity and stability of finished forms, and the easy stripping of forms from the hardened concrete.

The concrete for the Ingalls was mixed on the site, as it always was until the development of the self-propelled mixer. Four large bins were built up on the basement floor of the building, one each for cement and sand and two for the aggregate, the four together providing a capacity sufficient for the pouring of two stories. The mixer was first charged with a measured quantity of sand and aggregate, followed by the proper load of cement, proportioned by number of bags, and finally by the water. The mixer was small by contemporary standards, having a capacity of only 18 cubic feet or two-thirds of a cubic yard. The wet concrete, mixed with an extra quantity of water in order to render it sufficiently fluid to fill all the spaces around the bars, was discharged into the 1-cubic-yard hoisting bucket without stopping the mixer. The wheelbarrows, mixer, steel bucket, and automatic hoisting controls were all designed and patented either by Ransome or by the engineering staff of the Ransome Concrete Machinery Company. The mixer was operated by an electric motor and the hoisting bucket by a 20-horsepower Lidgerwood engine.[29] A 10-horsepower Columbia gas engine was located in such a way as to be belted to the mixer or the hoist as a source of auxiliary power. Except for mixing and hoisting, all construction activities were carried out by hand with a total work force of sixty men.

The process of pouring concrete was maintained continuously for any one floor level in order to avoid joints at that level between the partially set concrete and the fresh pour. The columns were poured in one-story heights, and the concrete was rammed down around the reinforcing rods in 12-inch lifts. All bearing and wind rods in the col-

[29] A steam-powered hoisting engine manufactured by the Lidgerwood Manufacturing Co., Brooklyn, N.Y. I am indebted to Professor Eugene Ferguson for this information.

umns were left with lengths projecting above the finished pouring level in order to bind the new pour as tightly as possible to the hardened concrete. The exterior concrete surfaces were covered with waterproof asphalt paint up to the third-floor level to prevent the stains that inevitably appear in concrete work from discoloring the marble veneer. All the techniques of construction used in erecting the Ingalls–building formwork, installing reinforcing bars, pouring concrete, preparing concrete surfaces, and applying veneer materials–have remained standard since the Cincinnati project was undertaken. Indeed, except for some increase in mechanization, the methods of construction have changed surprisingly little in the intervening period of more than sixty years. And the same may be said, consequently, of the time of construction, which was about eight months for the Ingalls (from September 1902 to the spring of 1903).

What has changed far more than the structural forms and the processes of construction is the architectural design and the external finish of concrete-framed buildings. In its architectural character the Ingalls is a conventional high-quality office tower of the time, designed in the mode of the late classical revival dominated by McKim, Mead, and White in New York and D. H. Burnham in Chicago. Elzner, however, saw the possibilities of a new approach to design in concrete structures, in which all these embellishments of veneers, entablatures, cornices, rustication, and pseudo-colonnades might be done away with, so that a truly organic architecture could come in their place. He concluded his description of the Ingalls with a prophetic passage:

> It is not incumbent on us to face the concrete with marble, or brick and terra cotta, as was done in the Ingalls Building, for reasons of momentary expediency, for as the state of the art advances, the architectural forms, mouldings and what not, will be incorporated with the moulds of structural work, and upon removing the formwork, the surface of the exposed concrete will be given the desired finish of rubbing or tooling, as the case may be. Thus we will have a truly rational architecture, in which there is no sham, no deception, a solid thing, no joints, every member incorporated with and a part of the living body; living because it is straining every particle of its substance in the performance of a great work, in its own self-preservation; a living architecture, indeed, and a rational one in every sense of the word, which will rise far above criticism and endure as long as the hands of man shall not be raised to its destruction.[30]

[30] Elzner, p. 544. Two years after the completion of the Ingalls, Elzner and Anderson had an opportunity to come closer to the realization of this ideal in their design

24 *Carl W. Condit*

IV. *The Coming of Age of Concrete Framing, 1903–5*

The rapid progress in the development of reinforced-concrete frames, which reached a culmination in the Ingalls Building, provided a secure basis for the subsequent evolution of the art. Ironically enough, however, a new invention appeared in the year the Ingalls was completed that was for a short time to eclipse all others in popularity but then to pass quickly into history without leaving any progeny. In 1903 Julius Kahn of Detroit founded the Trussed Concrete-Steel Company to provide a complete designing and contracting service for the erection of concrete buildings and bridges embodying Kahn's patented system of reinforcing. He worked closely with the architectural-designing empire established by his brother Albert, and between the two of them they quickly expanded the scope of their operations far beyond anything that Ransome could command.[31] Julius Kahn's system of reinforcing was expensive, redundant, and awkward to handle, but the principle was sound, and there is no question that it worked very well.

The system differs from all others in that it incorporates a multiplicity of shear bars that constitute an integral part of the tension bars lying in the same vertical planes (Fig. 4).[32] This was accomplished by rolling

---

of the Terminal Warehouse, Broadway near Twenty-fourth St., Kansas City, Mo. (1905). The warehouse is a complete work of reinforced-concrete construction: foundations, columns, girders, beams, floors, walls, roof, penthouses, and gravity water tank on the roof to supply the automatic sprinkler system. The three-story warehouse was remodeled into an office building in 1959–60. Elzner wrote of it, "In this way the entire building is practically a monolith, and has tremendous rigidity. The building stands unique in that the use of wood has been entirely avoided throughout the building" (quoted in Donald L. Hoffman, "Early Concrete Construction in Kansas City," *Skylines and Midwest Architect*, XV, No. 1 [December–January 1965], 22). The Ferro-Concrete Construction Co. closely followed the structural system of the Ingalls Building in a work of straightforward functional design when they built the huge Terminal Warehouse of the Louisville and Nashville Railroad at Atlanta, Ga., in 1906–7 ("The Terminal Warehouse of the Louisville and Nashville Railroad at Atlanta, Ga.," *Engineering Record*, LV, No. 9 [March 2, 1907], 312–13).

[31] Albert Kahn has rightly been called the architect of the American automotive industry, and one might add that he was in good part the architect of the Soviet Union's tractor industry in the 1920's. The point is that the Kahn brothers were prepared to master the new scale of mass-production industry, whereas their competitors for the most part were still living in a nineteenth-century world. See George Nelson, *Industrial Architecture of Albert Kahn, Inc.* (New York, 1939).

[32] See the following booklets issued by the Trussed Concrete-Steel Co.: *Kahn System of Reinforced Concrete* (Detroit, n.d.); *Tests and Other Facts concerning the Kahn Trussed Bar* (Detroit, 1906).

the bar in a square section with a pair of flat wings running along the opposite edges (the cross-section of the girder at the top of Fig. 4 gives some idea of the sectional appearance of the bar). By cutting the wings for short lengths along the edges of the main bar and bending them up at an angle of 45°, tension and shear bars could be united rigidly in a single piece. With the addition of upper bars to take the tension in a cantilever or continuous member, or simply to resist buckling, the intersecting patterns of bent-up wings formed a kind of truss, which, although far from being exactly scientific in the arrangement of its elements, worked effectively in resisting tensile and shearing stresses. A beam reinforced in this way was tested at the St. Louis Exposition of

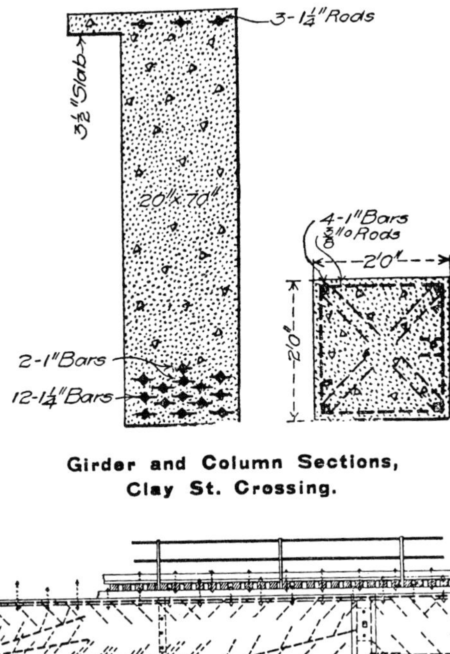

Girder and Column Sections, Clay St. Crossing.

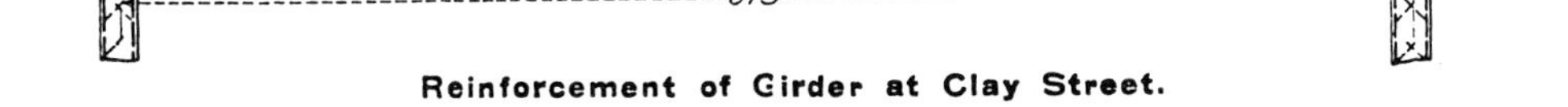

Reinforcement of Girder at Clay Street.

FIG. 4.—An example of the Kahn system of reinforcing for girders and columns. Longitudinal and transverse sections of a long-span girder in the viaduct of the Richmond and Chesapeake Bay Railway, Richmond, Va., 1906–7. (*Engineering Record.*)

1904 and showed extraordinary capacity to resist extreme loads without failure.[33]

Kahn's success is revealed not only by the number and variety of buildings whose concrete structures his firm designed but also by the number of reinforced-concrete bridges for which he was responsible. Many of these involved novel forms such as long-span simple girders and cantilever girders that appear to have been introduced by Kahn himself, at least as far as American practice is concerned. The building that commanded the greatest attention for structural virtuosity is the Marlborough Hotel Annex in Atlantic City, New Jersey (1905–6).[34] The architects were Price and McLanahan; the structural engineers were the designing staff of the Trussed Concrete-Steel Company, with M. Goldenberg as chief engineer; and the general contractor was Edwin Gilbert and Company of Philadelphia. The only precedent for the Marlborough was the Imperial Palace Hotel at Nice, France (1900), designed and built by Hennebique, but the Atlantic City project outdistanced it in size, being, in fact, the largest reinforced-concrete building in the world at the time. Much more than sheer volume was involved in the construction of the hotel: the building represented the triumph not only of Julius Kahn but of the art in general, since reinforced-concrete techniques were applied successfully to a great variety of large structural forms, including frames, slabs, gables, towers, domes, and girders designed to carry offset columns under multistory loads. The only criticism one can make, and that is possible only by virtue of a historical ex post facto judgment, is that everything in this extraordinary work could have been accomplished by simpler means.

The dominant techniques of reinforcing were to remain those associated with the Ransome and Monier-Wayss systems, although many others were still flourishing around 1905.[35] Ransome's most advanced

[33] The 16-inch-deep beam was simply supported for a span of 16 feet. It carried a total uniformly distributed load of 94,962 pounds, or 5,935 pounds per lineal foot, without failure (Trussed Concrete-Steel Co., *Tests and Other Facts . . .*).

[34] "Reinforced Concrete and Tile Construction in an Atlantic City Hotel," *Engineering Record,* LII, No. 26 (December 23, 1905), 719–21, and LII, No. 27 (December 30, 1905), 743–45; Collins, p. 87, Pls. 22–24.

[35] For a brief but comprehensive review of reinforcing technology at this time, see Emile G. Perrot, "Reinforced Concrete in Building Construction," *Engineering Record,* XLIX, No. 22 (May 28, 1904), 670–72. Typical of the pioneer phase of any major technical innovation, construction in reinforced concrete led to such a proliferation of patents in the period of 1885–1905, with many concentrated around 1900, that no engineer at the turn of the century felt free to design a structure without infringing on one of them (Ralph Modjeski, cited in Heidenreich [n. 15 above]).

work is the factory of the Kelly and Jones Company at Greensburg, Pennsylvania (1903–4), which embodies a complete framing system including several innovations beyond the square-twisted bars and the stirrups that Ransome had already developed.[36] Most important from the technical standpoint was the use of bars shaped in the form of continuous helixes wound around the tension bars of the columns (another of Considère's inventions). The innovation of greatest architectural importance was a spandrel girder in the form of a hollow box, the outer web and flange area of which projected beyond the outer column face as a belt course to support continuous or ribbon windows.

Concrete trusses in building frames came with the construction of Terminal Station in Atlanta, Georgia (1903–4), a work containing long-forgotten elements that reappeared in recent years as supposedly new inventions (Fig. 5).[37] The station building, with its associated bridges and trackwork, was designed and built under the direction of Walter H. Harrison, chief engineer of the terminal company, but the concrete structure was the work of the Baltimore Ferro-Concrete Company. The entire complex of station building, drives, and train facilities, except for the former balloon train shed, is built of reinforced concrete throughout–piling, footings, columns, floors, roofs, roof trusses, and the original platform canopies extending beyond the end of the train shed. The reinforcing system is essentially like that of Ransome, with peripheral tension bars and hoops used in the chord and web members of the trusses as well as in the columns. The gable roofs over the office floors and the waiting room are carried by triangular Warren trusses, the latter of which span 60 feet clear. Most remarkable, however, is that the roof over the so-called midway, the broad trapezoidal area between the waiting room and the train concourse, is supported by rigid-frame trusses, that is, trusses in which the diagonals are omitted, the web and chord members being arranged as a series of quadrilaterals (right end of longitudinal section, Fig. 5). Reinforced-concrete trusses are a European invention of the 1890's, having been developed chiefly by Franz Visintini in Austria and Considère in France. Rigid frame or Vierendeel trusses of concrete have been used in building construction from time to time since about 1930, and have recently appeared as the primary elements of bearing-wall construction in the office building of

[36] "The Kelly and Jones Company's Concrete-Steel Factory Building," *Engineering Record*, XLIX, No. 6 (February 6, 1904), 152–54.

[37] "Reinforced Concrete Work at the New Railway Terminal Station at Atlanta, Georgia," *Engineering News*, LV, No. 15 (April 12, 1906), 391–401; Baltimore Ferro-Concrete Co., *Ferro-Concrete Construction* (Baltimore, 1904).

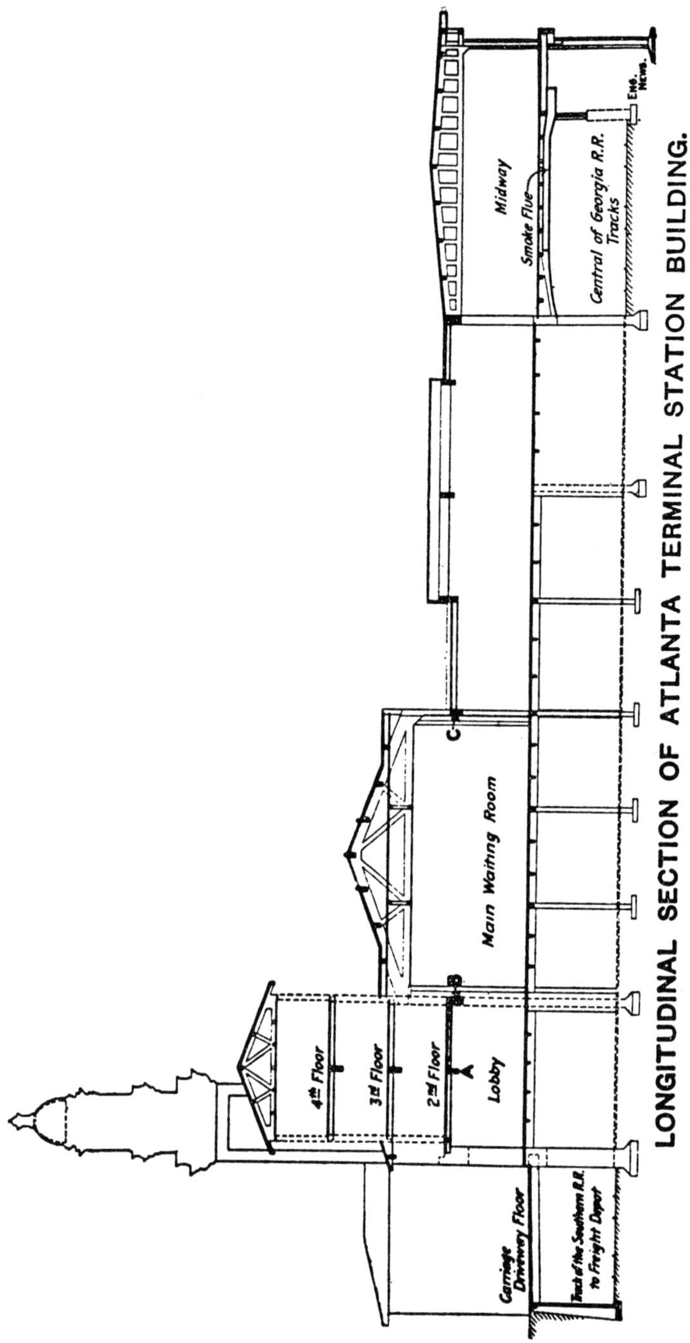

FIG. 5.–Terminal Station, Atlanta, Ga., 1903–4. Walter H. Harrison, chief engineer. Longitudinal section and truss and footing details. (*Engineering News.*)

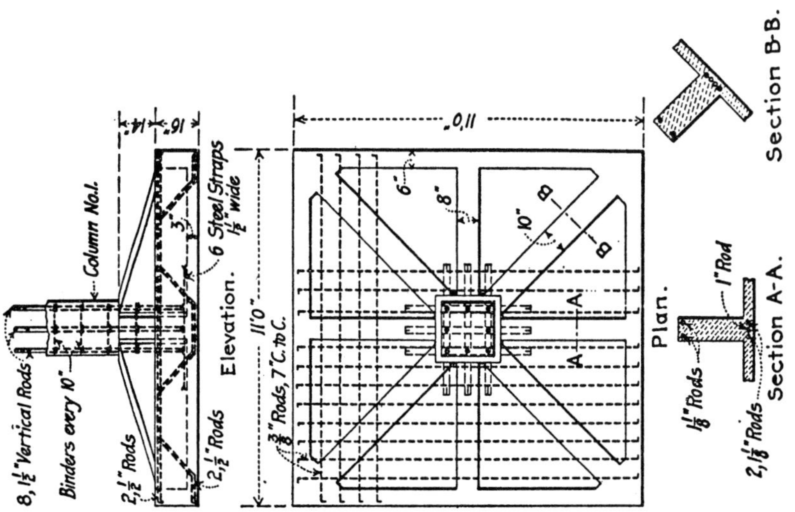

Ribbed Slab Footing for 20-in. Square Column Supporting Baggage Driveway.

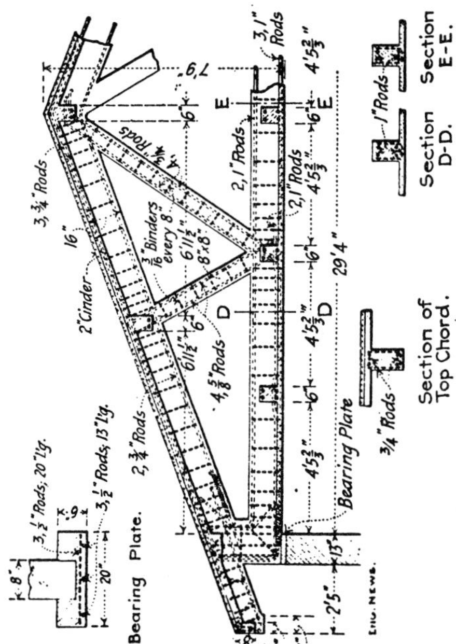

DETAILS OF MAIN TRUSSES IN ROOF BETWEEN TOWERS.

FIG. 5—*Continued*

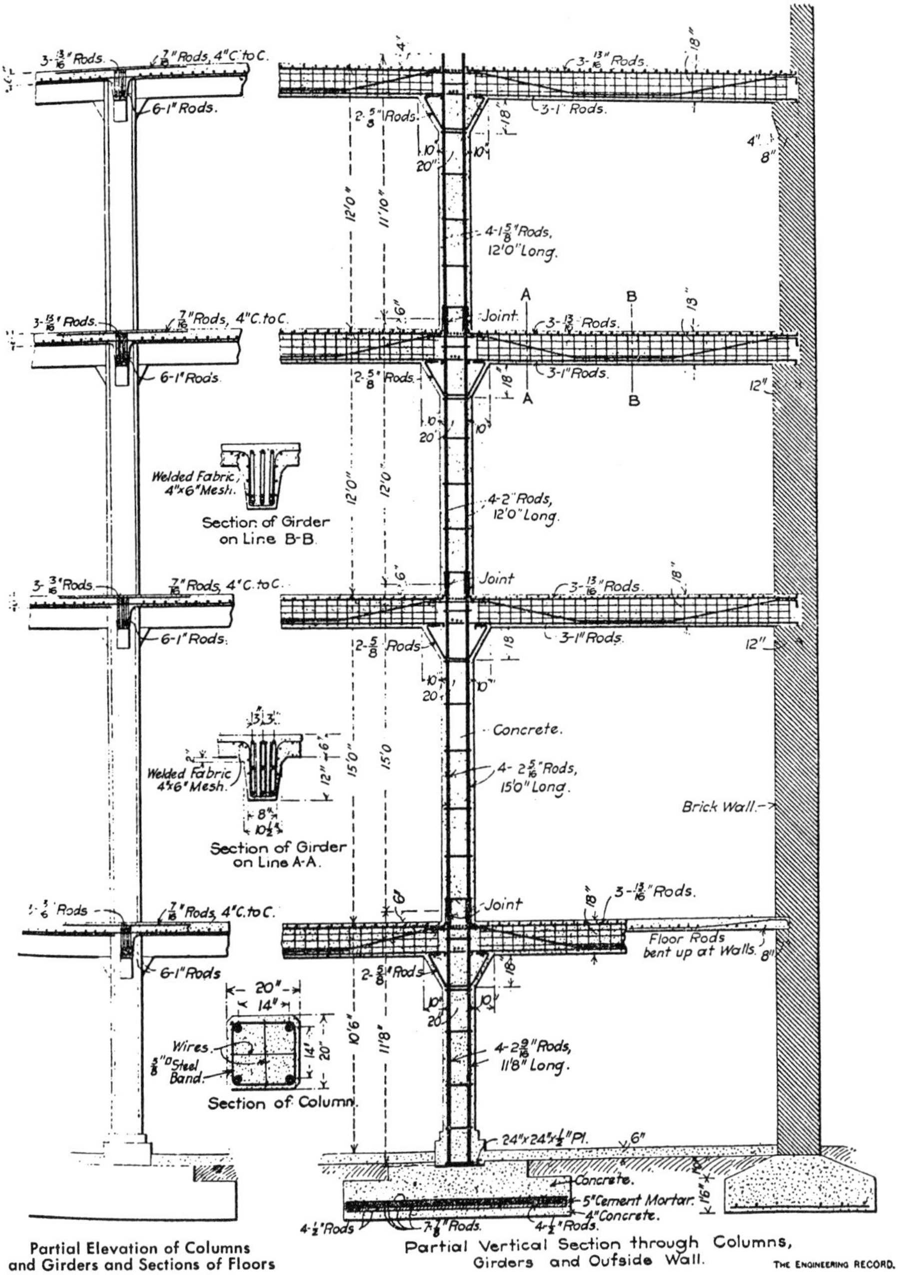

FIG. 6.–Winton Building, Chicago, 1904. James Gamble Rogers, architect. Vertical sections of the concrete frame. (*Engineering Record*.)

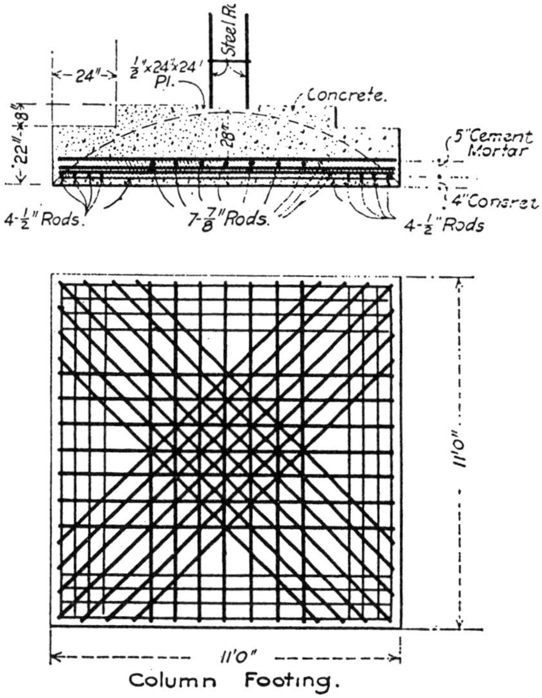

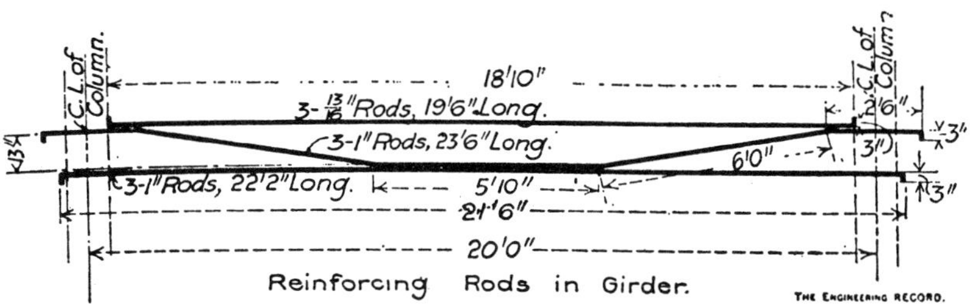

FIG. 6—*Continued*

the North Carolina Mutual Life Insurance Company at Durham, North Carolina (1964–65).[38] The evidence so far points to the conclusion that such trusses were originated for Terminal Station by the engineering staff of the Baltimore Ferro-Concrete Company, but why they were chosen in place of the traditional form is a matter of speculation. Another innovation in the terminal structure is the folded-plate construction of precast concrete that forms part of the supporting system for the sawtooth light monitors extending transversely over the flanks of the train-concourse roof. Most of the working drawings of the station have disappeared, along with the names of the engineers who created it, yet it remains in active use today, one of the most imaginative and most neglected achievements of American building art.

The progress of reinforced-concrete construction was rounded out by Heidenreich's continued development of the Monier-Wayss system in the United States. This European technique reached its maturity in the little Winton Building, erected in 1904 at Michigan Avenue and Thirteenth Street in Chicago.[39] The structural engineer was Heidenreich and the architect, James Gamble Rogers, who was later to achieve fame as the creator of the Gothic expansion of the Yale and Northwestern campuses in the 1920's. The now demolished Winton, designed as a seven-story building but erected only to the level of the fourth floor, was a typical combination of brick bearing walls along the sides and rear with an internal column-and-beam frame poured as a monolith. The reinforcing was another variation on the now-common practice of binding the steel bars into a tight network (Fig. 6). The columns contained the standard vertical tension or wind rods, located in the corners, bound by hoops set at intervals in the horizontal plane. The girder reinforcing was more elaborate, the rods having been arranged in what Heidenreich, like Kahn, thought of as a kind of truss. Each girder contained three of these so-called trusses, built up of the following components: three tension rods near the bottom; three shear rods, their ends bent up into the diagonal position at the one-third points; and three compression rods near the top, that is, close to the surface of the floor slab. The three rods lying in any one vertical plane were bound together by an envelope of welded wire mesh which extended throughout the length of the girder and functioned as the equivalent of a set of stirrups. The floor slab was

[38] "Friction Supports 14-Story Building," *Engineering News-Record*, CLXXIII, No. 8 (August 20, 1964), 46–49.

[39] "A Reinforced Concrete Store Building in Chicago," *Engineering Record*, XLIX, No. 23 (June 4, 1904), 713–14.

reinforced with the usual two-way grid near the undersurface. Except for the vertical wire mesh, which appears to have been introduced in the United States for this purpose by Heidenreich, the reinforcing closely parallels the forms developed by Wayss and Koenen in Europe and by Ransome in America. The fact that it does provides an example of the process of selection and convergence that nearly always characterizes the pioneer phase of any technological development.

By 1905 the structures erected by the large building and engineering contractors such as the Ransome, Kahn, Baltimore, and Cincinnati firms embodied all the essential features of modern scientific reinforcing systems. The Kahn system, as we have seen, offered peculiar solutions to the problems involved, and its needless intricacy and high cost eventually led to its abandonment. The other techniques, however, increasingly converged into a common body of variations derived from the principle of using steel to take tensile and shearing stresses, the variations arising from the stress pattern of any particular structural element. There were difficulties, of course, unexpected phenomena and unsolved problems that sometimes led to serious trouble, but for the most part the techniques were sufficiently well established to warrant their application to every kind of structural system. The builders were not slow to take advantage of the possibilities. By 1910 they had extended reinforced-concrete construction to cantilevers and simple girders of long span, continuous girders and trusses, Vierendeel trusses, ribbed vaults, shells, and plates. The steady progress of better than half a century led to such astonishing exhibitions of this new technical virtuosity as a hyperbolic-parabaloidal shell 221 feet on the long dimension but supported at only two points (Edens Theater, Northbrook, Illinois, 1961–63) and a column-and-flat-slab frame seventy stories high close to the very shore of Lake Michigan (Lake Point Tower, Chicago, 1965–68).

PLATE 1.–Ingalls Building, Cincinnati, 1902–3. Elzner and Anderson, architects. General view of the completed building from the southwest. (*Architectural Record*.)

PLATE 2.–Ingalls Building. Construction view showing the framing system. (*Architectural Record.*)

PLATE 3.–Ingalls Building. Exterior view of the concrete work during construction. (*Architectural Record.*)

PLATE 4.—Ingalls Building. Construction view showing reinforcing in the floor slab. (*Architectural Record.*)

15
The consulting engineer and early concrete bridges in California

John Snyder and Steve Mikesell

To understand the place of reinforced concrete in the waning years of the 19th Century, it is necessary to look briefly at the material's developmental history. As early as the mid-19th Century, European engineers had experimented in the reinforcing of concrete. They understood that concrete not reinforced provided great compressive strength but lacked tensile strength. In 1849 Joseph Lambot of France designed a cement boat reinforced with wire mesh, patenting his work several years later in 1855. That same year the boat, credited by some as the world's first reinforced concrete structure, received international attention at the Paris Exposition.

Countryman Francois Monier followed in 1867 with additional patents, and Francois Hennebique applied reinforcement to columns and beams at about this time. In fierce competition, German engineers bought rights to Monier's patents in 1879, and continued the development of reinforcement. In England, American Thadeus Hyatt was experimenting with concrete, publishing his work in 1877. In the United States, the first major use of reinforced concrete appears to have been the work of William Ward, who built a house of the material in Port Chester, New York in 1873-76. Yet interestingly, Ward's work was somewhat of an anomaly: Most of the innovations that followed occurred in California, in a region usually viewed by historians as developmentally behind the rest of the nation.

The pioneer reinforced concrete work in California was concentrated in and around San Francisco. Peter H. Jackson, basing his work on that of Hyatt, began use of the material for sidewalks as early as 1877. In 1885 he tested the material for use in floors and walls, noting, "... the invention may prove of great value ... ." Through the 1880s and 1890s, California architects and engineers continued to discuss and develop the material, chiefly for its fireproofing potential. Most of the notable, early reinforced concrete building work in these years was that of engineer Ernest L. Ransome, and architect George W. Percy, working in collaboration from San Francisco.

Ransome arrived in San Francisco during the 1860s, fresh from apprenticeship in his father's ironworks factory in Ipswich, England, and knowledgeable of the senior Ransome's patent work in portland cement. In 1884 he received U.S. patents on cold-twisted square iron reinforcing bars, on concrete reinforced with those bars, and on a concrete mixer. Immediately following, in 1885, he incorporated a reinforced concrete roof in his design for the Arctic Oil Company warehouse in San Francisco. By the latter 1880s he collaborated with Percy in a series of commissions employing reinforced concrete: Greystone Cellars at St. Helena in 1888, with reinforced concrete floors; the California Academy of Sciences in San Francisco in 1889, with reinforced concrete floors; and, most notably, the Stanford Museum and Roble Hall at Stanford University in 1889, with reinforced concrete floors, walls, and roof. It was also in 1889 that Ransome designed the Alvord Lake Bridge in San Francisco's Golden Gate Park, the first reinforced concrete bridge in the United States (Fig. 1).

Given the architectural quality of the Alvord Lake Bridge, it is likely that this was another collaborative effort with Percy. Ransome designed the bridge to carry a busy carriage drive above a pedestrian pathway connecting the Haight Street park entrance with the newly completed Children's Playground — including Percy's Children's Playhouse. Unmodified, the bridge is still in use with three lanes of Kezar Drive located the park. Equally remarkable, it went almost unnoticed at the time of its building, and even Ransome failed to mention it in his reminiscences of his early years in San Francisco. Still, engineer and architect combined to give the children a fun and mysterious entrance to the playground. The arch face and wingwalls were molded and hammered to resemble ashlar stone masonry, and the concrete color mimics that of the nearby Playhouse, of ashlar

Fig. 1 — Alvord Lake Bridge.

sandstone construction. Further, the intrados — the interior surface of the tunnel-like arch—is covered with concrete stalactites and stalagmites. Its significance is threefold:

- it is widely recognized as the first reinforced concrete bridge in the U.S.
- it is the earliest bridge known to have been designed by Ransome
- in this structure Ransome's method of reinforcement — small bars as opposed to the alternative method of I-beam reinforcement found in the Melan system — pointed the way to the predominant 20th Century practice.

Following Ransome's departure for Chicago in the mid-1890s, there was a decade-long hiatus in reinforced concrete development in California. Yet though development slowed, it did not stop. In the years between 1889 and about 1905, concrete bridges were built in large numbers throughout California, especially in Northern California.

There are records of only a few of these, either in physical remnants or extant plans. Several broad generalizations characterize these early concrete structures. First, these bridges were quite small, most scarcely more than culverts over the smaller creeks in the state. Second, a high percentage of these were "plain concrete" structures, concrete without metal reinforcement. Third, both county surveyors and private consultants designed concrete bridges during this period. Finally, there was a great deal of experimentation among these pioneering designers, illustrating the newness of the technology, and the lack of standardized practices.

One anecdote illustrates the confusion that prevailed among the designers and builders of these early concrete bridges. The winding mountain and coastal roads of rural San Mateo County include the largest number of very old concrete bridges in the state.

In the late 1890s, the county of San Mateo contracted with a San Francisco architectural and engineering firm to design a small concrete bridge across Pilarcitos Creek on Main Street in the little coastal village of Half Moon Bay. The winning contractor refused to construct the bridge as planned, however, informing the Board of Supervisors that, in his opinion, "... the specifications were defective to such an extent that structure if built thereto would be utterly worthless." The specifications were redrawn, the bridge was built in 1900, and stands today, almost completely unmodified. As if to underscore the novelty of the occasion, the county installed a plaque on the railing of the new bridge announcing it to be the "First Concrete Bridge Erected in San Mateo County, 1900."

During these years, a few county surveyors made significant contributions to the evolution of concrete bridge design. Chief among these was John G. McMillan, county surveyor for Santa Clara County between 1890 and 1914. Like many early highway bridge engineers in California, McMillan learned his trade through an apprenticeship in railroad and mining work. Raised in California but lacking formal training, his first apprenticeship was in the mining industry. He subsequently worked for the Central Pacific Railroad, the Market Street Railway in San Francisco, and the Guatemala Central Railroad. He also worked as a structural engineer in the construction of the first buildings at Stanford University, where he must have been exposed to the reinforced concrete work of Ernest Ransome, and George Percy. Self-taught and extremely adventurous, McMillan epitomizes the pioneer generation of California engineers.

McMillan's turn of the century bridges in Santa Clara County illustrate how tentatively county surveyors moved into the use of reinforced concrete. Between 1891 and 1900, he designed 29 bridges. Of these, 18 (more than 60 percent) were timber and metal combination trusses. Another four were all-metal trusses. The remaining structures were concrete bridges in a general sense, although none represented what would be standard practice a decade later. Two, for example, were arched steel trusses encased in concrete. Four of the original 18 might best be described as compound masonry structures, using concrete in combination with stone or bricks. The final bridge was a plain concrete arch. Only after 1905 did McMillan design what is considered conventional reinforced concrete structures — concrete with metal reinforcement.

In a 1909 article in an engineering journal, McMillan looked back on nearly 20 years of bridge design and explained his reasons for preferring to work in concrete, as well as his unorthodox use of brick and stone in lieu of metal reinforcement. He had two very straightforward answers: esthetics and economy. Working in what was then an overwhelmingly rural county, McMillan obviously savored the pastoral setting and respected the common sense of local farmers. Wooden and steel bridges were, in his view, ". . . a conspicuous and not altogether attractive part of our county roads." He continued, "There is no field of construction in which the use of cement appeals more strongly to the practical business sense of the farmer than in the building of highway bridges. The flimsy and unsightly bridge structures of wood and steel are more expensive in the end than bridges and culverts of concrete, which are at once beautiful, permanent and fireproof."

As for his use of stone in lieu of reinforcement bars, the explanation again was a dual concern for esthetics and economy. He wrote:

> *There are in Santa Clara County between 35 and 40 concrete and stone bridges of more than 20 feet. The designs are made in harmony with the locality. In the foothills and mountains, where boulders and field stones are available, they are worked into the structure not only for harmony, but also, as a matter of economy. With the proper use of boulders and field stones a savings can be made in labor and cement of 25 to 30 percent.*

He went on to explain that, after about 1905, the cost differential between stone and reinforcement metal had equalized and that all of his post-1905 concrete bridges used reinforcing bars.

Very few of McMillan's bridges have survived, and most of the survivors are heavily modified. The one great remnant of his work, a 1902 concrete arch bridge with stone veneer, is located on a remote state highway between San Jose and Santa Cruz (Fig. 2). From an engineering standpoint, it is a bold structure particularly considering the date of construction. What is most striking about the structure,however, is that it is so appropriate for its setting. It is, to use McMillan's phase, in "... harmony with the locality."

Fig. 2 — 1902 concrete bridge with stone veneer.

In hindsight, it can be said that McMillan and the other early California concrete bridge engineers made few major technological innovations. McMillan's unorthodox reinforcement, for example, attracted few imitators. Their contributions were more subtle. First, the simple fact that they built so many concrete bridges helped establish an economically viable cement and reinforcement industry as well as a construction industry familiar with reinforced concrete. They also contributed to our understanding of what a concrete bridge should look like. McMillan helped establish a belief that was later upheld by the California Division of Highways. He thought that the concrete bridge was vastly superior to the metal or wooden truss when built in a sensitive natural environment. He established a precedent for site-specific designs for bridges in beautiful natural settings, designs that could take into account, or enhance such environments. The challenge for the pioneer concrete bridge designer was both technological and esthetic, and engineers like McMillan fully grasped both aspects of the problem.

John Leonard was born in Union City, Michigan in 1864, studied engineering, and arrived in San Francisco in 1889. During the 1890s Leonard worked for several firms as a draftsman, and a civil engineer. As a member of the Technical Society of the Pacific Coast, the young Leonard likely knew and associated with fellow members Ernest Ransome, and George Percy, who may have fostered Leonard's interest in the potential of reinforced concrete at this time. In 1904 he opened his own office as a consulting civil engineer.

During the next 20 years, John Leonard became Northern California's foremost designer of reinforced concrete bridges. In 1905 he won a competition for his first reinforced concrete bridge, spaning the Truckee River in Reno. Virtually unmodified today, it is a two-span, filled-spandrel arch, originally carrying two traffic lanes, two sidewalks, and a center streetcar track. Even in this first example the gracefully proportioned arch ring springs to a remarkably thin section at the crown, a feature that is now known as a hallmark of Leonard's bridges. In keeping with the bridge's urban setting, Leonard chose Beaux-Arts detailing in the form of decorative railing and lighting elements.

Setting out next to sell officials in California on reinforced concrete bridges, Leonard gained three immediate commissions: the San Joaquin River Bridge at Pollasky, near Fresno; the Dry Creek Bridge at Modesto; and the Stanislaus River Bridge at Ripon. The Pollasky Bridge incorporated ten 75-foot spans in gracefully spanning the bed of the San Joaquin River; its composite length caused Leonard to refer to Pollasky, the longest reinforced concrete bridge in the United States at that time. At Dry Creek and Ripon, Leonard's designs were noteworthy for their individual span lengths, 112 and 110 feet respectively, with the Ripon Bridge employing two spans.

In late 1907 Leonard won a competition for a group of five bridges in Ross, Marin County, using Beaux-Arts detailing to produce bridges quite in keeping with the architecture of what was, even then, an exclusive suburb of San Francisco. At this time he also designed the Gianella Bridge, one of only two steel bridges which can be credited to him. And even here his preliminary proposal had been for a concrete bridge. His occasional failure to sell the idea of reinforced concrete bridges was also seen in a stillborn proposal for a three-span arch bridge across the Feather River at Oroville. This structure would have employed a 199 ft center span and would have marked Leonard's departure from the filled spandrel arch to the transitional spandrel cross-wall type.

The year 1911 climaxed Leonard's design work with the filled spandrel arch bridge with three designs. The greatest of the these was Fernbridge, which crosses the Eel River south of Eureka in northwestern California with seven 200 ft spans (Fig. 3). These are the longest filled-spandrel arch spans ever constructed in California. Fernbridge has survived the worst

Fig. 3 — Fernbridge.

floods of the Eel River, as well as the April 1992 Ferndale earthquake, and continues to carry traffic today.

A second 1911 structure was a reinforced concrete railroad bridge across the American River near Auburn. The structure is composed of three 140 ft spans towering above the river. Leonard met the company's requirements with a bridge that proved to be 20 percent cheaper than a steel structure designed for the same site. Like Fernbridge, this bridge was designed for permanence. In the 1950s and 1960s, it was twice pressed into emergency service as a vehicular bridge when floods washed out highway bridges a few hundred yards upstream.

Finally, in 1911, Leonard met the requirements of civic officials of the Oakland suburb of Piedmont, who wanted a bridge out of the ordinary. Retaining consulting architect Albert A. Farr, Leonard's collaborative effort resulted in a bridge far more architectonic than any other he designed. The pair gave the town a bridge in the Mission Revival style, then at its height. Though the bridge is still in use, Farr's architectural details have been removed.

In 1913 Leonard and junior partner W.P. Day published *The Concrete Bridge*. In it, they stressed the esthetics of concrete bridges, citing, "... conformity with environment ... pleasing outline and appropriate use of ornament." Beautiful bridges, they wrote, "... are a sure indication of a progressive community." They also noted that reinforced concrete bridges used California products — cement, sand, gravel, and reinforcing steel — negating the often lengthy wait for Eastern materials associated with steel bridges. Profusely illustrating Leonard's bridges, and in the tradition of a builder's catalog, the book represented a unique step for a consulting engineer to have taken.

During the second decade of the 20th century. Leonard developed a bridge type which he termed "canticrete." Essentially utilizing a cantilever steel truss to provide sidewall and floor substructure, the entire structure was then encased in concrete. The cantilever was self-supporting during construction, keeping falsework costs to a minimum. Because of the strength of the truss, less reinforcing steel was required, and sidewalls, and floors could be thinner in section, using less concrete.

Of eleven "canticrete" bridges known to have been constructed between 1914 and 1921, only three remain. The Tuolumne River Bridge in Modesto, built in 1917 and known locally as the "Lion Bridge," was one of the largest "canticrete" bridges and was one for which Leonard again involved a consulting architect, this time Fay Spangler of San Francisco. While Leonard gave Modesto a monumental "canticrete" bridge, Spangler provided such details as the cast concrete reclining lions which give the bridge its local name.

After the "canticrete" period, Leonard returned to more conventional reinforced concrete bridge designs for the remainder of his career, exemplified by his Van Duzen River Bridge of 1922. When he died in San Francisco on February 16, 1945 at 81 years of age, John Leonard left a legacy which included 47 known bridges designed throughout California and Nevada, all but three of which were constructed of reinforced concrete.

By the second decade of the 20th century, much of the initiative in reinforced concrete bridge design passed from engineers in Northern California to those in the Los Angeles area. There gathered in Los Angeles during this decade a large group of skilled and innovative consulting engineers specializing in reinforced concrete design.

Arguably the most important thinker in this group was William Thomas. Born in St. Louis in 1876, Thomas trained as an architect at the Chicago Art Institute, and received private tutoring in engineering. Like most early California bridge engineers, he began his work with the railroads, first designing a railroad terminal in St. Louis, and later serving as a structural engineer for the Southern Pacific Railroad in Tucson, and as a bridge engineer for the Arizona Eastern Railroad. Moving to California in 1906, he began work for the Union Traction Company, a street railway in Santa Cruz. He moved to Los Angeles in 1908, where he opened the firm of Thomas & Post in partnership with W.S. Post.

His most significant contribution to the science of bridge design stemmed from his experiments with the three-hinged concrete arch, and the use of pre-cast arch ribs. The precast, three-hinge arch bridge, which Thomas referred to as the "Thomas System Bridge," was first used on a Union Traction Company Bridge in Santa Cruz. This bridge was constructed in 1908 and still stands, although in a much-modified condition. It is an interesting conglomeration today, with the original 1908 bridge in the middle, a 1912 Thomas System widening upstream, and an unusual 1967 slab and column widening downstream. From the roadway it is impossible to discern as a historic structure.

Thomas did not, of course, invent the three-hinge arch, a concept first utilized in the design of metal bridges and previously executed in concrete. He freely acknowledged that many European, and a few American, engineers had built three-hinge concrete bridges before him. Indeed, it is a measure of the cosmopolitan nature of California engineering in the early 20th century that Thomas was aware of and patterned his work after international precedents.

The ingenious aspect of the Water Street Bridge, and the "Thomas System" bridges which followed, was Thomas' decision to cast the arch rings and spandrel columns in forms located away from the bridge site. These precast members were hauled to the bridge site, hoisted into place, and assembled, leaving hinges at the springlines and at the crest of the arch (Fig. 4). Writing in 1914, after having constructed more than a dozen such bridges, he looked back on the advantages of the system over the prevalent cast-in-place fixed arch. He listed several attributes, some of which related to the use of the three-hinge arch, some of which related to the use of pre-cast members. For example, he argued that off-site casting of critical members ensured a higher quality than could be achieved on-site because the forms could be made more rigid and uniform. Precasting also eliminated the need for costly arch centering falsework. The three-hinge design allowed for more effective expansion joints and better adjustment to changes in temperature, and the hinges could adjust better to settlement in the abutments and piers. Finally, he argued that the system was cost-effective, chiefly through savings in time and material for falsework.

While at least 17 of these bridges were built in California, only two remain today: the modified Water Street Bridge in Santa Cruz, and a little bridge on a remote Forest Service road in San Diego County.

Fig. 4 — Water Street Bridge.

To some degree, Thomas overstated the advantages of his bridge: virtually no one copied the form. If the system were so advantageous, we should have expected greater use of the general techniques. Nonetheless, we can say that Thomas was very precocious in recognizing the advantages of using precast structural members, a fact that would be recognized decades later when prestressed and post-tensioned concrete girders became widely used.

Thomas was also precocious in another, completely unrelated manner. His bridges were extremely honest from an architectural standpoint, devoid of applied ornamentation. Although trained as an architect, Thomas, a prolific writer, rarely spoke of the esthetic qualities of his bridges, focusing instead upon his innovations in construction methods. When he did refer to esthetics, he praised the simple strength that came from straightforward design. In describing one of his largest bridges, for example, Thomas said that it was,"... a pleasing structure, obviously strong and permanent." In an age that valued Beaux Arts detail in its bridges, Thomas designed simple, unadorned bridges which were esthetically pleasing because of, not in spite of, their structural systems.

The firm of Mayberry and Parker helped popularize the concrete girder bridge in Southern California. Edward T. Flaherty, a Los Angeles engineer, introduced in California the patented "Mushroom Slab" system, patented by C.A.P. Turner, for use in bridges as well as warehouses (Fig. 5). Frank P. Allen, a Los Angeles-based engineer and architect, designed the lovely Cabrillo Bridge in Balboa Park, built as part of the 1914 Exposition. The noted architectural firm of Greene and Greene designed a bridge in South Pasadena, and Irving Gill, the famous San Diego architect, designed a little bridge in Torrance, which is today depicted on the logo of the city.

Fig. 5 — Cabrillo Bridge.

Ironically, the best-known bridge in Southern California, the Colorado Street Bridge in Pasadena, erected in 1913, is the one major structure in the area not designed by a Southern Californian. When the city of Pasadena sought bids on the design of this project, it received three credible proposals. One proposal was from Thomas and Post, one was from Mayberry and Parker, and another was from the Kansas City firm of Waddell and Harrington. The Kansas City firm was chosen (Fig. 6).

Concrete bridge construction in California, as with public works spending generally, came to a halt during American involvement in World War I. Following the war, California entered into an era of unprecedented economic expansion and followed a much different approach to public works expenditures.

The most important post-1919 development was the creation of a corps of highly-skilled public service bridge engineers, who largely displaced consulting engineers as the designers of concrete bridges in California. The most important group of engineers were those working at the State Division of Highways, although cities and counties attracted talented engineers as well.

Concrete bridge-building also took place within a much grander scale of highway construction. During the 1920s and 1930s, the Division of Highways built the great trunk highways that link regions of California — Highway 99 through the Cascades, Highways 50 and 40 over the Sierra, Highway 1 along the coast, U.S. 101 through the northern redwoods, and so forth.

Great highway projects called for great bridges and, not surprisingly, most of the really impressive bridges built in

Fig. 6 — Colorado Street Bridge.

California in the 1920s and 1930s were concrete. Some of the best-known American concrete bridges were built in California during these decades, including the incomparable Bixby Arch in the Big Sur (Fig. 7).

Although the scale of construction was far greater, the California concrete bridges of the 1920s and 1930s owe a great deal to the pioneering bridges from the earlier period. To a large extent, the expertise for building great concrete bridges existed prior to World War I. The circumstances of the 1920s and 1930s offered the opportunity to use that knowledge on a very grand scale.

The bridges of the later period also share two important characteristics with the bridges that came before: technological daring, and careful attention to esthetics. These two traits unite more than half a century of California concrete bridges.

There is no need to overstate the regional distinctiveness of California in this regard; surely engineers elsewhere in the United States and particularly in Europe built beautiful and technologically daring bridges. Californians did, however, make several important contributions. First, the concrete bridge received its American debut in California and Californians built an enormous number of concrete bridges in the years 1889 to 1919, when concrete bridge building was in its infancy. Second, because they were pioneers in the field, the early concrete bridge engineers in California set very high standards for themselves, demanding bridges that were attractive, functional, and innovative.

This was the legacy of Ransome, McMillan, Leonard, Thomas, Allen, Mayberry and Parker and the others. They made an important imprint on reinforced concrete bridge building technology and esthetics and left behind a patrimony of great concrete bridges. The tradition of excellence survives even today; hold and beautiful concrete bridges are still being built on, and along the highways of California.

References

1. Andrews, C.E., "Bridges on California State Highways," California Highways and Punlic Works, July - August 1928.

2. California Division of Highways, *The Paths of Humanity: A Chronicle of California Highway Development*, February 1966.

3. Condit, Carl W., *American Building: Materials and Techniques from the Beginnings of the Colonial Settlement to the Present*, University of Chicago Press, Chicago; 1968.

Fig. 7 — Bixby Arch.

4. Draffin, Jasper O. "A Brief History of Lime, Cement. Concrete and Reinforced Concrete," *Journal of the Western Society of Engineers*, Vol. 48, No. 1, 1942.

5. *Historic Civil Engineering Landmarks of San Francisco and Northern California*, American Society of Civil Engineers, 1977.

6. *Historic Civil Eningeering Landmarks of Sacramento and Northeastern California*, American Society of Civil Engineers, 1976.

7. *Historic Civil Engineering Landmarks of Southern California*, American Society of Civil Engineers, 1974.

8. Hool, George A., and W.S. Kinne, *Reinforced Cincrete and Masonry Structures*, New York, McGraw-Hill, 1924.

9. Israel, Paul B., "Spanning the Golden State: A history of Highway Bridges in California," M.A. Thesis, University of California, Santa Barbara, 1980.

10. McMillan, J.G., " Reinforced Concrete Bridge for Country Roads," *The Architect and Engineer of California*, June 1909.

11. Panhorst, F.W., "Century of Bridge Progress," *California Highways and Public Works, Centennial Edition*, September 1950.

12. Pursell, Carroll W., "The Technical Society of the Pacific Coast, 1884 - 1914," *Technology and Culture*, October 1976.

13. Ransome, Ernest L., and Alexis Saurbrey, *Reinforced Concrete Buildings*, New York, McGraw-Hill, 1912.

14. Snyder, John W., "Buildings and Bridges for the 20th Century," *California History*, Fall 1984.

15 Thomas, W.M., "The Thomas System of Three Hinge Arch Bridges," *Southwest Concractor and Manufacturer*, April 4, 1914.

16. Tyrell, Henry Grattan, "Bridges in Relation to the City Plan," *The Architect and Engineer*, August 1913.

17. U.S. Bureau of Public Roads, *Report of a Study of the California Highway System*, Washington, D.C., 1920.

18. Zucker, Paul, *American Bridges and Dams*, New York, The Greystone Press, 1941.

Selected for reader interest by the editors.

16
Early reinforced concrete structures in New South Wales, 1895–1915

D.J. Fraser

SUMMARY: Reinforced concrete construction began in Europe around 1870, soon after Joseph Monier took out patents for concrete flower pots reinforced by wire netting. The Monier System, as it became known, was extensively used in Germany where it was refined to a viable, economical structural material for a wide range of applications. It was not until the early 1890's that the Monier System was introduced into New South Wales and then with some initial hesitancy and controversy about its use. But after the success of its first major application, the sewer aqueducts in Annandale, Sydney, 1895-97, its development and use grew rapidly. Within the next twenty years, Sydney's structural engineers became as proficient in the theory, design and application of this new technology as their counterparts in Europe, England and America. This paper traces the history of reinforced concrete in New South Wales during those formative years.

1 INTRODUCTION

The use of reinforced concrete in New South Wales began in 1895 when Carter,Gummow & Co. were awarded Contract No.77 by the Public Works Department for the construction of two concrete sewer aqueducts across the valleys of Johnstone's and White's Creeks Annandale, Sydney. At the time, reinforced concrete was known by other names such as ferro-concrete , concrete iron and steel-concrete construction, and by the names of patented methods such as Hennibique, Thacher and Melan Systems. The Annandale aqueducts, figure 1, were built using the Monier System of reinforced concrete arches. These aqueducts are still in service after 87 years.

The first reference to the introduction of the Monier System into New South Wales was not recorded in

---

Dr D J Fraser is a Senior Lecturer, School of Civil Engineering, University of New South Wales. (Paper G1173 submitted on 19 April 1985).

2 REINFORCED CONCRETE IN NEW SOUTH WALES

Figure 1 Johnstone's Creek aqueduct, one of the two 1896 sewer aqueducts in Annandale, Sydney.

contemporary technical papers but in the unusual source of the Report, Evidence and Appendices of a Royal Commission (1896-97). The Inquiry was initiated by accusations in the Colonial Parliament that Robert Hickson, Under-Secretary for Public Works, had shown favouritism towards Carter, Gummow & Co and had allowed irregularities in the contractural procedures at a cost to the Crown.

Legally, nothing came out of the Inquiry. Charges were not proved and Hickson and his officers were exonerated. To this extent the Inquiry followed tradition and had been largely "a waste of time and money". But what did come from the Inquiry is a priceless record of the beginnings of reinforced concrete, specifically the Monier System, in New South Wales. Reported fully and contemporaneously, it is information that might not otherwise have been recorded, because subsequent technical papers concentrated on the theory and use of reinforced concrete and only briefly mentioned its origins.

When the two sewer aqueducts were completed in 1897, they were the first large-scale use of reinforced concrete in New South Wales and, in fact, in Australia. Together with John Monash's work in Victoria, the Fyansford and Anderson Bridges, and the 1896 Lamington Bridge in Queensland, these projects established reinforced concrete as the new building material. The emergence of the new technology coincided with the emergence of a new nation, a federated Australia.

Local engineers were quick to recognise the merits of reinforced concrete because the examples in

Europe, Britain and America clearly demonstrated its flexibility of application. During the first twenty years of use in New South Wales, a great variety of works were constructed using the Monier and other systems.

This paper presents an historical review of the origins of reinforced concrete and its use, with specific reference to New South Wales and the period 1895 to 1915. The quality of construction was such that a great many of those early projects are still in use and are accessible for inspection. With many of the surviving works 80 or more years old, and with reinforced concrete currently dominating most forms of construction, these early projects have become part of our engineering heritage.

2 HISTORICAL BACKGROUND

Strengthening (reinforcing) masonry and concrete constructions, so as to resist the internal tensions that cause joints to open and cracks to appear, is a very old concept. The Romans are known to have embedded wood, brass and iron within their structures for that purpose, and the practice of including granite keys, metal rods and cramps in stonework, figure 2, was in common use during Medieval and Renaissance times. One of the largest and best examples of the application of this concept is Sir Christofer Wren's use of iron chains around the base of the dome of St Paul's Cathederal to help the concrete resist the tension caused by the outward thrust (Jones and Lakeman 1913, Building Research Station 1956, Cowan 1977 and 1978).

However, the useful principle of combining the high crushing strength of concrete and the high tensile strength of iron into a composite structural mater-

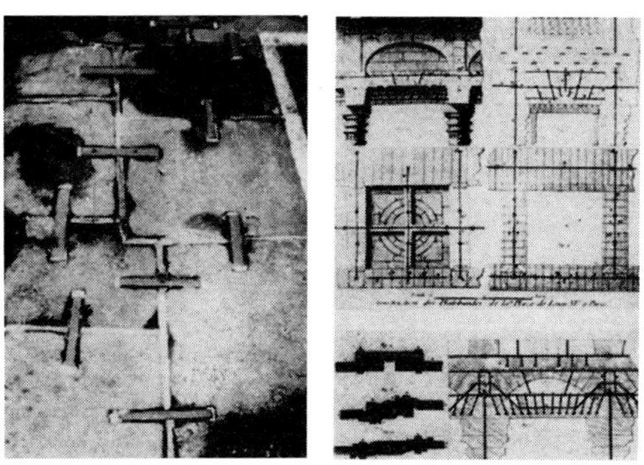

Figure 2 Examples of reinforcing masonry construction during the period 1650-1850.

ial, had to await the availability of cements, hence concretes, that were alkaline and reasonably waterprof thereby forming a reliable protective cover to the iron or steel. This did not occur until the nine teenth century. In 1824 Joseph Aspdin patented his method for making "portland cement" and this new material was used extensively in the construction of Marc Isambard Brunel's Thames Tunnel. But it was not in general use until some years later because it was confused with gypsum or "plaster of Paris" which quickly rusted the embedded iron.

By 1855 there were two specific examples of this composite form of construction that should have alerted others to the usefulness and range of its application. J.L.Lambot built a reinforced concrete rowing boat in 1848 and displayed it at the 1855 International Exhibition in Paris, and W.B. Wilkinson constructed suspended concrete floors reinforced by wire ropes, figure 3, in Newcastle-on-Tyne, which he patented in 1854. But it was the maker of flowerpots, Joseph Monier, who set in train the events that led to the development of reinforced concrete as twentieth century engineers know it.

Monier was a gardener in Paris and was manufacturing gardening tools and appliances such as clay flowerpots. When the substitution of concrete still left the finished product brittle, he added a skeleton of iron wires to the moulds and produced durable tubs that were shatter-proof. In 1867 he patented the invention and displayed examples at the International Exhibition in Paris the same year. During the next five years he took out patents for pipes, arch bridges, beams and reservoirs.

In 1879, Monier sold his inventions to G.A. Wayss whose firm of Wayss & Freitag of Frankfurt-am-Main, Germany, made widespread use of them in Germany and Austria-Hungary. A series of strength and durability tests was carried out for him by K. Koenen, fig-

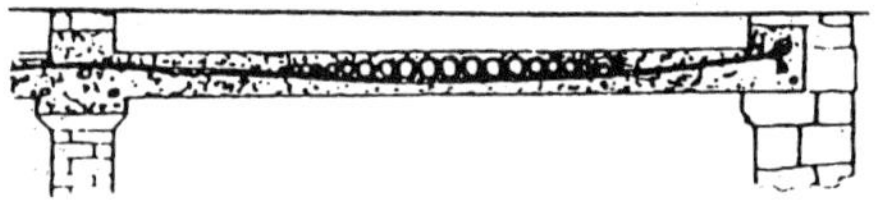

Figure 3 Wilkinson's 1854 patented floor. It incorpoated wire ropes for tensile strength and clay pipes to reduce self weight.

ure 4, and a manual was published in 1886 called "Das System Monier". There was now sound theoretical and experimental support for Monier's empirical designs, and Wayss succeeded in introducing the Monier system into public works. Between 1887 and 1891 the firm built 320 arched bridges, some of them with spans up to 40m (130 feet), figure 4.

This German connection is very important to the introduction of reinforced concrete to Australia because the early exponents, J. Baltzer, F.M. Gummow and John Monash were fluent in German and were ideally placed to exploit the new technology in New South Wales, Victoria and South Australia at the end of the nineteenth century.

Meanwhile, parallel developments had taken place in France where Edmond Coignet and Francois Hennibique made significant contributions to the theory and practice of reinforced concrete. Whereas the Monier system was based on the concrete arch in which tension was generally a minor problem, Coignet and Hennibique concentrated on beams and floor slabs in which high tensile stresses made the use of iron or steel reinforcement essential. Their patents of 1906 and 1897, figure 5, have all the characteristics of modern reinforced concrete construction. Subsequent changes over the next twenty years were mainly refinements to the theory and to the details by academics and engineers in Europe, Britain and America.

Figure 4
In Germany, a series of load tests was carried out on Monier arches. The success of the tests led to the construction of many large slender arch bridges throughout Europe.

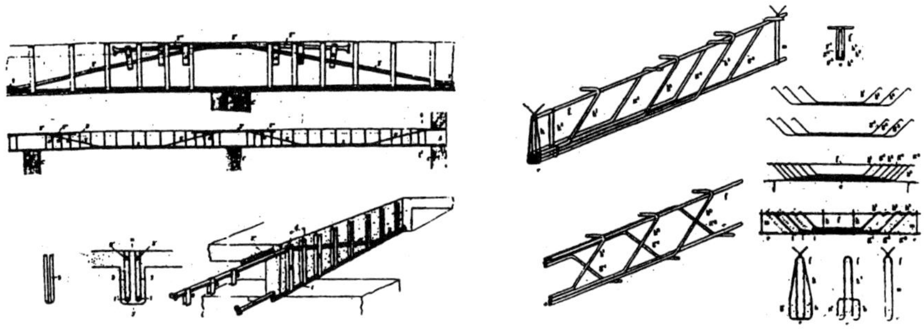

Figure 5 By the early twentieth century, reinforced concrete had acquired many of its modern features as these 1897 and 1906 patents show.

A more complete history of reiforced concrete (American Concrete Insitute 1976,1982) is beyond the scope of this paper because this brief review is sufficient background to its main theme.

3 PATENTED SYSTEMS

In the second half of the nineteenth century, technological changes were occurring at a rapid rate. Inventions were literall "pouring out of people's minds, off the drawings boards and out of the workshops". Any idea, no matter how trivial, was worth patenting because the subsequent expoitation might be a commercial "pot of gold". So it was with reinforced concrete.

After Monier took out his patents, the usefulness of the new composite material was immediately recognised, but the problem was how to use the new material without paying royalties or facing legal challenges over infrigements of the patent rights. The solution lay in the fact that the concept of reinforced concrete could not be patented, only the method by which it was achieved could be patented (the same applies to prestressed concrete today). Consequently, during the thirty-year period 1870-1900, a great many systems for reinforcing concrete were patented, Table I.
In each case the patent referred to the types of bars and their arrangements, figure 6, as the names Indented Bar and Triangular Mesh suggest. Companies were formed to market many of these patented systems

TABLE I
SOME PATENTED SYSTEMS FOR REINFORCED CONCRETE

Allen	Brannon	Clinton
Coignet	Considere	Corr Bar
Cottacin	Dennett	Dentile
Expanded Metal	Hyatt	Hennibique
Indented Bar	Johnson's Lattice	Kahn
Keedon	Klett	Koenen
Lock-woven Mesh	Melan	Moller
Monier	Paragon	Ransome
Scott	Shaler	Thacher
Triangle Mesh	Visintini	

and a host of Manuals and Catalogues (Mouchel 1909, American Steel & Wire Co. 1909, Indented Bar and Concrete Engineering Co. Ltd. c1916) were made available to all potential users. However, by the end of the period under review it was realised there was no valid patent preventing the use of plain reinforcing rods in any formation, consequently, specialist firms came to rely more on their skill and experience rather than their patents (again a parallel situation with pre-stressed concrete).

As for the early years of reinforced concrete in New South Wales is concerned, the dominant systems appear to have been Monier, Hennibique, Indented Bar and Clinton-Paragon.

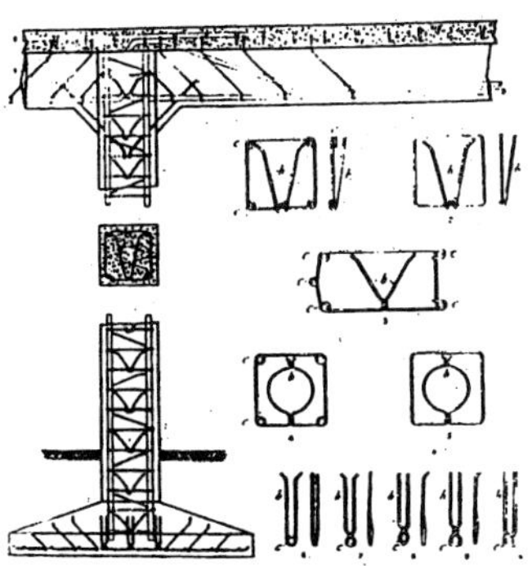

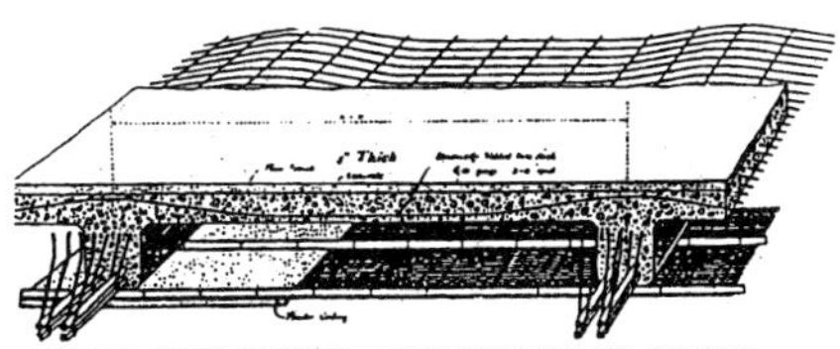

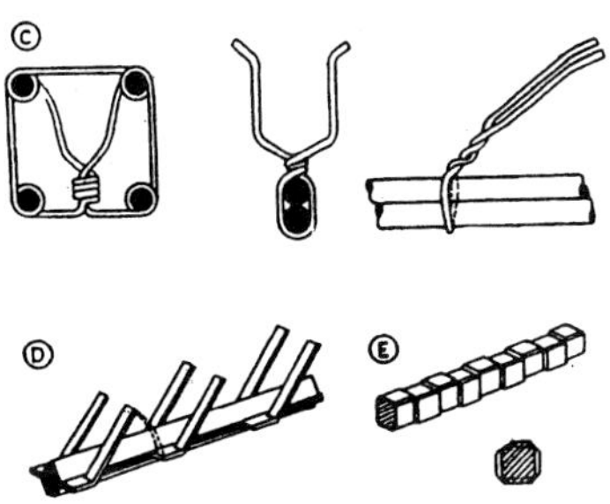

Figure 6 Early reinforced concrete was characterised by a flood of patents such as the Paragon System (top left) and the Clinton Floor (above), and (C) the BRC hoop and stirrup, (D) the Kahn Bar and (E) the Indented Bar.

8 REINFORCED CONCRETE IN NEW SOUTH WALES

4 THE START IN SYDNEY

4.1 The culvert at Burwood

Although the Annandale sewer aqueducts of 1895-97 were the first major use of Monier arches, a small Monier arch had been constructed in 1894 for a stormwater culvert, figure 7, located under Parramatta Road, Burwood between Wentworth Road and Philip Street. A drain of similar proportions is there at present, but the records of the Burwood and Concord Municipal Councils do not indicate whether it is the original culvert.

In 1894, the District Engineer for the Public Works Department was W.A.Smith. He stated in evidence to the Royal Commission that the culvert had been built as an experimental arch by Carter, Gummow & Co., and had cost about half the estimated cost of standard brick construction. The Monier culvert had been "tested by a severe strain" and the system had many advantages over conventional forms of construction. The Commission reported "that a moving load across an arch causes bending in other parts of the arch, sometimes flatteneing the curve of the arch, sometimes increasing the curvature of the arch. The iron adds strength to the arch by the added quality of a girder. Loads can be imposed at any point".

Robert Hickson, Commissioner for Roads and Bridges, and Engineer-in-Chief for Sewerage Construction,

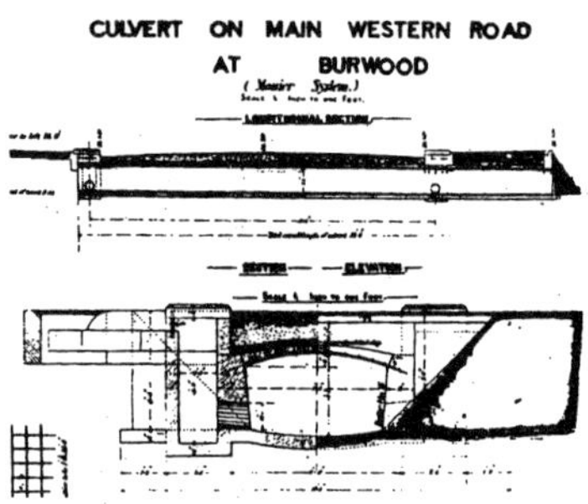

Figure 7

A modest stormwater culvert constructed in 1894 under Parramatta Road, Burwood, using the Monier System began reinforced concrete in New South Wales.

reported on 18 July 1894 "that the culvert was very satisfactory and that he had a firm belief in the system; schemes using the system would be successful, strong and graceful". He was therefore anxious to give it a fair trial and put into pratice this "new departure in engineering". Consequently, when the tenders for the Annandale Aqueducts were reviewed, he approved on the 16th May, 1895 the awarding of the contract to the agents of the Monier System, Messers. Carter, Gummow & Co. plus the payment of a 15% royalty.

4.2 The Inquiry

Hickson's actions led to a furore in Parliament. During construction of the aqueducts, Varney Parkes, MLA alleged "favouritism, defective work, incompetence, violation of terms of contract, and insufficient bond", was critical of the "speciality" nature of the design and expressed a lack of confidence in the ability of those involved to carry it out properly. The Royal Commission was appointed on 21 May, 1896.

The Commission heard evidence from a wide section of concerned parties, politicians, government officers, engineers of the Public Works Department, Contractors, Professor W.H.Warren and consulting engineers such as Norman Selfe. Sittings occurred during the twelve months after its appointment and its final Report was issued late 1897. Hickson and his officers were completely exonerated, indeed some of the original allegations were withdrawn early in the proceedings and the Commission "hinted" that the others were frivolous and precipitated by malice.

To that extent, the Inquiry could be seen as a waste of time and simply a standard ploy of politicians; but in calling all the expert witnesses to give evidence and receiving detailed technical submissions, the Report contains a priceless history of the beginnings of reinforced concrete in New South Wales.

4.3 Introduction of the Monier System

The Report and Evidence of the Royal Commission indicates that W.J Baltzer, a draughtsman/engineer with the Sewerage Construction Branch of the Public Works Department, initiated the use of Monier construction in Sydney, and, it seems, to Australia. He had received his engineering education in Germany and had joined the Department in 1885. He was regarded as a resourceful and capable officer who had designed some major works but, owing to the severe economic depression of the early 1890's, was retrenched in 1895 "with great regret" (Hickson's evidence).

Baltzer had a relative in Germany who was a Professor at the Technical School, Berlin, whom he had visited sometime around 1890. During the visit he toured Germany looking at engineering works and took photographs of many of the Monier arch bridges. When he returned to Sydney he had a letter from the Monier Company (Wayss & Co.) offering to afford him every information.

During 1892-95 Baltzer compiled a set of notes (Mitchell 1922) about Monier construction which subsequently became the 1896 Treatise issued by Carter, Gummow and Co. (see later). It is not clear from the Commission's Report if Baltzer was responsible for the design of the culvert at Burwood, but the inference is strong, however there is no doubt that he designed the viaduct of Monier arches for the Annandale sewer aqueducts.

4.4 The Annandale Aqueducts

Annandale is an inner western suburb of Sydney, located on the southern side of Rozelle Bay and bounded by Parramatta Road further south. Its best known street is the wide, gracious Johnson Street which runs straight from the bay to Parramatta Road along a ridge. To the east is the valley of Johnstone's Creek and to the west is White's Creek; both creeks flow northwards into Rozelle Bay.

During the 1880's Sydney was expanding rapidly due to the boom conditions and the large influx of immigrants. Public works, such as water supply and sewerage schemes, were a major feature of those hectic years. When the large sewer main was planned in 1893 to drain most of the suburbs on the southern side of the Harbour to the Bondi Outfall it was found necessary to carry a branch sewer across Johnstone's and White's Creeks, Contract No. 77.

The original PWD design , figure 8, proposed a series of brick arches for each aqueduct with a total length of about 610m (2000 feet) by 1m (3 ft 6 in) wide and at an estimated cost of £11,298 (Warren).

By 1895 the agency for the Monier System was held by Carter, Gummow and Co., a successful firm of contractors, who retained Baltzer as a consultant at a fee of 2½% of any Monier-related contract. He prepared an alternate design for them "out of office hours" based on 25.3m (82 feet 10 inch) Monier arches, figure 8, supporting a reinforced concrete flume at an estimated cost of £9,203. Carter, Gummow and Co. were awarded Contract No 77. Baltzer, of course, was still employed by the Department and so was seen to be in a compromising situation. This was another factor that precipitated the Inquiry.

The Monier arch design was seen by some as being a "speciality" and there were expressions of lack of confidence in the viability of the system and in the experience of all those involved in the project. But these "fears" were dismissed by the support from the overwhelming amount of evidence, for example, Baltzer and his credentials, Monier construction in Germany, the experimental culvert at Burwood, test arches loaded to destruction by Carter, Gummow at Forest Lodge, and the technical evidence by Prof. W.H. Warren and Norman Selfe. Hickson stated that "it was good for the Department to accept the substitute tender and so get valuable experience in this form of construction". Indeed it did, as later works described in this paper show.

Another important factor in the acceptance of the cheaper Monier arch design was the financial climate of the time. New South Wales was still suffering from the deep economic depression 1892-93, therefore, a project that offered substantial technical benefits with a saving in costs of about 20% was welcomed enthusiastically by those responsible for planning and constructing public works during those difficult years.

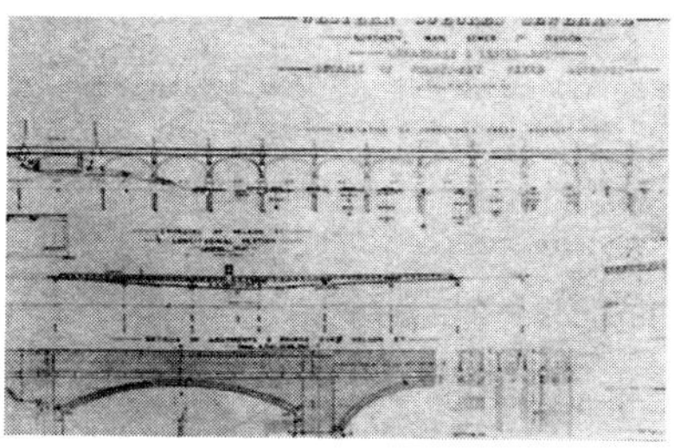

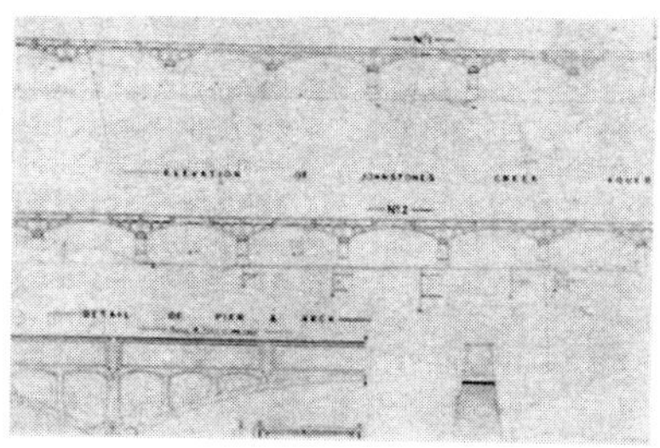

Figure 8 The PWD scheme for the Annandale aqueducts was a series of brick arches (left). The Monier arches (right) were 50% larger in span and their overall cost was 19% cheaper.

But did the Colony get value for its money despite the cost advantage of the Monier scheme? The unit cost of the aqueducts was £1.3 per ft^2 in 1896. When indexed to 1984 (Pope 1984) the unit cost is \$1060 per m^2 which compares favourably with current estimates of between \$1200 and \$1400 per m^2. So the Monier System appears to have been cost-effective and the decision to adopt the new technology was justified. However, it was not seen that way in 1896.

The contractors had to agree to £5,000 retention money and maintain and fully guarantee the completed aqueducts for three years. The onus of satisfactory performance of the structure was placed squarely with the contractors.

4.5 Carter, Gummow and Co.

The principals of this company were mainly contractors, Messers. Carter, Forrest, Snodgrass, Maddison and Ewing. F.M.Gummow and W.J.Baltzer (by retainer) were the engineers. The Company had been formed in 1892 and had successes with many projects including the Marrickville sewer scheme covered by Contract No.69. Later on, Cater, Gummow and Forrest became the only principals. In 1896, Baltzer brought to Australia the first machine for making Monier pipes and the Company produced a Treatise (Carter 1896) containing a complete review of the Monier System including photographs and drawings showing the scope of application of this form of reinforced concrete. Making Monier pipes soon became the dominant part of their activities (Gummow 1899) and the new firm of Gummow, Forrest & Co. established the Monier Pipe Factory, figure 9, at Alexandria. Eventually, the Government acquired the works in 1914 (PWD Annual Report, June 30, 1914) and continued business as the State Monier Pipe and Reinforced Concrete Works.

F.M. Gummow, figure 9, was a graduate in Civil Engineering from Melbourne University and by the time of constructing the Annandale aqueducts in 1896 he had some thirteen years of practical experience. Starting at 21 years old he had worked on the Bondi sewer, the Hyde Park sewer, excavations at the Potts Hill reservoir and other similar works in Sydney. In 1892 he was awarded his first contract for water supply works in Adelaide. From 1893 he was a principal with the companies noted earlier and their projects. One of their most successful departures from Monier construction was building the Dunolly Ford Bridge over the Hunter River at Singleton, figure 9, during 1904-05 (Dunolly Ford Br. Souvenir,1905).

Figure 9 F.M. Gummow was the principal pioneer of reinforced concrete in New South Wales. His company built the Annandale aqueducts and fabricated large quantities of concrete pipes. But he was a competent engineer in other fields as the erection of the steel trusses at Singleton in 1904-05 proved.

5 THE MONIER SYSTEM

The definition of Monier construction seems mundane today,but ninety years ago this new material was viewed this more interest. It had a degree of notoriety because it had figured prominently in the Inquiry, so much so that the popular newspaper "Town and Country Journal" carried a feature article about it in the issue of October 10, 1896. Extracted from Gummow's Treatise of January 1896, the article said,

> "The system of construction consists of iron rods placed longitudinally and laterally to form a mesh or lattice. The sectional area of the rods and the size of the lattice used, vary with the requirements of the structure, and are ascertained from formulae founded on pratical experiment. The lattice is erected on centring and cement mortar is well rammed through and around

it. In setting, the mortar shrinks and gets a firm grip of the iron. The lattice work is calculated to take up and distribute any tension the structure is likely to be subjected to, and it gives elasticity to the body".

The article continued in this flowery style as it mentioned the advantages and applications of the system and quoted some successful examples in Germany.

So new was this technology that it did not appear in Professor Warren's text book of 1894, and there seems to have been an air of mysticism and complexity about it among contemporary engineers because Baltzer introduced his 1897 paper on the subject thus,

"To make the paper more interesting and easy to follow, the author has refrained from entering into any abstruse calculations or intricate theory, but touching on the general principles, the opinions and experiences of leading engineers and scientists, and the results authenticated tests"

The theory of Monier arches was presented by Walter Beer in 1898. It is ironic, therefore, to find Baltzer, in giving evidence to the Inquiry, saying that "the line of pressure for the Annandale arches lies within the middle third hence there is no need for the grillage of iron". Indeed, the calculation books of his colleagues in the Public Works Department, particularly H.H. Dare and J.W. Roberts, show that for most Monier arches tensile stresses did not occur, or at worst were very small.

However, when the system was applied to free-standing water tanks, the lattice of iron rods was essential to resist the hoop tension, in fact the concrete merely acts as an impervious shell.

6 REINFORCED CONCRETE

In neither of the two early applications of the Monier System was composite action utilised structurally, but when beams and slabs (originally called plates) were to be constructed then a proper knowledge of the composite behaviour was necessary. The theoretical work was started in Europe by Coignet, Koenen who produced "Das System Monier" in 1886, Tedesco, Rabut, von Bach, Otto Gaf and Ritter just to mention a few (Cowan 1978).

Locally, Professor Warren conducted tests (1902, 1904) on cement mortars and reinforced concrete units to determine strength and elasticity, and a series of papers were published dealing with the theory of reinforced concrete beams (Cutler 1896, Woore 1902, Gummow 1904, Warren 1905).

Simultaneously, there was a flood of text books and manuals such as those by C.F. Marsh 1904, W.N. Twelvetrees 1905, F.D. Warren 1906, Prof. E. Marsh 1909, J.P. Brooks 1911 and, B.E.Jones and A. Lakeman 1913. By the beginning of World War I the books on reinforced concrete looked much the same as later books of the 1930's and 1950's, and reinforced concrete had achieved its "modern" image both in terms of design and practice.

7 SCOPE OF APPLICATION 1895-1915

With the success of the Annandale aqueducts, the engineers of the Public Works Department became the leading exponents of the design and construction of reinforced concrete work in New South Wales during the next twenty years. The Annual Reports for that period show the scope of application and the widespread use of the new technology. Other government agencies, such as the Water Board, the Sydney Harbour Trust (later to become the Maritime Services Board) and the Railway Department, embarked on large programmes of reinforced concrete construction which they found to be suitable, economical and durable for the long-term needs of the community.

As the success of these new concrete works became evident and with an increasing influx of information from overseas, engineers and architects in private practice began to include reinforced concrete components such as floor slabs, beams and stairs in their buildings.

The remainder of the paper is a summary of the scope of application of reinforced concrete between 1895 and 1915. Particularly useful references for identifying those early works are the papers by Bradfield (1900), the discussion to a Warren paper (1907), Mitchell (1922) and the 50th Commemorative issue of Constructional Review (1977). The Archives Section of various government agencies were also valuable sources of information. A very fortunate feature of this study was that nearly all of the examples are extant and accessible for inspection.

7.1 Arch bridges

Monier arches became the dominant form of concrete road bridge from 1896 to 1920, particularly over railways. At the time, the Railway Construction Branch was part of the Public Works Department and the design of Monier arch bridges, for both the road and railway works programmes, was performed by such prominent engineers as Harvey Dare, J.W. Roberts and J.J.C. Bradfield.

A semi-circular concrete arch was built over Black Bobs Creek in 1896 but its proportions are more like a Roman arch than the flatter Monier construction. The first of the new style was built over the railway cutting at Hill Top in 1897 followed by the much larger arch at Liddell in 1898 to carry the Great North Road (New England Highway) over the double track Main North railway, figure 10. A series of railway overbridges followed, over the Main North line between Hornsby and Cowan, the Main Soutn line from Bargo to Cootamundra, and along the Main West from Glenbrook to Lithgow.

In 1900 the Roads and Bridges Branch had their first Monier arch constructed over Moonbi Creek, north of Tamworth, figure 10 (Main Roads 1978) followed by the viaduct of thirteen arches across the Hawkesbury River at Richmond in 1905.

The Sydney Harbour Trust also showed a passing interest in Monier arches and built three over the rock cutting that is Hickson Road passing from Darling Harbour to Walsh Bay, figure 10. These 24 m (80 feet) arches support Munn, Argyle and Windmill Streets, and at the time of their construction 1910 -1914 were noted by the Daily Telegraph (1911) as the "largest reinforced concrete bridges in New South Wales.

7.2 Reservoirs

The Monier System had been successfully used in Germany for water tanks and pressure vessels, consequently the Water Board and the Water Supply Branch of the Public Works Department began building water tanks very soon after the Inquiry had officially endorsed Monier construction.

The first reinforced concrete water tank above ground was built on a hill above Kiama in 1899, figure 11, and soon after, a similar tank was constructed in the south-west corner of the Randwick Racecourse grounds. Beginning with the Mt. Pritchard (Liverpool) reservoir in 1901, the Water Board built

Figure 10 The success of the Annandale aqueducts and of the Monier arch bridges in Germany led to the design and construction of many such road bridges in New South Wales, for example, at Liddell (1898), at Moonbi (1900), at Richmond (1905) , and at Walsh Bay, Sydney.

a series of on-ground and elevated service reservoirs Howard Street Randwick 1910, Drummoyne 1910 and Bellevue Hill 1912 just to mention a few. The Public Works Department built the water supply tank for Corowa in 1906, and two Monier reservoirs were built for Pelow Main and Neath coal mines in 1908 and 1909 (Henson 1910).

But the introduction of the new technology was not without blemish. On January 22, 1909 the newly-built service reservoir at Mittagong collapsed, figure 11, and spilled 200,000 gallons down the hill (Town and Country Journal, January 27, 1909). E.M. De Burgh inspected the ruins and said "there was no theory as to the bursting, as a number of exactly similar tanks in other parts of the country are all giving entire satisfaction". Errors in construction and some inexperience by the contractors seem to have been the cause, however it did not halt the construction of the elevated service reservoirs that occupy many of the high spots in and around Sydney.

Figure 11 Success and failure in early reinforced concrete construction; the 1899 water tank above Kiama (extant) and the 1909 collapsed Monier tank at Mittagong.

7.3 Monier pipes

G.W. Mitchell (1922), who was Director-General, Department of Public Works in 1914, at the time the government acquired the Monier Pipe Works, claimed that in 1897 the making of Monier pipes was on a small scale because the pipes, made by the European system, had insufficient breaking strength. The design was carefully investigated and numerous tests were made to assess the best amount of reinforcement and its location. The new product "placed Australia ahead of other countries in this section of the use of reinforced concrete". From that time, the use of Monier pipes for sewerage and stormwater drainage became general, and the output of standard units increased rapidly.

Despite the thousands of pipe lengths used , the author has yet to see an example exposed in the ground, but an application of Monier pipes that has been easy to see is their use for bridge piers.

Around 1900, timber piles were still the dominant element of bridge piers, but in spite of the great density and durability of the Australian hardwoods, they were not immune to insect attack and wet rot. Both tended to be most troublesome at the ground line and in the length wetted by the tidal range. Sheathings of copper and Muntz metal had not proved satisfactory but the strength and durability of a reinforced concrete cover offered considerable improvements.

E.M. De Burgh (1900) recognised "that pipes constructed on the Monier principle, in which steel nettings and wires are introduced into the body of the cement forming the pipe, would be very suitable for pile coverings". The first application was in the construction of the road bridge over Cockle Creek, south of Newcastle, in 1901 (DMR archived plan No. 9001714). But Monier pipes, filled with concrete, were also used at many other sites to form the main pier unit, figure 12, instead of timber piles.

7.4 Sea walls

A feature of the introduction of any new technology is that some of its uses are determined by social rather than technical factors. The first use of reinforced concrete by the Sydney Harbour Trust was just such a case. At the turn of the century the shipping and commercial areas of Sydney Harbour were unpleasant waterways due to the presence of dead carcasses, offal and other foul flotsam. The residential strip, between the wharves of Darling Harbour and the City, was a squalid slum and ships' crews did little to control the influx of rats. Consequently, Sydney had a rat plaque, figure 13.

The Sydney Harbour Trust was established to bring some order to and planning for the Harbour and to clean up the mess. One of the problems was the ease with which rats could swim at will and climb up wooden sea walls or run along drainage/sewer pipes into the slum areas. The success of the eradication programme and of the remodelling of the Kent Street-Rocks areas is a very interesting social story in its

Figure 12 Monier pipes with in-fill concrete were used as piers for road bridges at many locations in New South Wales.

Figure 13 One of the methods for solving Sydney's rat plaque was to build smooth-faced concrete sea walls using precast reinforced concrete units.

own right, but its involvement with the early years of reinforced concrete stems from the use of precast reinforced concrete to form a smooth face on the tidal side thereby effectively restricting the free-roaming habits of the rats.

During the recent reconstruction of the wharves of Darling Harbour north of Pyrmont Bridge, the early precast concrete sea wall was either demolished or abandoned, but south of Pyrmont Bridge and in Walsh Bay, sections remain and are clearly visible at low tides.

7.5 Other harbour works

The Harbour Trust engineers were quick to see other uses for the new material; just three are mentioned here and illustrated in figure 14. Navigation beacons were constructed of reinforced concrete at many locations in the Harbour and on its foreshores.

Figure 14 Examples of the scope of application of reinforced concrete for harbour works early this century; a navigation beacon, a pontoon, and the Woolloomooloo wharf.

Wharf decks formed by reinforced concrete slabs were superior to wooden planks, consequently the large programme of wharf construction prior to World War I included such decks as standard practice. When the Trust decided in 1910 to build a new pontoon for Nos. 6 and 7 Circular Quay, timber and iron were rejected in favour of reinforced concrete. The unit was 33m long x 20m wide x 2.4m deep, weighed 600 t and had 44 watertight compartments. The Evening News of November 13, 1913 claimed that it was the largest of its type in Australia.

7.6 Canal linings

The Water Board extended the use of reinforced concrete from service reservoirs to lining the water supply canals, figure 15, during the years 1905 to 1913, and the lined sections are easy to see at the Prospect Reservoir outlet and where roads cross the canals in many places.

7.7 Buildings

From around 1910, architects and structural consulting engineers began to incorpoate reinforced con-

Figure 15 The canals of Sydney's water supply scheme were liner with reinforced concrete during the period from 1905 to 1913.

crete floor slabs, beams, stairs and walls into their building designs. The columns were either cast iron or fabricated steel and the main walls were solid brick; the all-concrete building did not appear until the 1920's. Four existing buildings from the period are the Australian Gas Light building in Parker Street, Ushers/Carlton Hotel in Castlereagh Street, Commonwealth Bank on the corner of Martin Place and Pitt Street, and William Adams' old workshop in Alexandria. Figure 16 shows three early examples of reinforced concrete building construction.

7.8 Slab and beam bridges

By 1914 the theory and practice of reinforced concrete in bending was well understood overseas and and the bridge design engineers of the Public Works Department were ready to extend the application of reinforced concrete from Monier arches to beam and slab construction.

The first of these was a slab bridge,figure 17,over American Creek near Fig Tree, Woolongong, completed in 1914 but recently replaced. In the same year a small slab bridge was built over Tooles Creek south south of Wagga Wagga, which has been widened using beam and slab construction.

There followed in quick succession a series of beam/slab bridges, over Mullet Creek near Dapto in 1916, over Throsby Creek, Newcastle, figure 18, also dating from 1916 (both extant) and over Shark Creek, Maclean. The latter was replaced in 1937 by the present concrete bowstring arch.

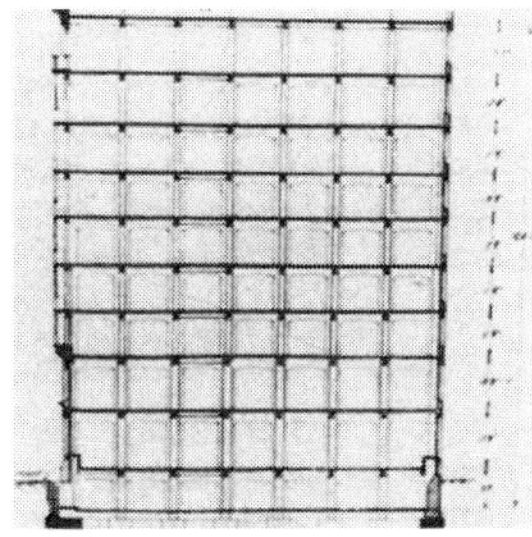

Figure 16 Three examples of reinforced concrete building construction before World War I; the Coogee Beach pavillion and a coal hopper in Parramatta, and the line drawing for No. 49 Wentworth Avenue - the building has reinforced concrete floors.

Figure 17 The 1914 slab bridge over American Creek, Fig Tree (replaced); the 1916 beam bridge, Wickham, Newcastle.

With the success of these bridges and the end of World War I, a whole new era of road bridge construction began, so much so that for the next fifty years reinforced concrete was the dominant form of construction.

7.9 Other works

The research into the early use of reinforced concrete is continuing and it is most likely that other applications will be revealed. For example, the author believes that some sections of reinforced concrete road pavement were built in New South Wales around the time of World War I, but has yet to find the evidence.

A reinforced concrete chimney was built at the Ryde Pumping Station in 1916 but demolished in 1982;and some hoppers and silos were built for private companies.

The examples quoted in this paper are far from exhaustive.

8 CONCLUSION

The development and use of any engineering technology is part of the total history of a community. The contemporary social attitudes, politics, economic factors and personalities all influence the success, or otherwise, of that technology. So it was with the introduction of reinforced concrete into New South Wales. Although technical papers from 1898 to 1915 illustrated the developments in theory, design and application, it was the result of a political row and the subsequent Royal Commission that revealed the origins of reinforced concrete in New South Wales.

The experienced historian is well aware of this technical-social interplay and so too must be the engineering historian, because, without identifying the impact of those other factors, the technical story is incomplete.

Reinforced concrete has become, after ninety-five years, a common part of our built environment,and we are fortunate that its durability has allowed a great many of the original applications to be extant. Most of these are actually still in use, and nearly all are readily accessible for inspection. Collectively, they form a significant body of hard evidence of our engineering heritage.

9 ACKNOWLEDGEMENTS

The author gratefully acknowledges the assistance of many people in researching this topic, particularly Noel Thorpe (M W S & D B), Bob Mayall (Archivist D M R), Ross Best (S R A), John Forsyth (Archivist S R A) and Alison Hair (M S B). Also, the staffs of the State and Mitchell Libraries.

10 REFERENCES

1 American Concrete Institute., "History of concrete, 30BC to 1926AD, Annotated", ACI Bibliography No. 14, 1982.

2 American Concrete Institute., "A selection of historic American papers on concrete 1876-1926" Publication SDP-52, 1976.

3 American Steel & Wire Co., "Handbook and catalogue of reinforced concrete", Feb. 1908.

4 Baltzer,W., "The Monier System and its uses", Jnl. Engg. Soc. of NSW, Vol. 12, 1897, pp81-94.

5 Beer, Walter., "The Monier System of construction", Proc. Inst. Civ. Engrs, Vol. 133, 1898, pp 376-393.

6 Bradfield, J.J.C., "Some notes on Monier construction", Jnl. Syd. Univ. Engg. Soc., 1900, pp 55-62.

7 Building Research Station (UK)., "A note on the history of reinforced concrete buildings", Special Report No. 24, Building Research Station, Herts, England, 1956.

8 Carter, Gummow & Co., "A treatise on improvements in concrete and cement mortar building, construction and manufacture, combined with iron rods", North Sydney, January 1st, 1896; copy held at the Archives Office of State Rail Authority, Sydney.

9 Cement and Concrete Association., "Concrete in Australia 1862-1977", Constructional Review, Vol. 50, No. 4, Nov. 1977, pp 9-71.

10 Cowan,H.J., "The Master Builders" and "Science and Buildings", John Wiley and Sons, New York, 1977 and 1978.

11 Cutler, A.E., "Monier structures", Proc. Engg. Soc. of NSW, Vol. 11, 1895-96, pp 51-70.

26 REINFORCED CONCRETE IN NEW SOUTH WALES

12 Daily Telegraph.,"Largest reinforced concrete bridge in NSW", Daily Telegraph, Oct. 30, 1911.

13 Dare,H.H., Calculation books c1900, Dept. of Main Roads NSW, Archives Office, Sydney.

14 De Burgh, E.M. "On the use of Monier pipes as a pile covering", Proc. Inst. Civ. Engrs., Vol. 142, 1900, pp 288-291.

15 Department of Main Roads., "Old Monier bridge near Tamworth", Main Roads, Sept. 1978, pp 10-11., historical notes, photographs and calculation books., Archives Section and Bridge Section, Sydney.

16 Dunolly Ford Bridge., "Souvenir of the opening of the Dunolly Ford Bridge, Singleton, 9th September, 1905", Copy held at Mitchell Library.

17 Engineer,The., "A reinforced concrete pontoon for Sydney Harbour", The Engineer (London), Vol. 118, Aug. 7, 1914, p149.

18 Evening News., "Reinforced concrete pontoon for a new wharf at Circular Quay" and "New bridges, reinforced concrete used, work of the Harbour Trust", Evening News, Nov. 13, 1913.

19 Gummow, F.M., "The manufacture of Monier pipes" Jnl. Royal Soc. of NSW, Engg Section, Vol. 33, 1899, pp xxxiv-xxxvii; and "Notes on the theory and practice of concrete-iron constructions", Jnl. Royal Soc. of NSW, Sept 7, 1904, pp 105-129.

20 Henson, J.B., "Reinforced concrete reservoirs, Pelow Main and Neath, NSW", Northern Engg. Inst of NSW, April 1910, pp1-17.

21 Indented Bar & Concrete Engineering Co. Ltd., "Reinforced concrete in Australia", undated brochure, but all examples are pre-1915.

22 Jones, B.E. and Lakeman, A., "Reinforced concrete", Historical notes, and Sytems described Cassell _ Co. Ltd., London, 1913.

23 Mitchell, G.W., "Genesis and development of reinforced concrete in Australia", Trans. I.E. Aust., 1922, pp 172-174.

24 Mouchel, L.G. & Partners Ltd. "Hennibique Ferro-concrete", 1900.

25 Pope, David., "Gross domestic product price index 1870-1984", Dept. of Economic History, Univ. of NSW, 1984.

26 Public Works Department,NSW., Annual Reports 1895-1917.

27 Roberts, J.W., Calculation books c1907, State Rail Authority Archives Office, Sydney.

28 Royal Commission. "Public Works Inquiry Commission, Report, 1896-97", Votes and Proceedings, Legislative Assembly, Vol. 6, 1897.

29 State Rail Authority,NSW. Plans of early reinforced construction, Way and Works Branch, Sydney.

30 Sydney Harbour Trust. Annual Reports 1901-1915, Library, Martime Services Board, Sydney.

31 Town and Country Journal. Many news items and articles, but particularly, "Monier Arches", Oct. 10, 1896, p 25; and "Collapse of Monier Reservoir", June 27, 1909, p 28.

32 Warren, W.H., "Engineering constructions in iron, steel and timber", Longmans, Green & Co, London, 1894.

33 Warren, W.H., "Investigation in regard to concrete with steel rods", Jnl. Royal Soc. NSW, Vol. 36, 1902, pp 290-303.

34 Warren, W.H. "Further experiments on the strength and elsticity of reinforced concrete", Jnl. Royal Soc. of NSW,Vol. 38, 1904, pp 140-169.

35 Warren,W.H., "Reinforced concrete", Jnl. Royal Soc. of NSW, Vol. 39, 1905, pp 49-64.

36 Warren, W.H., "Reinforced concrete constructions", Proc. Engg. Soc. of NSW, Vol. 21 , 1906-07, and the discussion in Vol. 22 , 1907-08, pp 75-98.

Name and Place Index

Please note: references to illustrations and tables are indicated by italic page numbers, and notes by the use of square brackets (e.g. 40 [n. 37] means page 40, note 37).

Subject Index

Please note: references to illustrations and tables are indicated by italic page numbers, and notes by the use of square brackets (e.g. 3 [n. 3] means page 3, note 3).